Conservation Biology
in Theory and Practice

Conservation Biology in Theory and Practice

Graeme Caughley
CSIRO Wildlife and Ecology
Canberra, Australia

Anne Gunn
Department of Renewable Resources
Government of Northwest Territories
Yellowknife, Canada

**Blackwell
Science**

Blackwell Science

Editorial offices:
238 Main Street, Cambridge,
    Massachusetts 02142, USA
Osney Mead, Oxford OX2 0EL, England
25
John Street, London WC1N 2BL,
    England
23 Ainslie Place, Edinburgh EH3 6AJ,
    Scotland
54 University Street, Carlton, Victoria
    3053, Australia
Arnette Blackwell SA, 1 rue de Lille,
    75007 Paris, France
Blackwell Wissenschafts-Verlag GmbH,
    Kurfürstendamm 57, 10707 Berlin,
    Germany
Feldgasse 13, A-1238
    Vienna, Austria

Distributors:

USA
Blackwell Science, Inc.
238 Main Street
Cambridge, Massachusetts 02142
(Telephone orders: 800-215-1000
or 617-876-7000)

Canada
Oxford University Press
70 Wynford Drive
Don Mills, Ontario M3C 1J9
(Telephone orders: 416-441-2941)

Australia
Blackwell Science Pty Ltd
54 University Street
Carlton, Victoria 3053
(Telephone orders: 03-347-0300)

Outside North America and Australia
Blackwell Science, Ltd.
c/o Marston Book Services, Ltd.
P.O. Box 87
Oxford OX2 0DT, England
(Telephone orders: 44-1865-791155)

Acquisitions: Jane Humphreys
Development: Debra Lance
Production: Tracey Solon
Manufacturing: Kathleen Grimes
Text Design by: Michael Granger
Typeset by: Best-set Ltd, Hong Kong
Printed and bound by: Edwards Brothers,
    Ann Arbor, MI

© 1996 by Blackwell Science, Inc.
Printed in the United States of America

5

**Library of Congress Cataloging-in-
Publication Data**
Caughley, Graeme.
    Conservation biology in theory and
practice / Graeme Caughley,
    Anne Gunn.
        p.   cm.
    Includes bibliographical references
(   ) and index.
    ISBN 0-86542-431-4
    1. Conservation biology.   I. Gunn,
Anne.   II. Title.
    QH75.C38   1995
    333.95'16—dc20   95-1616
                                                            CIP

Contents

Graeme Caughley 1937–1994
A Tribute

Graeme Caughley's publications are such a tribute to his academic achievements and intellect that no paean from another pen can improve on them. Known widely in biology today, he will still be referred to a century hence as one of the seminal minds that applied mathematical logic to comprehending biology. His laconic writing and wry, understated humor made easy reading and he had that gift, which comes with a first class mind, of making the seemingly complex simple.

While publications illustrate his originality, they throw faint light on the personality we knew as "Caul." In *The Deer Wars* he drops a few clues about an early career in his native New Zealand. Caul exhilarated in hunting deer and tahr high up mountains and in all weathers. He enjoyed the physical demands and while his trophies were not antlers but scientific data, professional deer hunters ranked him among the best. It pleased him, for he was competitive and proud of his achievements.

Though a first class theoretician, there was nothing otherwordly about his management recommendations. Being so competent an outdoorsman himself, they were models of common sense. From working in the Antarctic, Arctic, Himalayas and Africa as well as New Zealand and Australia, his breadth of outlook spanned academia and field management with rare completeness.

Biology may have been his profession, but Caul was a polymath. There were few subjects on which he could not comment sensibly. The product was sometimes unexpected and none more so than a discourse on fourteenth century brick-making in Umbria followed by another on Italian Renaissance music. We were in Assissi at the time and until then I had only heard him strum a guitar while singing Irish and Australian folk songs (not well). His erudition in early Italian composition was thus surprising, even though the mathematical mind and music often go together.

Both talks might have been leg-pulls as nobody present could gainsay him and, above all, Caul was an irrepressible humorist. Miss the initial eye-twinkle or that hint of a shy smile and his deadpan deliveries pulled many a leg. When recounting experiences, he was often the butt of his own humor. Recalling his first fire (he was a fireman in his student days) was typical. Overwhelmed by urgency and excitement, he charged into the burning house and amidst blinding, eye-watering smoke and flames found a person to rescue. That this man fought him off violently Caul put down to panic. After an epic struggle he got his victim across his shoulders in the approved fireman's lift and made for an exit. Bystanders were delighted when the diminutive Caul shot out of the smoke with another fireman twice as big as him on his back. Recounting the event Caul dwelt gleefully on his colleague's disbelief and indignation, unconscious of the corollary: while he may have been small, Caul was both tough and a man of action.

Graeme died on February 16th, 1994, well short of the hoped-for three-score years and ten. Yet in his life he achieved on a grand scale. And he was an even grander friend and personality to those privileged to know him. This book, co-authored by his partner, Anne Gunn, is Caul's last work: he was thinking originally to the end.

Ian Parker, Nairobi, Kenya
November 9, 1994

Preface

We have written this book for the people who want to stave off extinctions regardless of whether they call themselves ecologists, conservation biologists, wildlife managers or whatever. Our aim is modest. It is to demonstrate the need to understand the ecology of species in trouble and the necessity of applying scientific method to determine why species are driven into decline and how those declines can be reversed. We have used case-histories to examine how ecological evidence is weighed for and against competing explanations to account for declines. We term this step *diagnosis* and getting it correct is key to taking the necessary *recovery* action.

We have looked back at the major waves of extinction with the merits and shortcomings of their explanations (Chapter 2). Then for historic extinctions or near extinctions, we review the presumed reasons for their decline (Chapter 3). Tabulations, even of extinctions, tend to become only lists: we judge it important not to lose sight of what it means to forfeit a species. So we have tried to give enough details to "bring to life" the species whose extinction or near-extinction we describe. In more detail and admittedly with hindsight, in Chapter 4, we have looked at credibility of the diagnoses and recovery treatments as we can learn much from what has already been done. For that reason those case histories are early in the book as they are frequently referred to in the chapters that follow them. More on the logic and methods of diagnosing and treating problems are in Chapters 8 through 12.

Stemming declines and fostering recoveries have long been part of the relationship between wildlife and people. Those efforts became encapsulated within their own discipline—conservation biology— which came into being in the 1980s when it acquired a suite of theoretical concepts. Those prevailing concepts dwell on metapopulation dynamics (Chapter 5), small populations (Chapters 6 and 7) and island biogeography as applied to reserve design (Chapter 10).

Conservation biology takes in a cross-section of disciplines. Our focus is on ecology's role in helping species in trouble. Ecology by itself is not enough: we introduce scientific method (Chapter 1), paleontology (Chapter 2), population demography (Chapter 5) and genetics (Chapter 6). People's economic aspirations are part of the problem and part of the answer, which led us to introduce economics mostly as it influences harvesting (Chapter 11).

A word on our choice of examples: We did not attempt a comprehensive review but selected examples to illustrate and emphasize. Our choice may occasionally be idiosyncratic and biased toward mammals and birds but reptiles, amphibians, fishes and invertebrates are not neglected. We make no apologies for this bias, which reflects the relative amounts of information published and our own leanings and experience. Plants would be a volume unto themselves and we have left that to others.

Conservation's success stories almost inevitably have one or a handful of individuals behind them. More often than not their names are little known but it is their drive, tenacity, foresight and ecological skill that made the difference in saving species. We humbly acknowledge their efforts.

Many people gave freely of their time and expertise to help with the book and we thank them gratefully. David Grice (CSIRO) was this book's text manager and produced the graphs. Robin Barker (CSIRO) ferreted out references and compiled information. Both of them helped in many other ways and their efforts were unsparing. They both checked the nomenclature: For names of species, we used Howard and Moore (1980), Edwards (1982 and 1986), and Nowak (1991); place names on the figures were as used in Anon (1990).

CSIRO library staff, especially Barbara Staples and Inge Newman and Alison Welch of Renewable Resources library (Yellowknife) could not have been more helpful. Fleur Sheard prepared most of the maps and Natalie Barnett and Soussanith Nokham produced the remaining maps. Kathie Hollis drew the sketches and Liz Jones did the scraperboards that grace each chapter, as well as the cover illustration. Trish Boekel (Greenwords Pty) and Raylee Singh worked wonders with copyediting. Ray Bethke's computer skills saved the day more than once. Jane Humphreys (Blackwell Science, Inc.) was always encouraging.

Our colleagues helped by reviewing individual chapters: Nick Abel (University of Canberra), Doug Cocks (CSIRO), Michael Kingsley (Fisheries and Oceans, Canada), Chris Margules (CSIRO), Steve Morton (CSIRO), Roger Pech (CSIRO), Steven Sarre (Australian National University), Simon Stuart (IUCN) and Mike Young (CSIRO).

Brian Walker and Nick Nicholls (CSIRO), Tony Underwood (University of Sydney), Chris Shank and Mitch Taylor (Department of Renewable Resources, Yellowknife) and Ian Parker (Kenya) all took the time to read the first draft. Ian Parker contributed in his inimitable way to Chapters 3, 4 and 11. Their efforts made a difference and we thank them.

With our emphasis on practice as well as theory, we took a look at endangered species in their own settings. Chuck Stone and Howard Hoshido (US National Parks Service, Hawaii) introduced us to the Hawaiian goose *Branta sanvicensis*; Tony Friend and Paul de Tores (Department of Conservation and Land Management, Western Australia) gave one of us a delightful time with the numbats *Myrmecobius fasciatus* and Bruce and Kerry McFadyen (Lord Howe Island) revealed secrets of the woodhen *Tricholimnas sylvestris*.

These people freely supplied information: Jim Beckett (Fisheries and Oceans Canada), Steve Brechtel (Alberta Fish and Game), John Calaby (CSIRO), Albert Caton (Bureau of Resource Sciences, Canberra), Sue Campbell (CSIRO), Fiona Gell (CSIRO), Bob Harden (New South Wales National Parks and Wildlife Service), Jim Hone (University of Canberra), Michael Hutchinson (American Association of Zoological Parks and Aquariums), Bob Johnson (Metropolitan Toronto Zoo), Michael Kingsley (Fisheries and Oceans Canada), Don Merton (Department of Conservation, New Zealand), Bob Oakleaf (Wyoming Fish and Game), W.L.R. Oliver (Jersey Wildlife Preservation Trust), Ian Parker (Kenya), Bob Raupach (CSIRO), Carl Safina (National Audobon Society), Gwen Shaughnessy (Australian Nature Conservation Agency), Peter Shaughnessy (CSIRO) and Simon Stuart (IUCN).

The book was started after Graeme was diagnosed with cancer. We made the progress that we did with the support of Dr. Stuart May and the staff at Calvary Hospital, Canberra especially Pam Jones, Robyn Stanhope and Kay de Smet (Palliative Care Unit). Dr. Julia Hewitt, Sue Killion and Dr. Kerry MacFadyen helped in innumerable ways. We warmly thank them all.

Graeme died in February 1994. Brian Walker led the way for CSIRO Division of Wildlife and Ecology to provide the means to complete this book while I was in Australia. His constant encouragement was unstinting. I am grateful to the Department of Renewable Resources, Yellowknife for accepting my prolonged absence and to Mark Williams and Chris Shank who covered my job during that time. To all those friends who stepped in and helped, my heartfelt appreciation. And to Ian Caughley and his erstwhile prodigy, T. Caughley, a particular and warm thank-you.

Graeme's humor, courage and vision drove the book forward but any errors and incongruities in its completion are mine.

Anne Gunn, Yellowknife, Northwest Territories
November, 1994

CHAPTER 1

Introduction

Conservation has long been part of wildlife management. It mostly took the form of practical steps to reduce factors driving species toward extinction or to translocate species to safer refuges. In the last two decades, under the influence more from universities than wildlife management agencies, interest in preventing extinctions shifted to theory. Genetics was introduced and the theoretical consequences of being a small population and the implications of island biogeography for reserve design took center stage and became formalized as a new field—conservation biology. The split origin of conservation biology shows in its two paradigms that deal with different aspects of the same thing but tend to be treated as distinct. They are briefly described in this chapter.

A recurring theme in conservation biology is that, since time and resources are often critically short and data are scarce, simple models can be used to generate management recommendations. Computers that can run through a dazzling array of possibilities in seconds are readily available and have stimulated the development of theoretical models. But testing those models against reality through the rigor of scientific method is in short supply. This lack of science is a basic problem which has been with us for a while. President Theodore Roosevelt recognized the essential role of science in conservation in 1910. Aldo Leopold (1933) in his seminal book on game management later explained Roosevelt's particular contribution was the concept of "conservation through wise use" and that science was the tool to achieve this.

In this chapter we introduce science's tools and briefly outline the steps of experimental design to test hypotheses efficiently. These steps are as applicable to determining a problem—e.g., a species is declining—as to the diagnosis and treatment of factors causing a decline (Section 9.2). We use the term "diagnosis" to describe how ecological evidence is weighed for and against competing explanations to identify the factor or factors driving a decline. Limiting factors are those that affect births or deaths, and we use the term without prejudice as to whether they operate in a density-dependent or independent manner. The steps taken to halt a decline and rebuild a population imply "recovery."

1.1   THE TWO PARADIGMS OF CONSERVATION BIOLOGY

The declining-population paradigm deals with why the population is declining or has already declined to low numbers, what caused it, and what might be done to reverse the decline. The antecedents of the declining-population paradigm are found in wildlife management, and, as such, the paradigm has been around for decades. It is a loosely organized and less tangible set of ideas focused on detecting, diagnosing, and reversing a population decline. Wildlife management is about the practical aspects of the day-to-day realities of managing species to head off extinction, which may be why it has not yet acquired a theoretical framework. We suspect that progress in developing theory may come as much from experimentally forcing a common species' decline as from tracing the population trajectories of those in trouble. An alternative source might be from pest control—where an animal is deliberately driven to, at least, population extinction.

In stark contrast is the small-population paradigm. These are the ideas dealing with the effect of low numbers as such upon a population's persistence (Chapter 6). Questions within this paradigm tend to be restricted to stochastic influences on the population's current dynamics: demographic and environmental stochasticity, inbreeding depression, and genetic drift. How or why the population became small is not the issue. A flavor of the approach is encapsulated in this quote: "Modern conservation biology focuses on managing small populations, and this requires estimating their viability (or vulnerability to extinction) so that appropriate management can be taken." (Lindenmayer et al 1993).

Both paradigms have much to give each other, and in this book we have tried to bring them together. To this end, we have described the theory, laced with examples from the practice of conservation biology. Our starting premise is that the signal of a conservation problem is the sustained decline rather than the final stage of low numbers. Identifying a decline precedes diagnosing the factors driving it, and this, in turn, has to precede its management—halting and reversing it.

1.2 SCIENTIFIC METHOD

We think that Leopold's (1933) observation—"So far we have the scientist, but not his science, employed as an instrument of game conservation."—is as apt today as it was in 1933, especially for conservation biology. We understand the urgent feelings when faced with the last individuals of a species, but we do not accept that that is reason to throw science to the wind. To save a species, results should be unambiguous and defensible. In the absence of designed experiments, futile arguments can cloud issues such as whether forest fragmentation does cause increased predation or whether corridors should link reserves (Section 10.2.6). Frequently, hard choices have to be made and they increasingly lead to litigation. Controversies over the northern spotted owl, *Strix occidentalis caurina*, ended up in court with arguments over the validity of research results. In the courts, as elsewhere, weak inferences tumble before the evidence rigorously arrived at through sound experimental design.

The logic of scientific method starts with the formulation of a testable hypothesis. Experimental design is the means by which the hypothesis is most efficiently tested and its major purpose is to produce an unambiguous result in the presence of confounding factors. In contrast to data from experiments are observations of ecological phenomena that we have an incomplete picture or understanding of and little or no control over. The critical difference is that inferences drawn from statistical analysis of observational data are weak because uncontrolled variables have confounding effects. We cannot determine which variables are causing the effects that we observe. It is not uncommon, for example, to see, for endangered birds, reports that predation, food shortage, or inbreeding depression are singly or in concert depressing fledging success. Only through an experimental approach can we control for the effect of each variable and determine its effect.

In Section 9.2 we describe a fictional experiment to treat a decline in parrots, in which our observations lead us to suspect that either rat predation or a nest shortage is the cause. In the following text, we refer to this experiment. The discrete series of steps for experimental design start with observations about which we construct at least one or several models (model equates with theory or explanation) (Underwood 1990, 1991). The model describes as simply and as plausibly as possible an explanation for the observations. The second step is to use the model(s) to predict testable propositions—one or more hypotheses. Boldly stated, these two steps seem reasonable, but they conceal traps for the unwary. Underwood (1991) uncovers those pitfalls which relate to collecting observations unbiased by theories or preconceived notions—alternative models can help with this one. It is worth remembering that there are advantages to having several models and hypotheses that are tested simultaneously (Underwood 1991).

1.2.1   Null hypothesis testing

Given the ease of disproving rather than proving (as elegantly explained by Underwood 1991), the hypothesis is inverted and becomes a null hypothesis. All statistical testing begins with asserting no difference, a null hypothesis (designated H_0), such as "there is no difference between the density of rats on this island and that island." The alternative hypothesis, that the densities differ (H_a), is accepted if the null hypothesis is rejected by the test. The null hypothesis is chosen for testing because it is unique—no difference is no difference. For more complex problems the hypothesis will identify a single form which is either true or not true. If this specific hypothesis is not true, the relevant null hypothesis (which includes many other possibilities) must be true. The test's power can be determined (see below) if the magnitude of the alternative hypothesis is stated. So for the density of rats on the islands, a specified alternative hypothesis could be that the difference is at least 10 rats per hectare. The null hypothesis is that there is no difference. If the difference is at least 10 rats per hectare, we should reject the null in favor of the alternative hypothesis. In this case, if the difference is, say, 2 rats per hectare, we should not wish to reject the null—that a difference of 2 rats per hectare is not consistent with our hypothesis.

Statistical tests deal in probabilities, thus we may be wrong. We could have made one of two errors: *Type 1 error* (also known as an alpha error), in which the null hypothesis is rejected even though true, and *Type 2 error* (beta error), when the null hypothesis is accepted even though false. The probability of committing a Type 1 error is the specified significance level—the chosen possibility of making such a mistake. The probability of committing a Type 2 error is related inversely to the significance level for rejecting the null hypothesis. Both errors can be reduced simultaneously by increasing the sample size. Standard statistical tests concentrate on minimizing Type 1 errors. Minimizing Type 2 errors is called the *power* of the test, and it is 1 minus the probability of Type 2 error (or 1 − beta).

In conservation biology, avoiding saying there is not an effect when there is (Type 2 error) may be more important than ensuring the warranted rejection of the null hypothesis. This is most likely when the treatment itself is a benefit and the lack of treatment, a cost. Again, using our fictional parrot experiment as an example, our experience is reason enough for us to suspect that the rat predation will force a decline. The null hypothesis is that rat predation has no effect on the fledglings and, thus, population size. In this case the failure to reject the "no-effect" null hypothesis is insufficient reason to assume that rat predation is having no effect. At least, one would want to know about the power of the test. In this case the cost of making a Type 2 error (accepting that rat exclusion has no effect when it does) greatly outweighs the benefit of getting it right. Not controlling the rats could precipitate the loss of the parrot, whereas unnecessary rat control is, by comparison, only costly. This is the *asymmetry of risk*, and it is the consideration as to whether experimental design should minimize Type 1 or Type 2 errors.

Underwood (1993) gives the logic to reduce the probability of committing a Type 2 error while investigating environmental impacts. Sampling power can be estimated from preliminary sampling, or, if enough is known about the scale of variation, a simulated data set can be used. The consequences of Type 2 errors for detecting environmental impacts from pollution are serious; to conclude, there is no effect when there has been one (if the null hypothesis was that there is no effect). Likewise, the same logic applies to conservation biology, be it detecting trends in populations or the effects of treatments, and it comes down to using an appropriate sampling design (Section 8.3.1).

There are, however, dissenting views. Simberloff (1987) argued that Type 1 errors should be avoided in uncertain situations. He used corridors connecting reserves for the Florida panther, *Felis concolor coryi*, as an example. Rejecting the null hypothesis that corridors will not delay the panther's extinction when it is true (i.e., Type 1 error) could lead to creating corridors at considerable expense but without helping the panthers. But we do not really know if corridors are the answer (Section 10.2.6), and the money spent on corridors could, so the argument runs, be more efficiently spent on other aspects of panther conservation. Another point was that biologists would lose credibility, which could lessen their influence on other conservation decisions. On those premises Simberloff (1987) argued that the experimental design should reduce the probability of Type 1 errors. The trade-off is thus a reduced risk of inefficient spending and losing credibility, against an opportunity to reduce probability of the panther's extinction. In contrast to Simberloff (1987), Shrader-Frechette and McCoy (1993) expand the logic to minimize Type 2 errors.

Once we reject the null hypothesis, the hypothesis and the model from which it was derived are supported. Alternatively, we fail to reject the null hypothesis, and then the hypothesis and model are rejected. This is not the end of the matter even if we accept the model. We can make the model more specific or generalized and then retest it to attempt to demonstrate that it is wrong (Underwood 1990). This elimination of alternative models is the way to close in on ecological relationships.

Standard experimental designs

In conservation biology, we are likely to have to treat a population to foster its recovery or to determine if and why a species is declining. The answers to those questions often have a similar logical structure and lead to standard experimental designs. Details on these are set out elsewhere (Manly 1992; Eberhardt and Thomas 1991; Underwood 1991; Hurlbert 1984; Green 1979). Examples of published experiments whose aim was to answer more generalized ecological questions with comments on their design are found in Hairston's (1989) review of experiments in forests, deserts, and other habitats. Green (1979) succinctly summarizes points to remember as guiding principles (Table 1.1). Some points are worth reiterating and they follow, but first we must give an explanation of terminology.

We envision an area (study area) holding a population in trouble. The area is subdivided into plots (which may be grouped into blocks) where treatments will be applied. The management treatment is applied as a factor whose level we will be manipulating, such as supplementary feeding or predator removal. As a minimum, a factor has two

Table 1.1. Ten principles of sampling and experimental design

1. **State hypothesis:** State precisely the question to be answered, then restate the question as the hypothesis to be tested.
2. **Replication:** Take replicate samples within each combination of controlled variables so that differences among them can be compared with differences within treatments.
3. **Random sampling:** Take randomly allocated samples.
4. **Control (zero treatment):** Collect samples when the experimental treatment is present or absent and all other conditions are similar.
5. **Preliminary sampling:** Collect samples to evaluate sampling design.
6. **Verify sampling:** Determine that the sampling is effective and efficient across the range of sampling conditions.
7. **Stratify:** Divide large-scale environmental pattern into similar strata before sampling.
8. **Determine sample size:** Estimate the number of samples to achieve the required level of precision.
9. **Data transformation:** Test the data to determine if their variation is homogeneous, normally distributed, and independent of the mean.
10. **Adherence to design:** Do not abandon a statistical method once it has been selected, even in the face of unexpected or undesirable results.

Modified from Green RH. Sampling design and statistical methods. New York: Wiley, 1979.

levels: presence or absence (zero). The factor will affect the response variables, which might be fledging success or density.

A change in a response variable after a treatment can only be demonstrated by comparison with the response variable subjected to zero treatment—the control. Easily forgotten is that control does not mean undisturbed. Everything done for the other treatments has to be applied to the control, other than the manipulation that is formally the focus of the treatment. If the treatment is rat trapping with daily visits to the traps, then the control sites have to have an identical routine of daily visits. The control and treatment sites do not have to be identical, but the greater their similarity, the more they account for environmental variation.

Differences among responses to treatments (including zero treatment—the control) are demonstrated by comparison to differences within them. To measure variation of the difference among the treatments requires several plots. Suppose in the experimental supplementation of nest cavities for red-cockaded woodpeckers, *Picoides borealis*, from Walters et al (1992) (Box 9.2), artificial cavities had been fitted to trees on a single site, which by coincidence lacked suitable foraging opportunities. Woodpeckers would not have settled in the area and would not have used the artificial cavities. The treated site and control site would not have had significant differences in the response variable. This would have led to the erroneous conclusion that supplementing nest cavities was not a treatment to foster recovery (and that of nest shortage was not a factor in the population's decline). Without replication, a comparison of one treatment with another is a test of whether the combination of the treatment and the intrinsic characteristics of a single area differ from another treatment combined with that area's intrinsic characteristics. The area and treatment are *confounded*. Replicates are chosen randomly to sample the natural range of variability. The environment's natural variation is not only spatial. When the experiment is started relative to seasonal, annual or longer-term variation, this should be included in the design.

A common problem in experiments is using samples taken within a replicate as treatment replication. Observations taken within a replicate are not independent of each other, as they have been exposed to the same conditions including treatment. To return to the red-cockaded woodpecker experiment, increasing trees fitted with artificial cavities per site would increase the precision of the measure of the response variable (nesting success) for each site, but that would not help in measuring the variation between sites. That requires an increase in the number of sites (replicates).

How many replicates are needed? The answer is at least two per treatment (including the zero treatment, or control), but it depends on the number of treatments to be compared, the average variance among replicates within treatments, and the magnitude of the expected differences estimated from a pilot experiment or from a previous experiment in the same area.

The simplest design has one factor with two levels, but more useful is a design that examines two management treatments simultaneously. In addition to the effects of each level for the two factors on the response variable, we also have the interaction between them. Their individual effects on the response variable may be additive (i.e., independent of each other). Or the effects may be interactive: The level of one factor may change according to which level of the other factor is combined with it. We determine if there is an interaction by calculating whether the combined effect is greater or less than the addition of both effects when the factors are evaluated separately. Each level of the first factor is combined with each level of the second, and we call that arrangement mutually *orthogonal* (the factors are at right angles to each other).

Box 9.1 offers an experimental design for such a two-factor experiment. Each level of the first factor is combined with each level of the second; therefore there are $2 \times 4 = 8$ cells or treatments. Each treatment has two replicates and the number of replicates per treatment is the same for all treatments.

The design layout for treatment and control blocks has to have the treatments interspersed to avoid lack of independence between them. Hurlbert (1984) illustrates designs contrasting interspersed with partial interspersion. Random assignment of treatments eliminates the likelihood of any bias from the experimenter, but that is often difficult without a large number of replicates. When only a few sites are available, random assignment of treatments may not be possible and the treatments are then deliberately interspersed by the researcher. A common compromise is randomized blocking. The experimental plots are grouped as pairs, or more, within blocks that individually have less environmental variation than the area as a whole. Inglis and Underwood (1992) offer a practical way to design an experiment to test the effectiveness of corridors (Section 10.2.6), which illustrates randomized block designs.

In a *randomized block experiment* the treatments are randomly assigned to a plot in each block. When not enough blocks are available to include all the treatments, a *balanced incomplete block design* may be the solution. The balance is that each possible pair of treatment levels occurs together the same number of times. A caution for using block designs is that the treatments should not interact. The Latin square is another block design to test the effects of three factors, and it does not require many replicates.

An alternative approach to comparing levels of treatments (factors) is not to compare each level of a factor with each of another factor but to nest them instead—that is, to apply the levels of one factor within the levels of another. As described in an example by McKone and Lively (1993), a nested design with one usually random factor nested within another usually fixed factor is a possible solution to designs where only a few replicates are possible.

There are situations where replication is next to impossible and only two sites are available. The two treatment levels are unreplicated. The best that can be done is to choose the sites to be as alike as possible. Either the response variable can be monitored on both sites to establish that it is similar before the treatment or the two treatment levels can be reversed. The outcome of a weak design is a weak inference.

1.2.2   *Design of sampling*

General points for sampling design apply regardless of what and how it is being sampled. Statistical tests have underlying assumptions, and if they are ignored their validity will be compromised. Many tests are relatively robust to violations, with one exception. Samples have to be independent; that is, the sampling has to be random (Cochran 1963). Successive sampling over time often violates the assumption of independence. It is a common problem with using individual locations to determine home ranges or habitat preferences. McNay et al (1994) reviewed lack of independence in location data and recommended sampling animal locations systematically throughout time, depending on the behavior being measured. Kremsater and Bunnell (1992) give an example of testing for independence to determine the appropriate interval.

Sampling has to be scaled to the level of variation expected to avoid both inefficiency and the possibility of inadequately measuring the variable (Green 1979; Underwood 1993). Preliminary sampling can reveal inadequacies before they cause embarrassment, or worse. Inadequate sampling leads to imprecise estimates (wide confidence limits), which increase the probability of Type 2 errors, and that is something we want to avoid. The preliminary sampling can be tested on a surrogate using an ecologically similar species.

1.2.3   The problem of small samples

Small sample sizes plague investigations in conservation biology. Small populations occupying only a few sites do not lend themselves to experimental designs and sampling. The problems are not entirely solvable: in the final analysis there is no substitute for adequate sampling, and statistical tests on small samples will always be less powerful than those on large samples. Certain palliative measures are available, however.

We note, in passing, the need to avoid the temptation to increase apparent sample size by illegitimate stratagems. For example, the survival of individual litter mates is a tempting alternative to mean litter survival, offering the chance of sample sizes several times larger. The chance, however, of survival of one litter mate is not independent of others in the litter, because survival is ruled by conditions in common. So the true sample size is not increased, and using an artificially inflated sample size tends to generate spuriously significant results, not by increasing the estimated treatment difference but by reducing the estimated standard errors.

Manipulating statistical levels of significance between Type 1 and 2 errors is one way to boost sample size (Forbes 1990). When sample sizes are small, a test's power is low and a difference may not be detected even if it is real (Type 2 error). Conventionally we set the probability of a Type 1 error at $\alpha = 0.05$ or 0.01. But no graven tablets exist to say that all statistical tests in biology must be carried out at 95% to 99% levels of significance, or even higher if to do so means that power is reduced and that real effects of importance may be dismissed. We are often working with situations where the "balance of evidence" is more appropriate than "proof beyond reasonable doubt" (Section 1.2.1). In such cases, 90% or even 80% significance levels may be considered given that the probabilities of Type 1 and 2 errors are inversely related, and changing alpha increases the test's power, which correspondingly decreases the test's sensitivity to sample size. The nub is to report actual probabilities so we can see what was done and draw our own conclusions. Just as important, though, is to round up in advance all the available information on the probable variability of the experiment, the tolerable balance of error, and the sampling opportunities available and make a prior decision on the test's critical level.

Tests on large samples are somewhat proof against failures of assumptions, such as the normality of error distributions or the independence of sample mean and variance. Tests on small samples are less robust, and small samples are not a reliable means of testing whether the assumptions are valid. The standard response is to fall back on nonparametric tests. Those tests demand few assumptions, and for the common nonparametric tests exact critical values have been tabulated for small samples. But this often amplifies the original lack of power, especially when the test selected depends on ranking the data, the quantitative values being left aside.

Freely available computing power is a way to improve the performance of assumption-free tests. Tests based on ordinal statistics, such as the rank-sum test, may be replaced by their interval equivalents, in this case (Box 1.1) a permutational (more properly, combinational) test attributed to Fisher. In such tests the values of the data are retained and not discarded as they are in ranking tests. This provides a more powerful test, and, at the same time, the permutational test is less dependent on assumptions than the classic parametric tests (Box 1.1).

Permutational and resampling tests have never been easy to use until now. The critical levels depend on the actual data values in the samples and therefore cannot be looked up in standard tables as rank statistics like sums and correlations can. Consequently we calculate the distribution of possible values for the test statistic such as t or F by permuting or resampling the experiment's quantitative results and evaluating the probability that the observed value of the test statistic is exceeded. Those tasks are an impossibly heavy task without a computer.

For samples that are only moderately small, the number of permutations becomes large, not to mention tedious even with a computer. In our example (Box 1.1), two samples each of 5 members generate 252 possible combinations; two samples each of 10 members would generate 184,756. If we were rash enough to decide to use a permutational test on a correlation between two samples of 10 values, the complete set of permutations would be 3.5 million correlations. Fortunately in such cases we can achieve an adequate level of precision by sampling randomly from the complete set of permutations. It is worthwhile to repeat the sampling to ensure that it is reliable and to check that the "sampling" is truly random, as repeatability is no guarantee of this (a program check with a random number generator will do the trick). *Monte Carlo* is the general term applying to many situations that use a random

Box 1.1. A comparison of the *t* test, rank-sum test, and Fisher's "permutational" test

---

Sample 1: 32 47 33 44 56   Sample 2: 41 43 46 65 60
Ranks: 1   7   2   5   8   3   4   6 10   9

t test

The standard Student's *t* test between means—42.4 for Sample 1 and 51 for Sample 2, with an mean error square of 109—gives a *t* value of 1.303. Probability ($t_8 \geqslant$ 1.303, under H_0) = 0.1145. Do we know that the *t* test is valid for these samples?

Rank sum test

The ten values for the two samples are ranked together from 1, lowest, to 10, highest. If Sample 1 has lower *values* than Sample 2, it will also have lower *ranks*, but if the two samples are drawn from populations that have the same central tendency, the ranks associated with them will also be similar. The *sums of the ranks* of the two samples can therefore be compared to test whether their central tendencies differ. In this example, Sample 1 owns 5 of the 10 ranks from 1 to 10; these are 1, which belongs to the value 32, 7 (47), 2 (33), 5 (44), and 8 (56); the sum of these ranks is 23.

Altogether, there are 252 different sets of five ranks that might have belonged to Sample 1, with sums from 15 (the five lowest ranks, 1–5) to 40 (the five highest); evidently many of the sets duplicate the same sums of ranks. Twenty-three is on the low side, but not very small. Of the 252 possible sets of five ranks, 53 have a sum that is equal to or less than the observed value, 23; i.e., probability ($\Sigma R \leqslant 23$, under H_0) = 53/252 = 0.2103. For all experiments using two samples each of five members, these 252 sets are the same, so standard tables can be prepared, giving the probability level for any rank sum of ranks.

Permutational test on value

Sample 1 owns 5 of the 10 values—32, 47, 33, 44, 56—with a sum of 212, and Sample 2 owns the other 5 with a sum of 255. We can compare these sums to test whether the central tendencies of the parent population differ.

Among the 10 values, there are 252 different sets of 5 that might have belonged to Sample 1. With a personal computer it does not take long to calculate all 252 sums. They range from 193 (the five lowest values: 32, 33, 41, 43, 44) to 274 (the five highest). The observed value, 212, is on the low side, but not very small. Of the 252 possible sets of values, 32 have a sum $\leqslant$212; i.e., probability ($\Sigma R \leqslant 212$, under H_0) = 32/252 = 0.127. This is not so very different from the probability obtained by the parametric Student's *t* test. The sums of the 252 sets are not always the same but vary from experiment to experiment. No standard tables can be prepared.

(Modified from M. Kingsley, personal communication, 1994.)

number generator to produce results that reflect uncertainty in our knowledge of a system.

Despite good intentions, statistical tests are often chosen after the data have been collected. In that case other computer-intensive techniques, such as jackknifing and bootstrapping, are a help for small samples and non-normal distributions. More on these topics can be found in books such as Edgington's (1987) account of randomization tests, which covers analyses of variance (ANOVAs), *t* tests, and factorial designs, or a biometrician can be consulted.

1.3   REPORTING RESULTS

We feel a trifle sheepish in having a section on something so obvious as reporting results. While compiling information for this book, however, two problems kept cropping up, which may be why we might have inadvertently shortchanged some examples. The first is the lack of published results. Research had been done, but its results were unpublished or tucked away in some difficult-to-acquire institutional report. Second, partial reporting was not uncommon. Sharing knowledge of failures is as important as successes, given the urgency of problems facing species in trouble.

1.4   EFFECTS OF TECHNIQUES

Undesirable effects of experiments such as injury during marking of individuals are always a consideration, but even more so when dealing with the last few individuals of a species. Most effects are predictable and not insurmountable, but they do require planning and then modifications or accommodations. In the rush to save a species these can be overlooked.

Testing techniques on surrogates can be useful, particularly on animals difficult to rear or capture. Parrots are in that latter category, which is why Meyers (1994) tested different combinations of mist nets and decoys on introduced parrots in Puerto Rico. We give other examples on using surrogates in captive breeding (Section 4.9), translocating small birds (Section 4.7), and release techniques (Section 4.6).

Experimental treatments should take into account any deleterious effects of techniques and the separate but not unrelated problem of public perception. Support is essential for the recovery of endangered species, but not all people accept the handling or marking of wildlife. Discussions and explanations preceding the research may pay dividends in obtaining that support.

The effects of capturing and handling animals to mark them are now more often determined experimentally, and, although it depends on the species and the techniques, effects can influence results and undermine assumptions in analyses. Chatham Island black robins, *Petroica traversi*, caught for blood sampling, subsequently survived and behaved about the same as a control group who were not caught or handled (Ardern et al 1994). On the other hand, experimental marking and recapture of butterflies showed that "handled" butterflies were recaptured less and the effect varied between species (Morton 1984). Reduced survival is not the only problem and what animals learn from the experience can affect the technique's effectiveness and contravene assumptions. Wild turkeys, *Meleagris gallopavo*, not only avoided drugged corn once they had been previously caught after eating it, but "experienced" turkeys showed alarm and tried to drive other birds from the site (Davis et al 1994). Laurenson and Caro (1994) scrutinized effects of their nontrivial handling on cheetahs, *Acinonyx jubatus*, and concluded that, with "stringent precautions," effects were not detectable. Under the heading of "nontrivial handling" they included visits to maternal dens, attaching radio collars, and radio tracking (we could not resist wondering what is trivial handling and who would own up to it).

Treatment effects are not, of course, limited to handling animals. The presence of researchers and their monitoring (Section 9.5.2) sometimes have effects. Fledging success may be decreased when bird nests are monitored, but that effect can be minimized (Bart 1978; Bart and Robson 1982). We describe the effects of captive raising on the behavior and survival of individuals such as the nene (Section 4.11) and the black-footed ferret, *Mustela nigripes* (Section 4.6), and there are other examples (Section 9.5).

Less obvious, or perhaps more easily overlooked, are treatment effects on other species. Construction of a refuge for the Devils Hole pupfish, *Cyprinodon diabolis*—an endangered species—destroyed some of the last habitat for another endangered species, the Ash Meadows naucorid, *Ambrysus amargosus*, a creeping waterbug (Polhemus 1993). Unexpected effects nearly ended efforts to salvage the last population of kakapo, *Strigops habroptilus*. Cats had almost eliminated this last population of New Zealand's flightless parrot, and the survivors were transferred to four cat-free islands. Supplementary feeding of females started, but the food attracted Polynesian rats, who turned out to be effective predators of nestlings. The solution was ratproof food dispensers and rat poison baiting (Merton 1989; D. Merton, personal communication, 1994).

1.5   THE UNITS OF CONSERVATION

When we think about extinctions or wildlife in trouble, most of us tend to mentally picture a particular species or group of species. Likewise, most of us have little difficulty understanding that if a species is extinct it no longer exists. When a species is in trouble we have an idea what that means. When we talk with each other, however, we may discover that we do not all mean quite the same thing. Thus there are schemes for defining and categorizing the degree of risk faced by a species. Countries tend to develop their own definitions, and then there are international ones such as those of the International Union for the Conservation of Nature and Natural Resources (IUCN) and the Convention on International Trade in Endangered Species of Wild Fauna and Flora (CITES) (Chapter 12). We have been content to follow definitions in *Webster's New Collegiate Dictionary* for terms such as "extinct" and "endangered." And for that reason, we have resisted using the term "extirpated." Its current use in endangered species legislation in Canada and the United States restricts it to the extinction of a population. That is at variance with its etymology, which is to pull up by the root, and conveys complete eradication.

1.5.1   The species—a pragmatic unit of conservation

We favor the species as the unit of conservation because it is biologically a natural unit and because the idea of a species can be grasped intuitively by most people despite debate over the species concept (e.g., see Shrader-Frechette and McCoy 1993). It is the most important unit of evolution and the only taxonomic grouping that usually can be defined with little ambiguity. It also causes fewer problems in legal proceedings, and that is important. Biologically a *species* is a group of populations that are reproductively isolated. In practice we rely on an array of morphological, behavioral, or biochemical criteria that index ecological distinctions.

Subspecies (races) are local populations with genetically based traits grouped geographically or by habitat, but are not reproductively isolated. Occasionally and especially in the context of listings for endangered species legislation, the concept of biological species and subspecies becomes lost. We defer further discussion until Chapter 12, which deals with the listing of endangered species and subspecies.

The species, although the most important unit of conservation, is not necessarily the unit of management acted upon to achieve the desired result. That unit is the population. A few species such as point endemics, for example, constitute a single population, but most species have a number of populations separated from each other by unfavorable habitat or their behavior. These are called *local populations* or *subpopulations*, and they are largely closed breeding groups. When we consider the dynamics of populations and the gene interchange between them, we refer to the complex as a *metapopulation* (Section 5.6).

One objection to using species as the unit of conservation is that it ignores other species in the community. If a treatment for a species in trouble has unexpected effects on other species in the assemblage, then the problem was more a failure to think through ecological relationships. Using the species as the unit of conservation does not mean considering the species in isolation. In Section 12.3, we take up this topic again as it relates to recovery plans.

We caution against "ecosystem" or "community" as units of conservation because they cannot be defined unambiguously. Using these words usually reflects a desire to protect "habitat," but that is more easily achieved by defining the unit of conservation as the species whose habitat is threatened. However, we recognize that others have differing opinions on this (McIntyre et al 1992).

We use "ecosystem" for a multispecies assemblage without worrying too much about strict definition. Even a glance at the 16 possible definitions of ecosystem and then further defining terms implicit in them conveys the difficulties (Shrader-Frechette and McCoy 1993). We could equally have used "community" or "association." These assemblages are, according to what school you went to, either collections of species having the same environmental requirements and subject to the same environmental constraints and thus living in the same area, or they are coevolved taxa forming an holistic evolutionary unit. We will keep clear of that particular debate and refer instead to Underwood's (1986a) discussion of what constitutes a community.

The problem with declaring ecosystems units of conservation is the difficulty of knowing just what we are dealing with. A multispecies complex can range, on one hand, from a highly predictable set with relatively tight proportions (e.g., boreal forest) to, on the other hand, a mix of species which, although having a recognizable form, is highly variable in its species composition and their relative proportions (e.g., eucalyptus forest). At what disparity from type does such an ecosystem become another ecosystem?

At the other extreme is the championing of the gene as the unit of conservation and promoting *genetic diversity* as the number of distinct alleles in the population. Cases have been advanced for adopting the retention of rare alleles as the aim of conservation, the rationale being that they may be necessary for the future evolution species in a changing environment. Apart from this requiring the cancellation of normalizing selection, the notion is so transcendental that it has little practical application. More useful is *genetic variance*, which can be conceptualized as closely akin to heterozygosity (the proportion of loci that are heterozygous in an average individual; Section 6.3). The amount of genetic variance carried by a relatively small sample of the population will closely approximate the magnitude of the genetic variance of that population as a whole. We seek to conserve both the variance and the total number of distinct alleles within the population.

There is a more serious reason for declaring the species the fundamental unit of conservation. We can recreate a lost ecosystem from its component species shared by other ecosystems. We can reconstitute a subspecies from homologous genes hosted by its sister subspecies. But a species differs in kind. We cannot resurrect a species. Once a species is extinct, it is gone forever.

1.6   THE EFFECT OF POLITICAL BOUNDARIES ON CONSERVATION

The eastern barred bandicoot, *Perameles gunnii*, lives in Tasmania and Victoria, Australia (Figure 1.1). It is common in Tasmania but restricted to a single remnant population in Victoria, under whose state law it is declared endangered. This is clearly anomalous. The species, as such, is not endangered. The only thing endangered is one of its populations. Similar cases could be presented where a common species is declared endangered in a part of its range because its distribution is crossed by an artificial and arbitrary political boundary.

Such cases reflect a common characteristic of politics: the belief that the world stops at the border. Countries and states tend to view their biota in the context only of their particular patch of the earth's surface and enact laws accordingly. At the national level a country may seek to evaluate if a population is nationally significant because having only a single "representative" population is not necessarily serving the best interests of conservation but neither is listing every population. On the other hand, international listings such as those used by IUCN or CITES (Chapter 12) try to avoid split listings for individual species (some populations listed as endangered but not others), and species are evaluated on their global status.

Recovery actions have then to be cognizant of a hierarchy of scale from local to global. As an example, it may not be appropriate for Victoria to spend heavily on the eastern barred bandicoot's conservation because the species is not globally endangered. But some expenditure is appropriate to meet the wishes of Victoria's residents to see this species restored locally. Any system for the designating species as endangered needs thoughtful and informed application to take the question of scale into account. We can otherwise waste time devising rigid definitions and then contorting them to deal with all possible cases, be they global, national, or whatever.

Figure 1.1. Map of Asia and the western Pacific including Australia.

1.7 SUMMARY

Conservation biology as an independent field is the progeny of wildlife management on one hand and a particular set of theories of population genetics and small populations on the other. Our emphasis in this book on diagnosing and treating a population in trouble is an amalgam of those two paradigms.

The detection of a species in trouble, the diagnosis of its problems, and their treatment should follow the steps inherent in scientific methods if the results are to be unambiguous and defensible. The most efficient way to test the hypothesis is through an experimental design. A tight design has replicated levels of treatment to avoid confounding site and treatment effects, and the minimum is two levels of treatment, of which one level is zero, to determine changes in the response variables. The sampling of the response variable should be random and scaled to the expected pattern of variation.

When the null hypothesis is that a decline has not occurred or that a treatment has no effect on a species in trouble, the probability of accepting a null hypothesis when it is false (Type 2 error) is more serious than rejecting a null hypothesis that is correct (Type 1 error). Sampling design, especially increasing sampling intensity, will contribute to reducing the probability of a Type 2 error. Preliminary sampling and determining the precision needed to increase the power of the statistical test will reduce the probability of incorrect support for or rejection of null hypotheses. Difficulties from small sample sizes can be partially compensated for by experimental design and computer-intensive sampling techniques. The final step is to ensure the results, whether successful or not, are published.

We chose the species as the most important unit of conservation because it is the unit of evolution and the least ambiguous to define. Most species occur as populations—groups of freely breeding individuals. Individuals moving between those populations may link them together as a metapopulation. As the specics is the unit of conservation, the population is usually the unit for management action. Practical problems of defining ecosystems and communities impede their use as definitive units in conservation. Species are often distributed across political boundaries, which can lead to incongruities—a species rare in one jurisdiction is common in another. How species are ranked according to the degree of risk facing them then becomes a question of scale. What is appropriate at one level of jurisdiction (state or national) may not be at the next level (regional or global).

CHAPTER 2

Extinctions in prehistory

Although the themes of this book are diagnosing and treating modern species in trouble, we approach both in the context of extinction as an evolutionary phenomenon. We will examine the evidence, asking whether there is enough to draw firm conclusions about the processes and causes of extinctions. Table 2.1 gives the geological time scale. With data becoming progressively less abundant as one goes back in time, the overview is unavoidably biased by extinctions of the past 10,000 years, for which evidence is most abundant.

2.1   THE BACKGROUND RATE OF EXTINCTION

When dealing with extinctions in the remote past, we emphasize species extinctions that are components of mass extinction episodes, for the simple reason that their causes should be easier to diagnose. General effects are seldom the product of special causes. To identify such a phenomenon one must show that the extinctions occurred at a rate far above the background rate of extinction. By the background

Table 2.1.   The geological time scale

Era	Period	Epoch	Millions of years to beginning of interval
Quaternary		Recent	0.010
		Pleistocene	2.5
Tertiary		Pliocene	7
		Miocene	26
		Oligocene	38
		Eocene	54
		Paleocene	65
Mesozoic	Cretaceous		136
	Jurassic		190
	Triassic		225
Paleozoic	Permian		280
	Pennsylvanian / Mississippian	Carboniferous	345
	Devonian		395
	Silurian		430
	Ordovician		500
	Cambrian		590
Precambrian			650

rate we mean the number of species disappearing per unit time, which for many reasons is decidedly difficult to calculate. First, it must come from the fossil record, which becomes progressively fragmented as we move back in time. It is so imperfect that it leads to the second point. While we would like the unit of extinction to be the species, species are notoriously difficult to identify from often poorly preserved and in-complete fossilized skeletons, so we have to use the genus to run up a reasonable tally. It removes us a good step away from the biological phenomenon of species extinction. Third, we need accurate timings, but dating geological strata, other than for recent volcanic events, tends to be rough. At best then, calculating the background rate of extinctions can only be crude. Yet it is worthwhile because it may help identify those extinctions that were human induced, evaluate the role people played, and indicate lines for treatment in those species close to extinc-tion now.

Some of the most reliable information on extinctions when humans were absent comes from European and North American land mammals from the last part of the Tertiary and into the Pleistocene, the last 9 million years. We have used Webb's (1984) compilation to estimate extinction rates of genera per geological stage and per million years (Table 2.2). The Rancholabrean stage (see Section 2.3.4) must be ex-cluded from the calculation of background rate because its end coin-cides with people from Asia colonizing the Americas. The stratigraphic stages spanning the period from the base of the Hemphillian to the top of the Irvingtonian return estimates of the proportion of genera dying out per stage ranging from 0.08 for the late Irvingtonian to 0.55 for the late Hemphillian.

Those raw extinction proportions, p_E, must be scaled to correct for the stages differing in duration. To do so we calculate an instantaneous rate of extinction, d, for each stage from the relationship.

$$p_E = 1 - e^{-dt}$$

where t is the duration of the stage in millions of years. Rearranging the equation gives

$$d = \frac{-\ln(1 - p_E)}{t}$$

The instantaneous rate of extinction on a million-year basis is thus calculated for each stage and converted back to a finite rate of extinc-tion per million years p_{Em} by

$$p_{Em} = 1 - e^{-d}$$

The final column of Table 2.2 gives those corrected rates.

Table 2.2. Rates of extinction of genera of land mammals in North America

Period	Duration (millions of years)	End (millions of years)	Genera of mammals present	Genera of mammals extinguished	Proportion extinct	Proportion extinct per million years[a]
Rancholabrean	0.5	0.01	129	43	0.33	0.66
Late Irvingtonian	0.5	0.5	114	9	0.08	0.16
Early Irvingtonian	1.0	1.0	103	10	0.10	0.10
Late Blancan	1.0	2.0	109	35	0.32	0.32
Early Blancan	1.5	3.0	92	9	0.10	0.07
Late Hemphillian	1.5	4.5	112	62	0.55	0.37
Early Hemphilian	3.0	6.0	115	27	0.23	0.08

[a] Rates corrected for duration of each stage (see Section 2.1).
Modified from Webb SD. Ten million years of mammalian extinctions in North America. In: Martin PS, Klein RG, eds. Quaternary extinctions. Tucson: University of Arizona Press, 1984:189 210.

Although these figures may well be some of the best available, they are rough. We see no utility in subjecting them to further analysis and simply present their mean, $\bar{d} = 0.1925$, which converts to a finite rate of extinction of $\bar{p}_{Em} = 0.175$ of the standing number of genera per million years. It implies a rate of loss of genera per thousand years, a period of time more closely aligned with our sense of history, of around 0.02%. We could take this further and, using an arbitrary conversion of ×3 for multiplying loss of genera up to loss of species, estimate the background rate as 0.06% per thousand years. Extinction rates for mammals were higher than those for the mollusks (Stanley 1979), which makes the point that estimated rates of background extinction should be viewed as specific to each major taxon.

2.2 MASS EXTINCTIONS IN GEOLOGICAL TIME

Extinctions seem to have occurred at a fairly steady rate throughout geological time. The number of genera that we know went extinct, per million years, rises as we approach our own time, but that probably reflects only the relationship between age and stratigraphic representation. The sedimentary record becomes exponentially less complete with time (Sadler 1981). These extinctions provide little or no information on why they occurred, but at least they inform us that extinction is a continuous and continuing phenomenon.

Superimposed upon that steady erosion of genera and species, the gaps being plugged almost immediately by new genera and species, are a set of mass extinctions. These are more interesting to ecologists because they suggest that a single large and widespread "something" was involved. It is these mass extinctions, both long ago and recent, that allow some chance of divining cause. The pre-Pleistocene mass extinctions have been reviewed in volumes edited by Donovan (1989) and Briggs and Crowther (1990). We have drawn our summary from Eldredge's (1987, 1991) and Stanley's (1987) overviews.

Mass extinctions tend to occur at the boundary between two geological time periods, simply because time–stratigraphic boundaries are defined by a sudden change in index fossils. Some 22% of marine families went extinct during the Ordovician–Silurian episode 440 m.y. B.P. ("before present" in radiocarbon terms in 1950). European strata are more complete than the North American strata near the boundary, and suggest that the episode came in two pulses separated by 1 or 2 million years.

The mass extinction at the Devonian–Mississippian (Carboniferous) boundary (360 m.y. B.P.) did not occur quite at the boundary of the Upper Devonian but between the last two of its subdivisions: 86% of marine brachiopods in the Frasnian were extinct by the Famennian. Corals were also hard hit. The record suggests that this episode may have stretched over 3 million years. A plausible cause is a worldwide drop in temperature. The surviving taxa appear to be mainly cold-water forms.

The most dramatic extinction episode was at the Permian–Triassic boundary. At the end of the Paleozoic (245 m.y. B.P.), 54% of all marine families and 83% of their genera failed to cross the boundary into the Mesozoic. The brachiopods were almost completely wiped out, as were the tropical coral reefs. Modern reefs are built by different corals. Extinctions on land were less severe (and more difficult to detect) but apparently still significant. The episode appears to have spanned several million years and may have been caused, at least in part, by the reduction of coastline as the continental plates coalesced into a single continent (Pangea). Erwin (1993) sifts evidence for this and concluded that it was only one of several factors which combined to have devasting effects on the fauna and flora.

The mass extinction at the Triassic–Jurassic boundary (210 m.y. B.P.) saw the end of about 20% of existing families. On land several reptilian lines came to an end and the mammal-like reptiles were severely pruned. In the sea the sponges, gastropods, bivalves, brachiopods, and ammonites were hard hit. The extinctions possibly spanned 10 or more million years.

The extinctions at the Cretaceous–Tertiary boundary (65 m.y. B.P.) are not the most numerous but they are certainly the most famous. They marked the end of the dinosaurs. Many other groups of plants and animals, on the land and in the sea, were severely thinned out. A number of those groups had been declining throughout the Upper Cretaceous; the dinosaurs of North America, for example, had declined from 30 genera 10 million years before to seven genera at the boundary; nonetheless a large number of plants and animals appear to have gone extinct right at the boundary. Theories abound and a much promoted cause of this catastrophic event was earth's collision with a comet or asteroid.

What do these mass extinctions teach us about the extinctions going on today? The answer is quite a lot and remarkably little. On the positive side they remind us that extinction is a natural process no less real than speciation. They show that the human hand is not a necessary appendage of the process. On the negative side, the evidence that can be extracted from rocks is so imperfect that the timing and duration of these extinction episodes can seldom be determined with any accuracy. If we cannot determine just what happened, and if we have no access to a tight chronology of the event or events, we can seldom do more than speculate as to the cause. We will run into that problem again when we look at the Pleistocene extinctions (Section 2.3).

2.3   LATE PLEISTOCENE MASS EXTINCTIONS

Over the last 50,000 years or so there have been a number of extinction events. They were restricted almost entirely to land mammals, a few birds, and even fewer reptiles. Large mammals were more vulnerable than small mammals (in contrast to early Pleistocene background extinction), and the occurrence of these events varied across the world in timing and duration. A monumental volume edited by Martin and Klein (1984) describes these extinction episodes and canvassed theories and models advanced to account for them. Here we summarize the phenomena and discuss the difficulty of reaching a consensus as to the cause.

2.3.1   *Extinctions in Africa*

In a sense Africa acts as a control on mass extinctions elsewhere. It lost and gained genera of mammals throughout the Pleistocene, entering the Holocene after 10 k.y. B.P. with the fauna essentially intact. There was no mass extinction as occurred elsewhere. Table 2.3 shows the

Table 2.3. Loss of genera of large (body weight >44 kg)
African mammals during the Pleistocene

	Lower Pleistocene (1,800–700 × 10^3 B.P.)	Middle Pleistocene (700–130 × 10^3 B.P.)	Upper Pleistocene (130–0 × 10^3 B.P.)
Duration	$1,100 \times 10^3$	570×10^3	130×10^3
Genera extinct	21	9	7
Loss per 10^3 years	0.019	0.016	0.054

Modified with permission from Martin PS. Prehistoric overkill: the global
model. In: Martin PS, Klein RG, eds. Quaternary extinctions. Tucson:
University of Arizona Press, 1984:383.

Table 2.4. Loss of genera of small (body weight <44 kg) and
large (body weight >44 kg) European mammals during the Pleistocene

	Villafranchian					Middle Pleistocene						Late Pleistocene			
	a	b	c	d	e	A	1(i)	B	1(ii)	C	2	D	3	F	4
>44 kg genera	2	2	0	2	5	2	0	0	3	4	2	1	0	1	13
<44 kg genera	1	1	1	1	2	1	2	0	5	3	1	4	0	0	4

a–e = Villafranchian phases; 1–4 = glaciations; A–F = interglacials.

Modified with permission from Martin PS. Prehistoric overkill: the global model. In: Martin PS, Klein
RG, eds. Quaternary extinctions. Tucson: University of Arizona Press, 1984:384.

large-bodied animals (body weight >44 kg) lost during the early,
middle, and late Pleistocene. One of the losses in the lower Pleistocene
(*Mammuthus*) and two in the upper Pleistocene (*Elephus* and *Camelus*)
were of genera that continued to survive elsewhere. In addition, 10
genera of smaller mammals were lost in the lower Pleistocene. There
seems to have been few losses from that weight range thereafter,
although that may be a sampling artifact. In summary, the loss of
mammalian genera in Africa appears to have been relatively low and
relatively steady throughout the Pleistocene.

2.3.2 Extinctions in Europe

Table 2.4, showing the loss of mammalian genera from Europe, is taken
from Martin (1984), which itself is an adaptation of Kurtén (1968). It
uses the European classification of Pleistocene phases. The loss of the
larger mammals is fairly uniform across the Pleistocene, but with a
modest spike of extinctions at its end. Loss of genera of smaller mammals shows no time trend.

2.3.3 Extinctions in Australia

Australia suffered a massive spike of extinctions: 86% of the genera of large animals, including the large extinct reptiles and the giant birds, went extinct in the late Pleistocene. The chronology is loose because many deposits have been mixed and reworked (Murray 1984), and the bones are beyond the range of ^{14}C dating for collagen. A few smaller species hung on until about 16,000 B.P., but the bulk of the megafauna seems to have died out about 30,000 B.P., with a few forms persisting until 25,000 or even 20,000 B.P. The extinction spike may have been delayed by 5,000 years in Tasmania (Murray 1984).

Few reptiles disappeared, but among those that did some were impressive in size. The largest varanid lizard, *Megalania prisca*, was, at 5.5 m, double the length of the extant Komodo dragon, *Varanus komodensis*. The large python, *Wonambi*, and the horned turtle, *Meiolania*—whose shell was 1 m long—also disappeared 30,000 to 50,000 B.P.

2.3.4 Extinctions in North America

As shown in Table 2.2, the North American mammalian fauna suffered extinction of 33% of its genera during the Rancholabrean phase of the upper Pleistocene. It converts to a rate of 56% of genera per million years, the highest rate for any division of the upper Tertiary or Pleistocene. The rate of extinction is higher than those figures indicate because of 43 genera (Webb 1984) that died out, 37 extinguished in the Wisconsinan period (a subset of the Rancholabrean), which lasted only about 100,000 years. Tighter still, the vast bulk of that die-off occurred in the interval 10,000–12,000 B.P. (Mead and Meltzer 1984). Four of those 37 extinguishing genera were mammals weighing less than 44 kg. The pattern of weight distribution is entirely different from that of the extinctions earlier in the Pleistocene (Martin 1984). Only three reptiles disappeared during this time: a large tortoise, *Geochelone* sp., a large rattlesnake, *Crotalus* sp., and a horned lizard, *Phrynosoma* sp. (Case et al 1992).

2.3.5 Extinctions in South America

The late Pleistocene mammalian fossil record and its dating are less well developed in South America. Nonetheless, it appears to follow the pattern of the North American record of extinctions. Table 2.5 shows that the frequency of extinctions through the lower and middle Pleistocene is low but accelerates in the upper Pleistocene. That sudden burst of extinctions was limited to genera weighing more than 44 kg.

Table 2.5. Loss of genera of small (body weight <44 kg) and large (body weight >44 kg) South American mammals during the Pleistocene

	Uquian (2,000–1,000 × 10^3 B.P.)	Ensenadan (1,000–300 × 10^3 B.P.)	Lujanian (300–0 × 10^3 B.P.)
>44 kg genera	7	3	46
<44 kg genera	14	5	0

Modified with permission from Martin PS. Prehistoric overkill: the global model. In: Martin PS, Klein RG, eds. Quaternary extinctions. Tucson: University of Arizona Press, 1984:372–373. Also modified from Marshall LG. The great American interchange: an invasion induced crisis for South American mammals. In: Nitecki MH, ed. Biotic crisis in ecological and evolutionary time. London: Academic Press, 1981:133–229.

2.3.6 Causes of the late Pleistocene extinctions

There are two major interpretations purporting to account for the late Pleistocene extinctions. The first postulates that they were caused by man, either directly or indirectly; the second reckons that a change of climate was responsible. We do not adjudicate between them or their combined effect because the evidence available at present does not lead to an unequivocal verdict.

The climatic interpretation (for)

The North American extinction episode occurred at the end of the Pleistocene when the temperatures were rising. According to Graham and Lundelius (1984) and Guthrie (1984), this led to a change in plant communities which differed from their form throughout the Wisconsin (Würm) Glaciation and indeed from their form today. That disrupted the plant–herbivore grazing systems, resulting in the extinction of many herbivores, particularly the large ones, and the predators that depended on them. In a similar vein, Horton (1984) postulated that the Australian extinctions were caused by a change in climate, this one being an increased frequency of droughts.

The climatic interpretation (against)

The climatic interpretation requires a set of special events to account for a general effect. Thus, separate cases must be made for the noncontemporaneous extinction episodes in North America and Australia and, to a lesser extent, to explain the apparently synchronous extinctions in both Americas over several climatic zones. (Note, however, that Horton [1984] advances this as a strength of the climatic interpretation, suggesting that climatic asynchrony is to be expected.) Further, it accounts poorly for the lower rates of extinction in Eurasia and Africa except in terms of highly regionalized climatic change.

The climatic interpretation requires that species which survived the numerous marked climatic fluctuations of the Pleistocene glacials and interglacials were ambushed by climatic changes of lower magnitude during the Wisconsin Glaciation (Australia) and at the end of it (the Americas).

The human-impact interpretation (for)

By this interpretation people caused the late Pleistocene extinctions by hunting populations, or by changing their habitats, or both. Thus, the expected pattern is for numerous extinctions to coincide with the initial colonization of a landmass or by a more modest rate of extinction where people and animals evolved together. The extinctions must necessarily be asynchronous between the Americas and Australia because their colonization by people was asynchronous.

The human-impact interpretation (against)

The human-impact interpretation depends principally on a tight association between the arrival of people and the departure of species in Australia and the Americas. So far that tight chronology has not been forthcoming.

The North American chronology of extinction has tightened up over the last 20 years to a generally accepted 12,000 to 10,000 B.P. That for South America still rests on too few dates, seldom replicated by strata, from too few species, but it is generally recognized as not differing greatly from the more rigorous chronology of North America. Mass extinctions in the Americas lie just within reach of ^{14}C dating of bone collagen, thereby allowing direct estimating of date of death rather than having to use associated charcoal or plant material. However, most, if not all, of the collagen has dissipated from the bone after 10,000 years and the residue extracted from the bone by the standard acid treatment is likely to be other organics less amenable to dating. False dates, usually too young, often result from dating "collagen" out beyond about 8,000 years (Taylor 1980).

The chronology of colonization has a continuing history of emergent dates, suggesting colonization well before the extinctions, rudely challenged, stoutly defended, often subsequently invalidated, and seldom replicated. The latest dates, anomalous with respect to the human-impact interpretation, are mainly from South America but lack verification from farther north.

The Australian chronology is in a worse state. The colonization and extinctions are around three times older than the American equivalents, and it shows. The stratigraphy is primitive, the dating is indirect and often based on methods of dubious utility, and replication (as opposed to pseudoreplication) is seldom used. The deduced dates of colonization range from 120,000 to 35,000 B.P. and the dates of extinction from >30,000 to 6,000 B.P. (Murray 1984). A consensus as to precise timing has not yet begun to emerge for either. The only identifiable agreement is that megafauna and people overlapped in time, but there is no direct evidence, such as would be provided by midden material or spear points in *Diprotodon* ribs, that the people hunted the large extinct mammals.

2.4   LATE HOLOCENE MASS EXTINCTIONS

Mass extinctions occurred during the second half of the Holocene (i.e., after 5,000 B.P.) on Madagascar, New Zealand, the Pacific oceanic islands, and on the western Pacific fringing archipelagoes. Because they are relatively recent, they present fewer problems of stratigraphy and dating than do the Pleistocene extinctions. They are thus ideal candidates for testing the logic of diagnostic procedures that will be used on extant species currently under threat.

For reasons of space and the coverage of information available, we have neglected some geographic areas. In Section 4.17, we touch on the loss of mammals within the last few thousand years when Amerindians reached the Caribbean. Otherwise we leave it to others to trace the chronology of species loss and their purported causes in the Caribbean, the Mediterranean, and elsewhere.

2.4.1   *The problems of dating*

Section 2.3.6 touched briefly on the problems of dating extinctions and colonizations. The technical difficulties increase exponentially with time (Sadler 1981). A standard radiocarbon date looks like this: 740 ± 35 B.P. One could be excused for assuming that the dated event occurred around 740 years ago and certainly between 705 and 775 years before present.

That ±35 years confidence interval relates to the statistical sampling problems faced by a physicist estimating the rate of radioactive decay. It is the standard error of the number of β-emissions per unit time per unit of carbon mass translated into a number of years according to the half-life of ^{14}C (about 5,730 years). The bigger the specimen the smaller

Table 2.6. Radiocarbon dates from the main occupation layer of the Tairua moa-hunter site in New Zealand

Material	Laboratory number	Estimated age B.P.[a]
Moa-bone carbonate	NZ 558	0
Moa-bone carbonate	NZ 751	0
Moa-bone carbonate	NZ 765	0
Moa-bone collagen	NZ 766	393 ± 37
Marine shell	NZ 749	485 ± 32
Moa-bone collagen	NZ 559	503 ± 32
Moa-bone collagen	NZ 578	503 ± 32
Moa-bone collagen	NZ 752	543 ± 32
Charcoal	NZ 750	831 ± 33

[a] Dates determined in 1965 and 1966 are given in terms of the "old" half-life uncorrected for secular variation.
Modified from Trotter MM. Tai Rua: a moa-hunter site in North Otago. In: Anderson A, ed. Birds of a feather. British Archaeological Reports 62 and New Zealand Archaeological Association Monograph II, 1979:203–230.

is the standard error; the longer it is monitored, the smaller the standard error. The conformity between the estimated date and the timing of the event it purports to date has nothing much to do with those things.

Table 2.6 shows a fairly typical set of dates from an occupation layer containing human artifacts and the bones of extinct megafauna. They range between 0 B.P., as determined from bone carbonate, and 831 B.P., as determined from charcoal. Yet these nine dates came from a single thin layer that apparently accumulated over a short period of time (Trotter 1979). This example makes the important point that dates must be replicated. A single date is useless.

The explanation for the spread in the dates in Table 2.6 is that the carbonate dates are obviously wrong; in fact, bone carbonate is no longer used for dating. At the other end of the range the charcoal date is suspiciously old. Charcoal does not date the fire but the death of the plant cells. If it came from driftwood or the heartwood of an ancient tree, it could be significantly older than its date of deposition. Box 2.1 gives the most common sources of dating errors.

The precision of radiocarbon dates is best investigated by examining the spread of dates from an event of negligible duration. An explosive volcanic eruption is an obvious candidate. If the standard errors attached to the dates by a dating laboratory measured the intrinsic precision of those dates, then the mean of the standard errors would, within the limits of sampling variation, be the same as the standard deviation of the replicate dates. For such comparisons we work with variances rather than with standard errors or standard deviations. The variance

---

Box 2.1. The most common sources of dating inaccuracies

---

1. False calibration of the test results to absolute years (common in new methods)
2. Contamination of the sample by old (i.e., radioactively dead) carbon (important only when the sample is <1,000 B.P.)
3. Contamination by modern or atomic bomb carbon
4. False association, when the date of death of the specimen differs from the date of the event that specimen is supposed to date:
   a. Redeposition (common with wildfire carbon washed into streams and swamps)
   b. Inclusion (including human burial, deposition in holes in an older stratum, the treading into a stratum of a specimen lying on its surface, and material carried into the stratum by burrowing animals)
5. Mixing in place
6. Contemporaneous deposition but different radiocarbon ages (e.g., driftwood and heartwood used to date a cooking fire; these date not the fire but the death of the plant cells)
7. Assuming that residual extracted by acid from bone is in fact collagen (often leads to underestimates for specimens around 10,000 B.P.)

---

of each date is the square of its standard error, and so the mean of their variances V_e, called the statistical error variance, is

$$V_e = \frac{\sum_n SE^2}{n}$$

where n is the number of dates in the sample.

The total variance V of the dates, each date being labeled x, is

$$V = \frac{\sum x^2 - (\sum x^2)/n}{n-1}$$

V_e equals V within the limits of sampling variation if there is no contamination, inbuilt age, false association, interval of deposition, or fumble-finger in the dating laboratory. In the event of $V > V_e$ the difference between the two is the "nonstatistical variance," which

Table 2.7. Dating of the essentially instantaneous Taupo (New Zealand) and Thíra (Santorini) (Mediterranean) pumice eruptions

Statistic	Taupo	Thíra
Number of dates	29	17
Unweighted mean age B.P.	1,944	3,435
Total variance of dates V	57,754	130,344
Statistical error variance V_e	10,083	8,833
Technical error variance V_T [C–D]	47,671	122,013
Technical error SD [$\sqrt{E}$]	218	349
Ratio [F/B]	0.11	0.10

Modified with permission from Caughley G. The colonisation of New Zealand by the Polynesians. J. R. Soc. N. Zealand 1988;18:270. Also modified from Hammer CU, Clausen HB, Friedrich WL, and Tauber H. The Minoan eruption of Santorini in Greece dated to 1645 BC? Nature 1987;328:517–519.

may itself be divided into technical error variance V_T and interval of deposition variance V_I.

Table 2.7 summarizes a tale of two volcanic eruptions which were essentially instantaneous. The nonstatistical error variance (E) contains only technical error because there is no contribution from an interval of deposition. In both cases the technical error standard deviation is about a tenth of the mean of the unweighted dates. If that holds generally it tells us something important about the spread of radiocarbon dates. Suppose we wished to date a stratum containing the skeletons of mammoths, *Mammuthus* spp., that (unbeknown to us) all died on the same day about 10,000 years ago. We take 100 samples at random and date them thoroughly so that each date has a negligible standard error. The tightest 95 of those dates would have a spread of 4,000 years (4 × 10,000 × 0.1), which might lead us to specious conclusions about what happened.

2.4.2 Madagascar

Extinctions

At the beginning of the Christian era, Madagascar hosted 17 genera of lemurs, *Lemuridae*. About 1,000 years ago seven genera had become extinct (Dewar 1984). Those lost genera were relatively large species exceeding—with one possible exception—the size of the species that survive to this day. The extinct species were probably all diurnal, and several were terrestrial quadrupeds. In contrast, the species surviving this episode of mass extinction are either nocturnal or small diurnal arboreal lemurs (Walker 1976).

Three nonlemuroid mammals were lost: a pygmy hippo, *Hippopotamas lemeriei*; a large viverrid, *Cryptoprocta spelea*; and an endemic aardvark, *Plesiorcteropus madagascariensis*. Two species of giant land tortoises, *Geochelone*, also went. The elephant birds of Madagascar (very likely the *roc* of *Arabian Nights*) died out at the same time.

Dewar's (1984) listing of 20 radiocarbon dates for the extinct megafauna shows a dearth of replication within sites. Some of the dates are from materials known to generate an erratic chronology. Nonetheless they suggest that at least a proportion of the megafauna species were extant during the first millennium A.D. and may have overlapped with people.

Human colonization

Madagascar was colonized not from Africa but from Indonesia. The language closest to Malagasy is found in Borneo (Dewar 1984). Thus, the arrival of people in Madagascar resulted from a late westward eddy of the Austronesian easterly dispersal wave that colonized the Pacific (Section 2.4.4). Radiocarbon dating suggests that the island was widely settled by A.D. 1000 (Dewar 1984). The dates before that are from archaeological sites, but are poorly replicated by site and sometimes of dubious association with artifacts. Initial colonization was probably not long before A.D. 900.

Causes of extinction

Dewar (1984) reviewed the hypotheses and their evidence for the Madagascan mass extinction. The climatic interpretation, whereby a shift in climate changed habitats to the detriment of many species, had some currency in the 1970s, but the critical evidence from pollen profiles is not available from Madagascar. The nearest useful profiles, from east Africa, show no significant climatic shift over the first millennium A.D. The only recent variant of this interpretation is by Eldredge (1991). He attributed the extinctions to habitat change wrought by "natural climatic shifts," which were exacerbated by the clearing of vegetation by people. The low rate of loss of reptiles, however, tells against the role of climate and habitat change.

The anthropogenic interpretation requires that the now extinct megafauna overlapped in time with the colonizing people. The six "mixed" sites, those where megafaunal remains and human artifacts are intermingled, individually did not provide knockdown evidence of temporal overlap (Dewar 1984). Their megafaunal remains could have been reworked subfossil bones. Together those sites are persuasive evidence, taken with the overlap of radiocarbon dates, that people and megafauna overlapped in time. As to the mechanism of extinction, if indeed it was anthropogenic, the current Madagascan data are too sparse to support extinction by hunting as against people altering habitats, or either of those against their combination.

2.4.3   New Zealand

New Zealand, in contrast to Madagascar, provides multitudinous sub-fossils and cultural artifacts with time–stratigraphic control. Since the mass extinction occurred there within the last 1,000 years, it allows a close examination of what occurred.

Extinctions

New Zealand is an isolated piece of continental crust 270,000 km^2 in area (Figure 2.1). Before the Polynesians arrived about A.D. 900, give or take a century (Anderson 1989), there were 125 species of land and freshwater birds (the precise number varying according to classification) in New Zealand. The number of species had declined to 89 by the

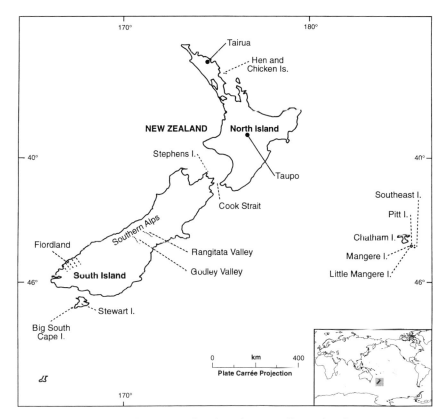

Figure 2.1.   Map of New Zealand and its satellite islands.

time the first European stepped ashore (*Endeavour*, Captain James Cook, 1769), and since then it has declined to 81 (Holdaway 1989). The losses comprised six families in their entirety, 32 genera in their entirety, and 44 species.

Evidence for the composition of the pre-Polynesian avifauna is from subfossil deposits (i.e., bones so young or soft that they are not mineralized, as are conventional fossils). Their age is Holocene (the last 10,000 years).

The pre-European extinctions included 11 species of flightless moas (*Dinornithiformes*) ranging in average weight from 42 kg (*Megalapteryx didinus*) to 178 kg (*Dinornis giganteus*), large individuals of the latter reaching 270 kg (Worthy 1988; Atkinson and Greenwood 1989). But the extinctions were not restricted to large birds. The three species of small flightless New Zealand wrens (*Xenicidae*) that went extinct about the same time would have weighed less than 100 g. This extinction episode included rails, ducks, three raptors, a goose, a swan, a crow, a nightjar, and at least three species of *Liopelma* frogs, which perhaps might have died out within the last 200 years, and hence represent post-European extinctions (Cooper et al 1993). Three of 38 reptile species disappeared

Table 2.8.   The extinct New Zealand megafauna (i.e., >10 kg body weight)

Species	Range	Deduced habitat	Weight (kg)
Moas			
Megalapteryx didinus	SI	Upland, montane/subalpine	42
Anomalopteryx didiformis	NI, SI	Lowland wet forest	42
Pachyornis mappini	NI	Lowland wet shrub-forest mosaics	38
P. australis	SI	Wet montane forest/subalpine	75
P. elephantopus	SI	Grass–shrub–forest mosaic	130
Emeus crassus	SI	Coastal dry shrubland and open forest	72
Euryapteryx curtus	NI	Dry lowland shrub/open forest	32
E. geranoides	NI, SI	Dry inland low shrub/open forest	90
Dinornis struthoides	NI, SI	Lowland to subalpine	83
D. novaezealandiae	NI, SI	Lowland to subalpine, wet areas	155
D. giganteus	NI, SI	Lowland shrub/open forest	315
Large "rail"			
Aptornis otidiformis	NI, SI	Forest	11
Eagle			
Harpagornis moorei	NI, SI	Forest and forest edges	12
Flightless goose			
Cnemiornis calcitrans	NI, SI	Forest–scrub–grassland mosaic	13

NI = North Island; SI = South Island.
Modified with permission from Cooper A, Atkinson IAE, Lee WG, and Worthy TH. Evolution of the moas and their effect on the New Zealand flora. Trends Ecol. Evol. 1993;8:435. Also modified from Holdaway RN. New Zealand's pre-human avifauna and its vulnerability. N. Zealand J. Ecol. 1989;12(suppl.):11–25.

completely, and nine species lost from the North Island and South Island are now found only on offshore islands (Case et al 1992).

Despite the abundance of subfossil deposits, no unequivocal evidence attests to a species dying out over the 70,000 years before the arrival of the Polynesians. Fossils and subfossils predating the last glaciation are uncommon and limited almost entirely to moa bones in marine shelf sediments of upper Pliocene and Pleistocene age (the last 2.5 million years). Nonetheless, there are 40 such moa fossils, all of which are referable to modern species (Worthy et al 1991). The extinction episode appears to have begun suddenly in the late Holocene with few, if any, precursory losses during the Pleistocene or lower Holocene. Table 2.8 lists the extinct megafauna.

Table 2.9 gives the figures allowing calculation of the rate of loss of New Zealand landbirds. We use the same technique to compare rates over different periods, as was introduced in Section 2.1. Correcting the proportions of species going extinct for the differing intervals of time (Table 2.9) reveals that the extinction rate of 0.0383 of the avifauna per century between A.D. 900 and 1769 is very close to the rate of 0.0411 from 1769 to present. In the sense that the rates are close, the extinctions from 900 onward can be viewed as a single event.

Human colonization

Radiocarbon dates of early Polynesian occupation terminate fairly abruptly backward in time at about A.D. 1000, with very few extending back into the first millennium A.D. (Anderson and McGovern-Wilson 1990; Anderson 1989). The Polynesian colonization of New Zealand is dated considerably tighter than, for example, the colonizations of Madagascar (Section 2.4.2) or Hawaii (Section 2.4.4). We have taken A.D. 900 as the date of colonization, but it may have been a little earlier or a little later. The voyagers brought with them plants and, signifi-

Table 2.9. Extinctions of New Zealand land and freshwater birds over time

Year A.D.	Centuries between	Species extant	Species extinct in interval	Proportion extinct	d (per century)	p_{Ec} (per century)
900		125				
	8.69		36	0.2880	0.0391	0.0383
1769		89				
	2.24		8	0.0899	0.0420	0.0411
1993		81				

d = instantaneous rate of extinction on a century basis; p_{Ec} = proportion of species lost per century.
Modified from Holdaway RN. New Zealand's pre-human avifauna and its vulnerability. N. Zealand J. Ecol. 1989;12(suppl.):11–25.

cantly, the Polynesian rat, *Rattus exulans*, possibly as a stowaway, and the Polynesian domestic dog, *Canis familiaris*.

Causes of extinction

Climatic changes have previously been advocated to account for extinctions of the New Zealand birds (e.g., Mills et al 1984 with respect to takahe, *Notornis mantelli*), but a lack of evidence over the critical period has largely eliminated such mechanisms from serious consideration.

That leaves the anthropogenic interpretation. Its validation requires, among other things, a demonstration that the extinction of species was closely associated temporally with the arrival of people. The evidence for a tight association between the extinction of many species and the colonization of New Zealand by the Polynesians is overwhelming. Anderson (1989) discussed the distribution of "mixed" sites where bones of extinct birds and human artifacts are intermingled. Of the 33 species known definitely to have died out during the Polynesian period, at least 30 (including all 11 species of moas) have been found in kitchen middens (Millener 1990). Most dated occupation sites rich in avian remains are 400 to 800 years old. Bird remains are scarce in younger sites.

Caughley (1988) used variance stripping on the dates of marine shell and moa-bone collagen to estimate the interval of deposition of mixed strata and hence the mean interval over which Polynesians and moas coexisted in a district. Anderson and McGovern-Wilson (1990) repeated the exercise on a larger collection of dates, estimating the interval of deposition separately from marine shell, moa-bone collagen, and charcoal. They got 124 years for collagen, but 466 years when the calculation was restricted to dates from shell and 625 years when restricted to charcoal. The samples are large enough to eliminate sampling variation as the cause of discrepancy. We accept the shortest interval as the most reliable, given that the influences on dating accuracy (Section 2.4.1) tend to expand rather than contract variances.

The lack of climatological evidence and overlap between people with the extinct fauna for over a century lead us to conclude that the New Zealand extinctions were anthropogenic, but those lines of evidence do not tell us what happened. People can cause extinctions in a number of ways: by hunting, by changing habitat, and by introducing alien competitors and predators. The demonstration that moas and people overlapped in time and that the people killed and ate moas is necessary but insufficient validation of hunting as the cause of the extinction of the moas.

The botanical evidence that habitat was unmodified over large areas in which moas lived, and the archaeological evidence that Polynesians penetrated those regions and hunted moas there strengthens the argument in favor of hunting as the mechanism for the extinction of the moas. Caughley (1989) estimated from the density of hunted moa remains in the lowlands of the southern half of the South Island that about 100 moas were killed per square kilometer and that this deduced hunting pressure was sufficient to cause regional extinction. Hence we tentatively accept hunting as a plausible mechanism for the extinction of the moas while recognizing that the evidence remains circumstantial. By extrapolation we are inclined to implicate hunting in the extinction of other medium to large birds such as the rails, ducks, goose, and swan.

That argument fails to explain the extinction on the main islands of very small species of birds such as three genera of New Zealand wrens. Similarly it fails to account for the extinction of frogs and lizards. Again, habitat modification was probably not sufficiently widespread to account for these extinctions, although it may have played a role regionally. The most plausible agent of decline for these species is the Polynesian's camp follower—the Polynesian rat. Bones of this rat are present in archaeological deposits of all ages, and so its spread through New Zealand was at least as fast as that of the Polynesians (Atkinson and Moller 1990). The rat lived in a wide range of habitats, including grassland and extensive forest ranging across climates from very dry to very wet. It was particularly prevalent in beech (*Nothofagus*) forest where it would erupt in numbers during years of high seed production. It is an efficient egg predator, an agile climber, and has been recorded as killing adult birds and chicks of several species (Atkinson 1985). We nominate the Polynesian rat as the most likely cause of the extinction of many small species of birds, reptiles, and amphibians in New Zealand, but the evidence, although plausible, is still circumstantial.

2.4.4   *The Pacific islands*

The recognition of a mass extinction on the oceanic islands of the Pacific Ocean is remarkably recent. If we exclude New Zealand, where a pre-European mass extinction had been recognized for well over a century, the bulk of the literature on Pacific extinctions is dated to within the last 10 years. A lot more literature is in press or in preparation, but enough is now published to sketch in broad outline what occurred.

Extinctions

The extinctions of species in the Pacific are limited mainly to birds and a few reptiles, because indigenous mammals are sparse in the Pacific and almost all of those are flying foxes (*Megachiroptera*) and smaller insectivorous bats (*Microchiroptera*). The Polynesian rat is ubiquitous throughout the Pacific, but it is commensal with people. It surfed along on the wave of human colonization. The extinctions are detected from the presence in recent subfossil deposits of species not known from the area today. Thus, it is possible to detect both extinctions (species that have died out entirely) and population losses (species gone from one island but still occurring on other islands). Our knowledge of the species of birds that died out is presently biased toward the larger forms because until recently the sieving techniques used mesh sizes greater than the 2 mm needed to recover the bones of small passerines. Milberg and Tyrberg (1993) tallied 151 species and subspecies of birds that had recently become extinct in the Pacific (including New Zealand), but that total is already out of date.

The historically known avifauna of the Hawaiian Islands comprised between 40 and 55 endemic species. An additional 35 species are known only from late Holocene fossils (Olson and James 1991; James and Olson 1991). The difference represents the magnitude of the late Holocene extinction episode. Many more specimens of extinct species await description. At the rate subfossils are being discovered on the Hawaiian chain, there is little doubt that the number of extinct Holocene species will soon outnumber those of the historically known avifauna. The estimate of the proportion of the avifauna that died out is ultimately likely to approach 50%.

The extinct forms on Hawaii were diverse. Of the several raptors (including owls), only one species survives today. Only one of 11 flightless rails survived but then died out in the mid-nineteenth century. The lost species include flightless gooselike ducks, flightless ibises, and a plethora of passerines belonging mainly to the honeyeater family and the endemic honeycreeper subfamily. Radiocarbon dates are sparse and mostly from associated land snail shells (which tend to produce overestimates of age), but nonetheless they indicate that most of the subfossils were deposited in the upper Holocene, only a few being of Pleistocene age (Olson and James 1991).

Steadman (1989) reviewed extinctions in eastern Polynesia, reporting that extinctions and population losses have been recorded in all eastern Polynesian archipelagoes in which bird bones form part of the archaeological record (Marquesas, Society, Pitcairn, and Cook groups). More species of eastern Polynesian landbirds have become extinct since human arrival than survive in the region today. Nearly all species that survive there, whether seabirds or landbirds, have been lost from most of the islands that were once included in their range of distribution. "Fossils from archaeological sites have extended the ranges of many extant species by hundreds to thousands of kilometers. Such species, traditionally regarded as endemic to one or several islands, were widespread until their disappearance . . . from island after island" (Steadman 1989). Petrels and shearwaters suffered the greatest losses. The landbird extinctions are dominated by flightless rails, but various pigeons, parrots, passerines, swifts, and kingfishers either went extinct or were lost from parts of their ranges. The most spectacular extinctions have been of the flightless rails, one or two species of which inhabited almost every island, large or small, at the time of initial colonization (Steadman and Olson 1985; Steadman 1987). Olson (1989) estimated that "hundreds" of rail species died out, leaving only 14 at European contact, which declined further to the three still extant today. Steadman (1991) thought the number of prehistoric extinctions of rails in the Pacific may have topped 2,000.

The fossil record of western Polynesia (Niue, Tonga, Samoa, Wallis, Futuna, Tokelau, Tuvalu) is limited in comparison to eastern Polynesia (Figure 2.2). Extinctions and population losses have been recorded from the few known late Holocene fossil localities (Steadman 1989). Fiji (ethnically part of Melanesia) is equally depauperate in subfossil sites, and a single site records an extinct megapode and pigeon (Steadman 1989). Micronesia has an even sparser fossil record, but a recently found cave deposit on Rota Island in the Mariana Islands chain has uncovered several extinctions (Steadman 1992).

New Caledonia, the largest Pacific island (16,750 km²) after New Zealand, has recently provided evidence of extinctions. Balouet and

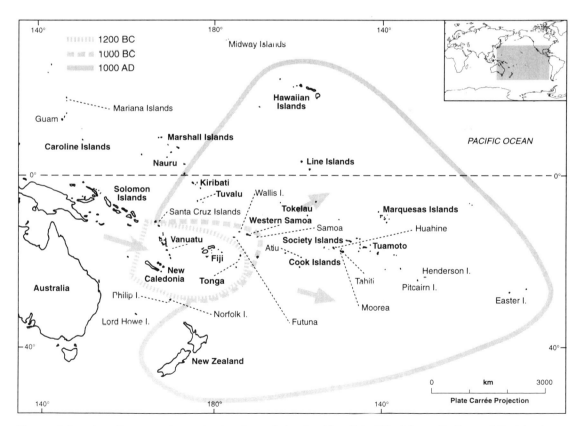

Figure 2.2. Isoclines on the colonization of the Pacific. (Modified from Bellwood P. Man's conquest of the Pacific. The prehistory of Southeast Asia and Oceania. Auckland: Collins, 1978.)

Olson (1989) reported only the nonpasserine subfossils in three sites. Radiocarbon dating is inadequately replicated, yielding one date of 3,450 B.P. and another of 1,750 B.P. Those sites attested to 11 nonpasserine species extinct prehistorically, which, together with the historical nonpasserine avifauna of 32 nonmigratory landbird species, implies an extinction rate of 26%. However, the subfossils sampled only half the species known historically and, therefore, presumably only about half of those species that died out. Correcting for this and for recent immigrants to the historical avifauna gives an estimated extinction rate of 40%. Also lost were large varanid lizards, horned tortoises, and crocodiles whose subfossils occurred in or near human artifacts.

We deviate here to speculate on the *Meiolaniidae*—the large-horned turtles known only as fossils from South America, Australia, New Caledonia, and Lord Howe Island. The South American forms were Eocene and Cretaceous fossils. As we have mentioned, the disappearance of the horned turtles from Queensland (Australia) and New Caledonia coincided with settlement by people. Lord Howe Island was settled by people in the 1700s (Section 4.2), long after the disappearance of its population of horned turtles. Perhaps that loss was through bad luck—Lord Howe Island is only a small island. The losses of the populations elsewhere left no dispersers to recolonize Lord Howe Island. Alternatively, Lord Howe Island was visited by people who took the turtles but left no conspicuous traces of their passage.

Human colonization

The colonization of the Pacific is documented by Bellwood (1978), and we follow his interpretation (Figure 2.2). The people were Austronesians, quite different physically, culturally, and linguistically from the people who colonized Australia and New Guinea many millennia previously. The initial Austronesian penetration of the western Pacific can be traced from their distinctively patterned pottery called Lapitaware. (Lapita is the type locality on New Caledonia.) The spread of the Lapita culture was recent and rapid. The earliest sites dated to around 1300 B.C. are known from New Caledonia, Santa Cruz Islands, Vanuatu (New Hebrides), Fiji, and Tonga. Samoa seems to have been colonized a little later, about 1000 B.C.

The second phase of colonization, peopling the huge extent of the rest of the Pacific, began about a thousand years later with Samoa as the launching point. The oldest radiocarbon dates of this phase (around the beginning of the first millennium A.D.) are incongruously from the Marquesas Islands group a long way to the east. Confirmation of this being the first settlement of eastern Polynesia awaits further dating from the intermediate island groups. From some point in eastern Polynesia within the region occupied by the Marquesas, Cook, and

Society groups, all the remaining Polynesian Pacific islands, from Hawaii to New Zealand and out to Easter Island, were colonized by the end of the first millennium.

Causes of extinction

The fossil record of the Pacific is patchy. The island of Atiu in the Cook group, for example, yielded 41 identifiable bird bones from three caves but without any accompanying clues as to where they fitted into Polynesian prehistory. In contrast, over 11,000 identifiable bird fossils have been collected at one site in the Marquesas Islands that spans much of the time of Polynesian occupation (Steadman 1989). However, it is possible to prove for most eastern Polynesian sites that extinct species overlapped in time with people. Western Polynesia has a sparser record, but Pregill (1989) showed that a Tongan site rich in early Lapita artifacts and dated to early in the first millennium b.c. contained three extinct species of pigeons, two extinct megapodes, and the butchered and charred bones of an extinct species of giant iguana.

Henderson Island in the Pitcairn group was discovered by Europeans in 1606, at which time it was uninhabited (Figure 2.2). It has not been settled permanently since. Yet excavations in 1970 turned up unequivocal evidence of an earlier Polynesian settlement, and identification of bird bones from a single archaeological site boosted the tally of landbird species by 33% (Steadman and Olson 1985).

The association and timing of the colonization of Pacific islands by people and the extinction or extirpation of indigenous species are too close to be passed off as an artifact. They provide strong inference that people caused the extinctions in some way or another. However, we need controls and offer two: the Galápagos Islands of the eastern Pacific (Figure 2.2) and the Mascarene Islands (Mauritius, Réunion, and Rodrigues) in the Indian Ocean (Figure 4.1). Neither group was colonized before European contact. The Galápagos Islands were first settled permanently from Ecuador in 1832. Despite a reasonable late Holocene fossil record there are no clear-cut cases of vertebrate extinction before that date (Steadman 1989), although common sense dictates that there must have been a few.

The Mascarene Islands were visited by the Portuguese in the 1500s and by the Dutch in the 1600s (Section 4.1). Cheke (1987a) and Cowles (1987) examined the present, historical, and subfossil records of Mascarene vertebrates. There is a long list of extinctions but, as yet, no unequivocal evidence of extinction prior to European discovery. Likewise, the currently known fossil record is for birds becoming extinct after Polynesians colonized islands and not going extinct over the same period on islands not colonized by Polynesians. But some species and populations would have become extinct through natural causes and have not come to light in the fossil record yet.

The evidence implicating people in the Pacific extinction spasm is overwhelming, but it is more difficult to deduce the mechanism or mechanisms by which the species were extinguished or locally lost. Hunting probably accounted for many of the species, particularly the large pigeons, flightless rails, and communally nesting seabirds. The effects of direct hunting may have been augmented by predation by the pigs, *Sus scrofa*, and dogs introduced by the Polynesians. Archaeologists are intrigued by those islands colonized and then abandoned by Polynesians before European contact—Bellwood's (1978) "mystery" islands—but evidence, particularly from Henderson Island, suggests that overexploitation of both landbirds and seabirds on islands not suitable for agriculture is the explanation (Steadman 1989; Diamond 1991).

Hunting is not a plausible explanation for extinction of some of the smaller passerines, as occurred in Hawaii and New Zealand. The Polynesian rat, which accompanied the Polynesians to every island they colonized, is a more likely culprit, perhaps aided on some islands by a clearing of the birds' habitat. Another contributing agent may have been avian disease carried by Polynesian chickens, *Gallus gallus* (Diamond 1991), but the history of the New Zealand extinctions (Section 2.4.3), where chickens (and pigs) were not introduced, suggests that it was not a dominant cause.

2.5 SUMMARY

Our brief examination for the causes of prehistoric extinctions suggests that the resolution of coincidences in timing between causes and effects is insufficient to draw firm conclusions. We are on firmer footing for events of the last 10,000 years where we can find a tighter relationship in assigning dates to compare the frequency of extinctions following first settlement by people. The postsettlement extinction rates in the Americas and Australia are much closer to those on the Pacific islands following Polynesian colonization, on the islands of the Caribbean following Amerindian colonization, and on the islands of the Mediterranean following European colonization. But here we run into a problem of interpretation. The direct and indirect effect of people on the island biotas is generally accepted even though it has been so recently revealed through descriptions and dating of subfossils. The same cannot be said with equal clarity for the American and Australian Pleistocene extinctions.

Historic extinctions and near extinctions

Smith et al (1993) reported that, since A.D. 1600, some 486 animals and 600 plants have become extinct and that another 3,565 animal species and 22,137 plants are presently threatened with extinction (Table 3.1). Many others have addressed the subject (Hornaday 1913; Allen 1942; Harper 1945; Greenway 1958; Wilcox 1980; Terborgh and Winter 1980; Soulé 1983; Frankel and Soulé 1981; Diamond 1984a,b).

Some of those authors merely catalogued extinctions, and others postulated causes or probed vulnerability to extinction. Together the volume of work alone imparts, rightly or wrongly, an impression of certainty about the causes that drove species to extinction. Yet many cases happened so long before the twentieth century and in the obscurity of such distant places that, at least, some of this certainty may be unwarranted. With this in mind, we sampled extinctions and near extinctions from the International Union for the Conservation of Nature and Natural Resources (IUCN) (1990) compendium and Collar et al (1992).

Historically, collecting endangered species took precedence over describing natural history, and we are left with little with which to speculate what happened to those species. The desert rat-kangaroo,

Table 3.1. Number and percentage of species extinct since A.D. 1600 or presently threatened

Taxon	Species extinct	Percentage extinct	Species threatened	Percentage threatened
Animals				
Corals	1	0.01		
Mollusks	191	0.2	354	0.4
Crustaceans	4	0.01	126	3
Insects	61	0.005	873	0.07
Fish	29	0.1	452	2
Amphibians	2	0.07	59	2
Reptiles	23	0.4	167	3
Birds	116	1.2	1,029	11
Mammals	59	1.3	505	11
Total	486	0.04	3,565	0.3
Plants				
Fern allies	4	0.3		
True ferns	12	0.1		
Gymnosperms	2	0.3	242	32
Monocotyledons	120	0.2	4,421	9
Monocots: palms	4	0.1	925	33
Dicotyledons	462	0.2	17,474	9
Total	600	0.25	22,137	9

Modified with permission from Smith FDM, May RM, Pellew R, Johnson TH, Walter KR. How much do we know about the current extinction rate? Trends Ecol. Evol. 1993;8:223.

Caloprymnus camprestris is an example. It was a sandy-colored marsupial that could flee speedily, hopping along with its body leaning forward and its tail straight out behind. "The European history of the desert-rat kangaroo was remarkably short and fatal" (Calaby and Flannery, in press). It was first collected in the early 1840s, and then in the 1930s the only 23 specimens found were collected. The species has not been seen since.

The starting point for analyzing causes of extinctions and near extinctions is three sets of species selected without regard to their habitats, habits, or numbers other than an IUCN classification as extinct or endangered. We chose mammals and birds because more is known about them.

We have, when possible, separated whether we are dealing with proximate or ultimate factors. Proximate means the "next before" and ultimate means "last, final, beyond which no other exists or is possible" (*Webster's College Dictionary*).

3.1 BIRDS EXTINCT SINCE A.D. 1600

In 1990, the IUCN listed 17 bird species presumed extinct (Table 3.2; Figure 3.1). There are grounds to suspect that, in the strict sense of the word, some may not be extinct, but we have not changed the IUCN's classification. We used Greenway's (1958) compilation of extinct birds as our primary source, supplemented by Collar and Stuart (1985), Collar et al (1992), and other references as cited.

The data on causes of extinction are so sparse that we divided them into three categories: unknown causes, guessed causes, and causes through reasoned speculation. We assumed that if an environmental change was recorded within the species range, its effect (decline in numbers, range contraction) was possible but not proven. Without that caution we can easily fall into the trap of assuming that because a grassland is plowed a grassland dweller will then decline, which leads to the conclusion that it declined because the grassland was plowed. We need more than coincidence for a secure diagnosis.

3.1.1 Unknown causes

So little is known about 8 of the 17 birds in our sample (44%) that it is impossible to say why they became extinct. Indeed, so little is known that some of them are possibly not extinct. The caerulean paradise-flycatcher, *Eutrichomyias rowleyi*, of Sangihe Island was known from one specimen collected in 1873 and then not seen again for 100 years. Possibly, despite the conversion of its forest habitat to coconut and

Table 3.2. Dates of first and last sightings and possible causes for birds extinct since A.D. 1600

Causes	Dates of first and last sightings, possible causes
Unknown	
Caerulean paradise-flycatcher	One collected in 1873; one seen 100 y later
Choiseul pigeon	Six collected in 1904
Forest owlet	Six seen before 1914; one photo in 1968
Himalayan mountain quail	Ten specimens collected before 1868
Jamaican pauraque	Three specimens collected before 1859
Marquesas fruit dove	Last seen in 1922
Ponapè Mt starling	Last seen in 1975
Woodford's rail	Few specimens collected before 1880
Guessed	
Snail-eating coua	Common in 1831; last seen in 1834
Canarian black oystercatcher	Declining by 1850; not recorded from 1913 to 1970
Guadalupe storm petrel	Abundant in 1906; last seen in 1912
Javanese wattled lapwing	Last seen in 1939
Pink-headed duck	Sold in markets in 1915; gone by 1936
Through reasoned speculation	
Colombian grebe	Declining by 1944; 300 seen in 1968; none seen in 1977
Glaucous macaw	Last seen in 1860; possibly one in 1970
Molokai creeper	Common in 1890s; unreported from 1940 to 1961; last seen in 1963
New Zealand bush wren	Losses monitored; causes: rats and, ultimately, human error
Stephens Island wren	Rats, cats, and collectors

nutmeg plantations or secondary forest (Whitten et al 1988), it may still exist. Blewett's forest owlet, *Athene blewetti*, from woodlands north of Bombay was known from six specimens, the last of which was taken in 1914. Thereafter the bird was not seen until one was photographed in 1968. The Jamaican pauraque, *Siphonorhis americanus*, was a nighthawk described from three specimens collected by 1859 and a possible sighting in the 1980s. Its disappearance before the introduction of mongooses, *Herpestes*, eliminates them as causing the loss (Greenway 1958). Caerulean paradise-flycatchers, Jamaican pauraques, and Blewett's forest owlets may well have been seen by ordinary people in the decades when they went unrecorded by ornithologists. In this context it is worth bearing in mind that new species are still being found and others are being rediscovered. Therein lies an important point: Scientific ignorance should be recognized for what it is and not treated as anything else.

The Choiseul pigeon, *Microgoura cyanopsis*, of the Solomon Islands was a large-crested ground pigeon known only from six specimens

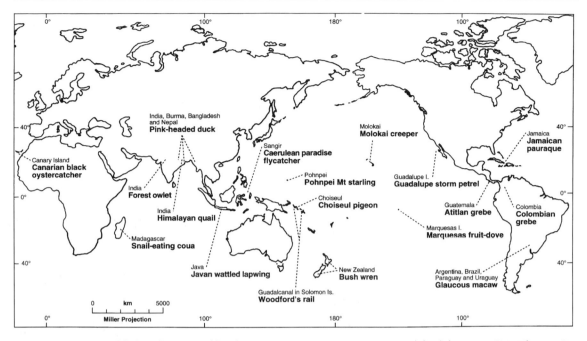

Figure 3.1. World distribution of birds extinct since A.D. 1600. (Modified from IUCN. The IUCN red list of threatened animals. Gland, Switzerland: IUCN, 1990.)

taken in 1904. Local people thought that introduced cats caused its disappearance (Greenway 1958). The Himalayan mountain quail, *Ophrysia superciliosa*, was known from some 10 specimens and was kept in captivity, but unsuccessfully. It was last seen in the wild in 1868. The Marquesas fruit dove, *Ptilinopus mercieri*, was last seen in 1922 and all we know is that it was believed to live in woodlands. The Ponapè Mountain starling, *Aplonis pelzelni*, was common this century but disappeared by 1975 for reasons unknown. Woodford's rail, *Nesoclepeus woodfordi*, was a large and flightless rail known from only a handful of specimens collected from the Solomon Islands and Fiji Islands before the 1880s. Ripley (1977) opined weakly that those on Fiji may have been lost to mongooses.

3.1.2   Guessed causes

The category of guessed causes differs from the last only in that the literature contains more speculation about the causes of disappearance and a tad more information on the species. The snail-eating coua, *Coua delandei* (Figure 3.2), from Ile de Sainte-Marie, northeastern Madagascar, was one of the world's largest cuckoos and a specialist snail eater. It was hunted for food and its handsome plumage. Day (1981) postulated that introduced rats were responsible, but Collar and Stuart

Snail-eating Coua
Coua delalandei

Figure 3.2. Snail-eating coua.

(1985) note the lack of evidence for this, although rats may have reduced snails. They favor forest clearance as the main reason for its disappearance. The rate at which the cuckoo disappeared—not rare in 1831, to the last confirmed reports 1834—seems, in our view, to be too fast for either case to be convincing.

The Canarian black oystercatcher, *Haematopus meadewaldoi*, was only known from the eastern Canary Islands. It was uncommon in the 1850s. By 1919, its extinction was predicted by the ornithologist who collected the last known specimen in 1913 with "a long and lucky shot" (Bannerman 1963). Subsequently he wrote that "from that day to this the bird has never been seen alive or dead" until his unconfirmed sighting on Teneriffe (Bannerman 1969). Lighthouse keepers and fishermen report that their last sightings were about 1940. Subsequent sightings of this conspicuous bird were on the African mainland in 1970 and 1975 and Teneriffe in 1981. Again this poses the question of whether it is truly extinct, or whether there has been an extension of the species range to mainland Africa, or whether its true range has been overlooked until the 1970s. Nonetheless, its numerical decline in the Canary Islands is indisputable. Collar and Stuart (1985) suggested that rats and cats played a role in the oystercatcher's extinction. Hockey (1987) offered an alternative view. The local name for the oystercatcher was "lapero," meaning limpet eater. People also gathered limpets and piled the shells in middens. The growing population together with decreases in agriculture over the past century led people to overharvest the limpets and this, in turn, triggered the oystercatchers' decline. Interesting as this is, it is only speculation.

The Guadalupe storm petrel, *Oceanodroma macrodactyla*, was abundant in 1906 but not seen after 1912. Feral cats killed some and were blamed by Greenway (1958). Collar et al (1992) reported that the petrels nested in the forests; goats, *Capra hircus*, destroyed all but one grove of the forests. The forest loss and the petrel's disappearance have not been shown to coincide, and the relationship may or may not have been causal. The Javanese wattled lapwing, *Vanellus macropterus*, was only known from a few records prior to 1939. Its habitat was deltas and marshes and they are now cultivated for rice. The birds were hunted, and so it is uncertain as to the effects of habitat loss or hunting in its disappearance.

The pink-headed duck, *Rhodonessa caryophyllacea*, lived in the swampy lowland grass jungles of northern and eastern India and was always rare except on the Ganga (Ganges) River (Philips 1986). It was available in the markets of Calcutta until 1915, which suggests it was then common enough. Hunting as the cause of its disappearance by 1923, despite an increase in the availability of guns (Bucknall 1926, in Day 1981), seems unlikely. The duck was widely but sparsely distributed in habitats not easily hunted, which leaves the explanation as tenuous. The duck was in captivity but never bred, and by 1936 the last one died.

3.1.3 Reasoned speculation

In this category we still lack proof of why the extinctions happened, but we have more support for speculation. The Colombian grebe, *Podiceps andinus*, is a large grebe known only from a few high altitude lakes in Colombia. By the end of the 1940s, grebes had disappeared from lakes on the Bogotá plateau. Until the 1960s, the grebe was relatively common; 300 birds were reported in 1968 on Laguna de Fúquené, Ubate Plateau, but only one bird was seen in 1972, two in 1977, and then none, despite searches (Fjeldså 1993). The decline's causes may have been one of several factors or a combination of them. Grebes were hunted. Some wetlands were drained, and soil erosion increased the water's turbidity. Waterweeds, which had sheltered the grebe's prey, disappeared, apparently as a consequence of the murky waters. In the 1960s, at Laguna de Fúquené, fertilizer and pesticide use was high and may have entered the food chain. Water levels dropped and the trout, *Salmo*, introduced in 1944, possibly competed for food with the grebes. As matters stand, it is not possible to say what killed off the Columbian grebe. With time and more information, it might be possible to analyze the data in greater depth and close in on the causes.

The glaucous macaw, *Anodorhynchus glaucus*, a splendidly attired bird, was widely but locally distributed along the middle reaches of major rivers in south central South America. It nested and roosted in riverside cliffs: Observations and bill structure suggest that this macaw depended on nuts of the chatay palm, *Butia yatay*. The palm's presence indicates good agricultural soil (Chebez, in Collar et al 1992), and it has been selectively cleared as agriculture opened up along the major rivers over the past 400 years. Cattle grazing prevented regeneration of the palms. Loss of its specialized habitat over four centuries would have set the glaucous macaw on a long decline, which was reinforced by birds being taken as pets. The macaw was disappearing by the mid-1800s, and after 1860 no specimens were taken until an unsubstantiated report from Brazil in 1970. Yet not all the chatay palms had disappeared by 1860, and the ultimate causes of the parrot's extinction are unknown.

The Molokai creeper, *Paroreomyzia flammea*, was common on Molokai Island, Hawaii, in the 1890s. By that time, forest clearing and browsing by stock were well advanced and had caused widespread habitat change. Introduced predators (mongooses, feral cats) and avian cholera were also present. Clearly, those agents had not reduced the creeper's numbers. Black rats either did not reach Hawaii until 1870–1890, or their numbers erupted then. After 1890 the Molokai creeper declined and reached such low numbers that ornithologists did not record it between 1940 and 1961. Subsequently it has not been seen after 1963. The timing of the disappearance coincided with the appearance of black rats, which implicates them (Atkinson 1977).

The disappearance of New Zealand's bush wren, *Xenicus longipes*, is associated with a greater volume of data. Subfossils of this genus are widely distributed on the North and South Islands (Turbott 1990). The wrens were rare by the 1850s on the North Island but common in dense bush until the 1880s on the South Island (Buller, in Turbott 1967; Oliver 1955). The Polynesian rat, implicated in the extermination of the flightless Stephens Island wren, *Traversia lyalli*, from the mainlands of New Zealand, disappeared before the last sightings of the bush wrens (Atkinson 1973). Thus, while it may have played a part, it could not have been the wren's ultimate cause of extinction.

Chronologically, other alien predators which arrived in New Zealand are Norway rats and cats, which probably arrived about the same time. The former probably had spread throughout New Zealand by the early 1800s. Cats spread more slowly and did not penetrate the thick bush until between 1830 and 1850 (Fitzgerald 1990). Stoats, *Mustela ermina*, reached into Fiordland after 1855 (Figure 2.1) and could have contributed to the demise of the bush wren (King 1990). Black rats arrived after 1860 on the North Island and after 1890 on the South Island (Atkinson and Moller 1990). The wren's range contraction overlapped in timing with the arrival of predators. The Polynesian rat probably started the wren's decline, and the next round of introduced animals continued it. The bush wren was a weak flier and often fed on the ground where it would have been vulnerable. The Polynesian rat itself fell prey either to the black rat or to competition with the house mouse (Atkinson and Moller 1990).

We know more about the final disappearance of the third subspecies of bush wren. Stead's bush wren, *X. l. variabilis*, occurred on Stewart Island and its outlying islands until the 1960s when it was restricted to Big South Cape Island (Figure 2.1). Black rats reached Stewart Island in 1911 (Atkinson 1973) and Big South Cape Island (911 ha) in 1962. The case for rats having exterminated wrens from other parts of their range is strengthened by the persistence of wrens where rats had yet to reach. This supposition is further supported when the rats reached Big South Cape Island. The rats disembarked down the mooring lines of fishing vessels and once ashore their numbers erupted. The island was too large (930 ha) and rugged to consider controlling the rats. The last-minute transfer of the few remaining wrens to a rat-free island in 1964 failed and the subspecies was lost. In this case the proximate cause was the presence of the black rats, and the ultimate cause was humans fumbling the final rescue.

Another New Zealand wren, like the above case, offers us a reasonable assumption as to what caused their extinction. Stephens Island wren was a flightless species whose widely distributed subfossil remains indicate that it was once widely distributed on New Zealand (Millener 1990). By the 1880s it was confined to a small wooded island in Cook Strait (Figure 2.1). Its disappearance from mainland New Zealand preceded European occupation and likely coincided with the arrival of the Polynesian rat, which had not colonized Stephens Island (Atkinson 1973). It became famous through its extinction by the lighthouse keeper's cat in 1894 (collusion between a collector and the lighthouse keeper may have inflated the cat's tally) (Turbott 1990). In this case, while still not incontrovertible, the evidence of predation by an alien—either feline or human collector—is stronger.

Collecting was often the last, and sometimes the only, record of a species. Rarity commanded a high price, and that was apparently what it took to end the species. Natural history museums mushroomed in the late 1800s (Sheets-Pyenson 1988), and their curators became competitively acquisitive in their dealings. Rivalry between taxonomists aided and abetted the collection of rare species, and rarity fed desirability. Collecting was the final straw for species already close to extinction, and we include here the example of the great auk, *Pinguinis impennis*, even though it was not included on the IUCN list (1990) (Box 3.1). Galbreath (1989) portrays the rapacity of ornithologists in the late 1900s, and, as mentioned, their zeal probably played a role in the extinction of the Stephens Island wren.

3.2 MAMMALS EXTINCT SINCE A.D. 1600

We took the sample of 42 mammals presumed extinct since A.D. 1600 from the IUCN list (1990). Our sources for what caused the extinctions were IUCN (1990), Nowak (1991), Allen (1972), and Hall (1981). We have broken the sample into three subsamples for ease of discussion. Our unit of conservation is the species, but our samples include sub-

Box 3.1. The great auk—an example of extinction driven by harvesting and collecting

The great auk was a large flightless seabird distributed along both sides of the Atlantic seaboard, but most breeding colonies disappeared before records were kept. The prehistoric populations probably numbered in the millions. They were hunted for subsistence since at least 3000 B.C. The birds and their eggs were taken by the tens and hundreds of thousands for food. By the mid-1700s, the harvest shifted to take the birds for their feathers. The feather trade wiped out colonies except those on a few isolated islets off the Canadian and Icelandic coasts by the early 1800s. The largest known colony was on Funk Island, Canada, and may have held some 100,000 pairs when first found by Europeans in the 1700s. Commercial exploitation for feather, oil, and fat quickly destroyed the colony, and it was gone by 1841. The effect of the "Little Ice Age"—the cooling of the Atlantic region from 1550 to 1850—is unknown, but the fate of the last known colony is not in doubt. It was on the island of Eldey, off the southwestern coast of Iceland, and was inaccessible until collectors drove the price of eggs high enough to fund a voyage dedicated to taking eggs and birds. The last two birds taken in 1844 from Iceland were for a museum, and, except for a sighting in 1852 and a dead auk picked up in 1857, that was the end of the species. The demands for museum specimens and the fashion of egg collecting in the late 1800s drove the price up. In 1819, eggs sold for £0.3, but that jumped to £20 by 1850 and £333 by 1946. Even until the 1970s, the eggs and skins were still changing hands.

(Modified from Bourne WRP. The story of the great auk, *Pinguinis impennis*. Arch. Natur. Hist. 1993;20:257–278.)

---

species. As we state (Section 12.2), there are reasons to include subspecies, and the IUCN list (1990) does so.

3.2.1 "Unknown" mammal extinctions since A.D. 1600

Too little is known about 17 of the extinct mammals to really have an idea of their fate (Table 3.3; Figure 3.3). Four species were defined from one or two skulls, and five are known only from a few specimens. Two species are known only as subfossils. Nothing seems to be written about the extinction of the Puerto Rican flower bat, *Phyllonecteris major*.

Table 3.3.   Dates and possible causes of extinction
for little known mammals extinct since A.D. 1600

Mammals	Initial status (no. specimens)	Date of last record	Causes
Central hare wallaby *Lagorchestes asomatus*	Unknown (1)	1931	Unknown
Darling Downs mouse *Notomys mordax*	Unknown (1)	1846	Unknown
Alice Springs mouse *Pseudomys fieldi*	Unknown (1)	1895	Unknown
Big-eared hopping mouse *N. macrotis*	Unknown (2)	1843	Unknown
Giant African forest shrew *Crocidura odorata goliath*	Unknown (3)	Unknown	Unknown
Chapman's fruit bat *Dobsonia chapmani*	Species?	Unknown	Unknown
Tanzanian woolly bat *Kerivoula africana*	Species?	1878	Unknown
Large Palau flying fox *Pteropus pilosus*	Unknown (2)	1874	Hunting?
Guam flying fox *P. tokudae*	Unknown (3)	1968	Hunting?
Pemberton's deer mouse *Peromyscus pembertoni*	Species?	1932	Unknown
Puerto Rican flower bat *Phyllonycteris major*	Subfossil	Unknown	Unknown
Sardinian pika *Prolagus sardus*	Subfossil	Unknown	Unknown
Nelson's rice rat *Oryzomys nelsoni*	Unknown (2)	1898	Rats, cats?
Bavarian pine vole *Pitymys bavaricus*	Unknown	Unknown	Unknown
Short-tailed hopping mouse *N. amplus*	Unknown (2)	1894	Unknown
Long-tailed hopping mouse *N. longicaudatus*	Unknown (4)	1894	Unknown
Gould's mouse *Ps. gouldii*	Widespread	1857	Unknown

The Sardinian pika, *Prolagus sardus*, lived until Roman times. Deforestation for agriculture and the arrival of the black rat, *Rattus rattus*, possibly signaled its end (Vigne 1992).

As with the birds, some of those mammals may not be extinct. An example might be the Tanzanian woolly bat, *Kerivoula africana*, which is known only from a single specimen collected in 1878. Most African

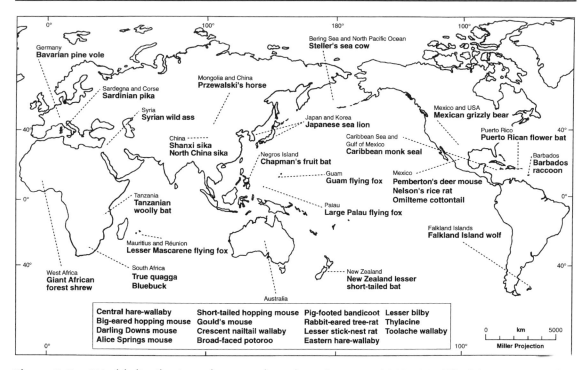

Figure 3.3. World distribution of mammals extinct since A.D. 1600. (Modified from IUCN. The IUCN red list of threatened animals. Gland, Switzerland: IUCN, 1990.)

species in this genus are known from very few specimens. As nocturnal forest dwellers they are particularly difficult to observe. High rainforest canopies and the air above them are, for practical purposes, out of human reach. Just how unknown those habitats are is illustrated by Chapin (1938), who collected two bats new to science from the crop of a bat hawk, *Macheiramphus alcinus*, which, in turn, had taken them from the forest canopy. The discovery and rediscovery of mammals are still happening. In Australia, for example, three mice and four marsupials, including Leadbeater's possum (Box 9.3), were rediscovered in the 1960s and 1970s. Again, the point is the limitations of attempts to list extinct species.

3.2.2 *Australian mammals extinct since A.D. 1600*

Extinctions of Australian mammals are disproportionately high compared to the rest of the world (Tables 3.3 and 3.4). With 10 Australian mammals, we are on firmer ground to find causes for their extinctions. Judging from the last verified sighting or specimen, half went extinct this century and five after 1930. Some declines went unnoticed, but others were observed and concerns were voiced that, unless something was done, that particular species would disappear—and it did.

Table 3.4. Dates and possible causes of extinction
for Australian mammals extinct since A.D. 1600

Mammals	Initial status	Period of decline	Date of extinction	Causes
Crescent nailtail wallaby *Onychogalea fraenata*	Local	<50 years	1930	Habitat, foxes, cats
Broad-faced potoroo *Potorous platyops*	Rare?	<50 years	1875	Habitat, foxes, cats
Pig-footed bandicoot *Chaeropus ecaudatus*	Rare?	70 years	1907	Habitat, foxes, cats
Desert bandicoot *Parameles eremiana*	Local	<50 years	1931	Habitat, foxes, cats
Rabbit-eared tree rat *Conilurus albipes*	Uncommon	Unknown	1875	Habitat, foxes, cats
Lesser stick-nest rat *Leporillus apicalis*	Common	Unknown	1933	Habitat, foxes, cats
Eastern hare wallaby *Lagorchestes leporides*	Abundant	50 years	1890	Habitat, cats
Lesser bilby *Macrotis leucura*	Widespread	<50 years?	1931	Foxes, cats
Thylacine	Widespread	100 years	1936	Hunting, disease
Toolache wallaby	Local	40 years	1923	Hunting, foxes

Hunting played the key role in the disappearance of the two largest Australian mammals to have become extinct since A.D. 1600—the thylacine, *Thylacinus cynocephalus*, and the toolache wallaby, *Macropus greyi*. Thylacines occur as subfossils over Australia and New Guinea up until around the time that people had introduced the dingo, *Canis familiaris dingo*. Presumably they competed to the thylacine's loss, so that by the time Europeans colonized Australia its last refuge was Tasmania, where the dingo was absent. Once sheep ranching started in Tasmania, a campaign was launched against the thylacine (Morgan 1992). In 1830 bounties were paid for all feral dogs, thylacines, and Tasmanian devils, *Sarcophilus harrisii*, killed, and thylacines were also hunted for their pelt. At the same time they would have had to contend with loss of their natural prey through habitat changes and competition with feral dogs. By 1905, thylacines had sharply declined, and their final collapse was rapid. Disease may have played a role, as an unknown disease struck domestic and native cats, *Dasyurus* spp., apparently about 1910. Collecting thylacines for zoos and private collections was common in the early 1900s. The last known thylacine was caught in 1933, and it died in captivity on September 7, 1936 (Paddle 1993).

The toolache wallaby was renowned for its beautiful markings and agility. It was an open country species and was locally numerous.

Clearing native vegetation between 1860 and the early 1900s probably started the wallaby's decline, which was accelerated by bounty and pelt hunting and also fox, *Vulpes vulpes*, predation in arid areas. By 1923 only one small band of 14 and scattered individuals were left. Of those, four died during attempts to capture them for transfer to Kangaroo Island. The remainder was killed as trophies and the last one, a female, died in captivity about 1939.

The other Australian mammals to go extinct make an unusual record, as they were ecologically similar enough to be considered a set. Their disappearances in the 1800s and into the 1900s is one of the most sweeping cases of mammal extinctions historically documented. All were medium-sized (150–5,000 g) vegetarians of arid and semiarid regions. Explanations of their collective loss involve introduced predators and herbivores together with changing human land uses. This diagnosis of multiple causes is based on coincidental timing between the declines and landscape modification and alien introductions in the last 100 years.

Initial and final stages of the decline were driven by different combinations of factors and timing, which suggest that species did not respond uniformly to the changes. The story varies regionally with the rate at which introduced predators and herbivores spread (Figure 4.18) and changes in land use occurred. The slide in the Australian mammals started by the 1860s before the peak in numbers of sheep and rabbits (Allen 1983). Five endemic mammals possibly disappeared by 1875 and four more by 1900 (Tables 3.3 and 3.4). A peak in extinctions came with the last records of six species in the 1930s. This would have been after grazing, fire, and fuel became locked together with severe droughts from 1895 to 1903 and again in the 1920s, by which time rabbits were a plague (Allen 1983).

Australia's pre-European vegetation in the semiarid lands was a mosaic maintained by patchy burning. Europeans changed that configuration to more uniform but unfavorable pioneer grass and then pine scrub. Endemic mammals became isolated in patches of their specific habitat. During droughts, rabbits and stock would have converged on those patch refuges and their grazing would have degraded the habitat (Morton 1990). Foxes and cats, especially if dingo, *C. familiaris*, control was present, probably were attracted to the refuges to prey on the rabbits and endemic mammals. Supported at high densities by the fecund rabbits, the predators could have effectively mopped up the small, isolated, and shrinking populations of indigenous endemics. The myxomatosis-induced crashes of rabbits in the 1950s may have accentuated fox and cat predation on the remaining endemics. Grazing competition reinforcing predation may explain the greater loss of mammals compared to birds and reptiles.

Short and Turner (1994) took another approach to attributing causes to the extinction of Australian mammals. They compared the abundance of three marsupials between areas of high and low diversity of vegetation on an offshore island which lacked introduced predators, herbivores, and aboriginal burning. Nonetheless the vegetation was a mixture of seral successional stages but through clearing for oil fields. Marsupial densities were independent of the scale of vegetative diversity. From that, Short and Turner (1994) argued that the persistence of marsupials on the island did not support the hypothesis that their loss on the mainland had come about through losing the vegetation mosaic created by aboriginal fires. Instead they identified the absence of predators as the reason why the marsupials persisted.

We do not accept this reductionist explanation. It ignores any interaction between vegetation (food and cover), competition, and predation by introduced animals. It also assumes the serial stages in vegetation (indexed by density) were similar to those produced by burning and that the habitat on the island is comparable to the historic habitat on the mainland.

A historical retrospective for the timing of predator and herbivore introduction and changes in land use practices takes us tantalizingly close to diagnosing the causes of extinctions in Australia. Combing all the historical accounts would provide more details to fill out the picture, and it is worth doing because the disappearances are not over. Many of Australia's surviving mammals are endangered, such as the numbat (Section 4.14). Many are confined to reserves or off reserves (Table 10.2) and the historical interactions of predators, prey, and patches of vegetation have implications for reserve design and management.

3.2.3 Non-Australian mammal extinctions since A.D. 1600

Once we leave the Australian continent, mammal extinctions in historical times are not such a tidy story. The subsample of 15 mammals whose accounts follow was selected from the list of 42 mammals because there are more data with which to speculate than those recorded in Table 3.3.

The greater short-tailed bat, *Mystacina robusta*, was one of only two species in New Zealand's endemic bat family, Mystacinidae. The bats are unusual because they hunt while scurrying on the ground and tree trunks, taking insects, fruit, and nestlings. Subfossil remains indicate the bat was widespread, but it had disappeared from North Island and South Island before European colonization. By 1840 it was confined to two small islands off Stewart Island (Figure 1.1). Big South Cape and its neighbor, Solomon Island, were both free of both Polynesian and black rats (Daniel 1990). The bat roosted in mutton-bird burrows where it would have been vulnerable to rats. In 1962 or 1963 black rats reached both islands. The several hundred bats seen in 1961 disappeared quickly, and by 1967 the last bat had been seen. Whether the rats killed the bats, disturbed their roosting, or competed for food is not known (Daniel 1990). The tight coincidence between the rat's arrival and the bat's disappearance (and bush wrens) (Section 3.1.1) convinces us that black rats exterminated the bats.

The Falkland Island wolf, *Dusicyon australis* (warrah), was a large fox with a thick woolly coat, about the size of a small coyote, *C. latrans*. The foxes were curious and bold when the first Europeans arrived in the mid-1700s. Fear of those bold animals and the ease of killing them led to many being destroyed. Charles Darwin noted their reduced numbers and predicted their extinction (Day 1981). FitzRoy, who captained HMS *Beagle*, commented, in contrast to Darwin's observation of its tameness, that it was fierce and fearless, feeding on seals and penguins. The last one was killed in 1876 after some 10 years of bounty and pelt hunting. The bounties were justified for protecting recently introduced domestic sheep.

The Falkland Island wolf is a footnote in the history of evolution (Bourne 1992). It was FitzRoy who commented on geographic differences in the foxes' appearance on the Falklands, and both he and Darwin speculated that foxes could have drifted the 480 km from mainland South America on ice. Fortunately, by the time Darwin had reached the Galápagos Islands, his appreciation for the significance of intraisland variation in species was piqued. The wolf's tameness, which contributed to its demise, may ironically have been from domestication. The white muzzle, tip to the tail and lower legs, and its skull shape attest to that. South American Indians had domesticated other members of *Dusicyon*, and the warrah may have drifted over to the Falkland Islands (Clutton-Brock 1977).

The fate of the Mexican silver grizzly bear, *Ursus arctos nelsoni*, was typical for grizzlies over most of their range in western North America. Sweeping habitat changes, competition with agriculture, and fear of bears led to concerted efforts to obliterate them. The Mexican grizzly bear hung on until only some 30 years ago despite being hunted, trapped, and poisoned out of existence from Arizona, New Mexico, southern California, and Texas. This subspecies was not recognized scientifically until 1914 when the bear's abundance was already dwindling. By 1937, the bear's final retreat was an area of about 15 km in diameter in the Mexican state of Chihuahua, where ranchers and sportsmen harried the last 20 to 30 bears. In 1961 prolonged drought may have driven a bear to kill stock, which triggered an all-out effort to destroy the survivors by shooting and poisoning. The bears had been declared as protected the previous year, but lack of funds stalled efforts to set up a reserve. By 1962 it was too late—the Mexican grizzly bear was extinct.

Steller's sea cow, *Hydrodamalis gigas*, was a large manatee (Sirenia) which numbered in the thousands in the Bering Sea. Steller, the

German naturalist who, in 1741, recorded the only description of the sea cow, gave a harrowing account of its demise (Day 1981). The Russian discovery of the Bering Sea and its islands had attracted traders whose crews used the sea cows for provisions. The hunting was wasteful and the sea cow's plight was quickly recognized. The authorities ignored a petition to protect the sea cows, and the last one was killed in 1767 after only 27 years of hunting.

The Syrian wild ass (onager), *Equus hemionus hemippus*, was the last of five subspecies of wild ass that roamed the Asiatic lowland deserts. It ranged the deserts of Syria, Palestine, and Saudi Arabia (Groves 1974), where it had long been hunted by desert people. Firearms became more available after World War I, thus, hunting increased, which coincided with the onager's rapid disappearance. The last one was shot in 1927.

The foregoing cases are unequivocal extinctions, with the causes relatively well defined. Man was the proximate and ultimate cause for the loss of the Falklands wolf, the Mexican grizzly, Steller's sea cow, and the onager. Black rats probably exterminated the New Zealand greater short-tailed bat, but whether through predation or competition, or both, is less certain. The next seven cases in our subsample of non-Australian mammals are less clear-cut when we try to assign causes of their disappearances. Hunting was part of the problem, but competition with domestic stock and habitat change may have played their part.

Przewalski's horse, *E. caballus przewalski*, lived in the arid and saline steppes of the Mongolian and Dzungarian mountain ranges. Hunting and competition for pasture with domestic stock apparently reduced its numbers by the 1930s. Despite protection from hunting, its numbers continued to decline. Exceptionally severe winters in 1948 and 1956, political disputes over the land, competition for grazing, and the fact that the high steppes of the Gobi Altai may have been an unsuitable refuge, all contributed to the decline. Horses were seen until 1968, but none thereafter. About 1,000 horses living in the world's zoos are descended from three or four of the surviving lines of the original 12 founders (the thirteenth founder was a domestic mare).

The quagga, *E. quagga quagga*, was the third equid to go extinct in the wild within the last 110 years. It was Africa's southernmost zebra, and the Boers encountered it in large numbers. They shot it for meat and hides, and by early 1820, quaggas had significantly diminished. Cattle competing for grass added to the pressures. In the 1850s an enzootic viral disease, African horse sickness, was rampant and killed 70,000 horses, *E. caballus*, in the Cape of Good Hope, an event that coincided with the quagga's decline. By the 1860s, few were left, and the last wild quagga was shot in 1878. The last specimen died in a zoo in 1883.

The quagga's demise was foreshadowed by the loss of the bluebuck, *Hippotragus leucophaeus*, a large scimitar-horned antelope, which occurred on the open veldt of Cape Colony, South Africa. It was hunted for its attractive pelage, described as "a fine blue of velvet appearance" (Pennant 1781, in Day 1981). It also had to compete with rising numbers of cattle. By 1799, the last ones were taken.

Hunting appears to have been a factor in the loss of two subspecies of sika deer, *Cervus nippon*, but forest loss has also been suggested. The diagnosis is not ensured in the absence of information. The Shanxi sika deer, *C. n. grassianus*, was already rare by 1918 in the forested mountains of west Shanxi, China, where it was hunted for its antler velvet (Sowerby 1918). The north China sika, *C. n. mandarinus*, ranged over the forested mountains of northeastern China. Its numbers decreased until finally it survived only in two reserves, but in 1911–1912 they were opened for hunting and the subspecies was lost.

Eared seals, true seals, and walruses have been commercially hunted over the centuries (Chapters 8 and 11). For two seals, economic extinction (when it was no longer profitable to hunt) preceded possible biological extinction. It is assumed, but not known for certain, that they are extinct, but what caused the loss of the last individuals is just as uncertain. We note, for example, the Juan Fernandez fur seal, *Arctocephalus philippi*, was assumed extinct by 1917 after years of commercial harvesting. Local fisherman were apparently aware of its continued existence, and then biologists rediscovered the species early in 1965 (Hubbs and Norris 1971).

The Caribbean monk seal, *Monarchus tropicalis*, was abundant and widely distributed until the late seventeenth century when they were taken for oil and pelts. The hunting apparently severely reduced numbers (Busch 1985), and by the 1850s the monk seal was rare. The final blow may have been fishermen who saw the seals as competitors. The last sighting was in 1952.

The Japanese sea lion, *Zalophus californianus japonicus*, had long been hunted for meat and oil, and apparently they almost disappeared in the 1900s (Nakamura 1991). In the Sea of Japan sea lions numbered about 200 in 1958 (Scheffer 1958). Although listed as extinct by IUCN, their current status is unknown, since the islands are contested by both Japan and Korea.

The last three mammals in the subsample were medium sized, and other than habitat change and hunting being possible factors, there are too few facts to support a diagnosis. The Barbados raccoon, *Procyon gloveralleni*, was formerly abundant on southern Barbados, and a hint to its possible fate lies in the mention (Nelson and Goldman 1930) of a "bounty for destruction" in the mid-eighteenth century. Habitat changes may have contributed, although elsewhere in the Caribbean introduced raccoons, *Procyon* spp., have proved adaptable to changes.

The lesser Mascarene flying fox, *Pteropus subniger* (Figure 3.4), was lost while unknown to science. Cheke and Dahl (1981) concluded that it disappeared from Mauritius probably between 1864 and 1873. That period spanned an acceleration in forest clearing, and by 1880 only

Mariana Fruit Bat
Pteropus marianus

Kathie Hollis '94

Figure 3.4.   Flying fox.

3.6% of the forests remained (Cheke 1987a). The flying fox roosted in hollow trees, and its disappearance as the forests shrank may have followed loss of both its food and shelter. Reasonable as these suppositions seem, they are only suppositions.

The Omilteme cottontail, *Sylvilagus insonus*, is a large cottontail rabbit only known from a possible recent sighting (Ceballos and Navarro 1991) and from specimens collected in the 1800s from forests in the Sierra Madre del Sur, of southern Mexico. Clearing for forestry and cattle grazing left fragments of forests, but in reality we do not know that the cottontail is extinct or what caused its problems.

3.3 EXTINCT OR ENDANGERED BIRDS FROM THE AMERICAS

Our third set of species is gleaned from Collar et al's (1992) compilation of threatened birds of the Americas. *Threatened* is IUCN's all-encompassing term for a species (and sometimes subspecies) recognized as being in trouble. We took the 96 species classified as extinct or endangered, which are essentially those taxa in danger of extinction but known not to be actually extinct (Collar et al 1992). These birds, then, are thought to be extant, even if only just so. Thus more information should be available. We dropped 13 species (Table 3.5) known only from a few specimens because the information about their ecology was too sparse to guess causes of their rarity or decline (Table 3.6). We grouped the other 83 species according to the causes of their decline as reported by Collar et al (1992).

Habitat change was the dominant cause either as a single factor (43 species) or in combination with hunting and trade (19 species), predation (6 species), or combined with several factors (15 species). For the

Table 3.5. Extinct or endangered birds known from only type specimen or a handful of specimens

Species	Last record
1. Magdalena tinamou *Crypturellus saltuarius*	1943, single male
2. Kalinowski's tinamou *Nothoprocta kalinowskii*	1900, 2 specimens
3. Jamaican pauraque *Siphonorhis americanus*	1860, 3 specimens 1980, possible sighting
4. Turquoise-throated puffleg *Eriocnemis godini*	1850, 6 specimens
5. Moustached antpitta *Grallaria alleni*	1971, 2 specimens
6. Tachira antpitta *G. chthonia*	1956, 4 specimens
7. Brown-banded antpitta *G. milleri*	1942, 10 specimens
8. Bahia tapaculo *Scytalopus psychopompus*	1983, 3 specimens
9. Kinglet cotinga *Calyptura cristata*	1828, 45 specimens
10. Tumaco seedeater *Sporophila insulata*	1912, 4 specimens
11. Hooded seedeater *S. melanops*	1823, 1 specimen
12. Cone-billed tanager *Conothraupis mesoleuca*	1938, 1 specimen
13. Cherry-throated tanager *Nemosia rourei*	1870, 1 specimen 1941, sighting

Modified from Collar NJ, et al. Threatened birds of the Americas. Washington: Smithsonian Institution Press, London: ICBP, 1992.

Table 3.6. The basis for diagnosis of the major factors
driving declines of endangered birds of the Americas

	Basis for diagnosis		
	Natural history	Associative	Experimental
Habitat	42	0	1
Hunting and habitat	10	1	0
Predation and habitat	5	0	1
Habitat, hunting, and trade	4	1	0
Habitat and other	4	0	0
Habitat and trade	3	0	0
Brood parasitism and habitat	0	2	2
Brood parasitism, habitat, and predation	0	0	1
Other and unknowns	6	0	0

Modified from Collar NJ, et al. Threatened birds of the Americas.
Washington: Smithsonian Institution Press, London: ICBP, 1992.

blue-eyed ground dove, *Columbina cyanopsis,* and the recurve-bill
bushbird, *Clyocantes alixii,* the cause is unknown. For only five species
was there certainty over the causes of the decline when the decline was
measurably halted or reversed by removing or reducing the presumed
agents—the red-cockaded woodpecker; Bermudean petrel, *Pterodroma
cahow;* yellow-shouldered blackbird, *Agelaius xanthomus;* Kirtland's
warbler, *Dendroica kirtlandii;* and the dark-rumped petrel, *P. phaeopygia,*
fitted this category. Coincidence between the measured rates of change
in the population and a factor underpinned another four cases; an
example is the live bird trade and the decline of the red siskin,
Carduelius cucullatu. For the remainder of the cases, assigning a cause to
the decline was based on no more than natural history observations.

We took at their face value Collar et al's (1992) estimation of the
threats facing birds of the Americas. We selected as a subset 22 families
to cover a range of life histories, body size, and habitats and whose
species distribution was categorized as island or continental. We tallied
169 continental versus 48 island species, and although both groups
faced similar rates of threat from habitat loss and hunting, introduced
predators were a markedly greater threat to island birds (Figure 3.5).
The percentage of other threats (pollution, hurricanes, diseases, and
human disturbances) was greater for the island species.

The categorization of causes and even the extent of declines lacks
certainty. Most were inspired guesses and suppositions. This is not a
criticism of our source (Collar et al 1992), which is breathtaking in the
number of facts compiled. It is that the amount of information available
even for extant species is insufficient for one to be confident of identifying declines and their causes for the majority of species.

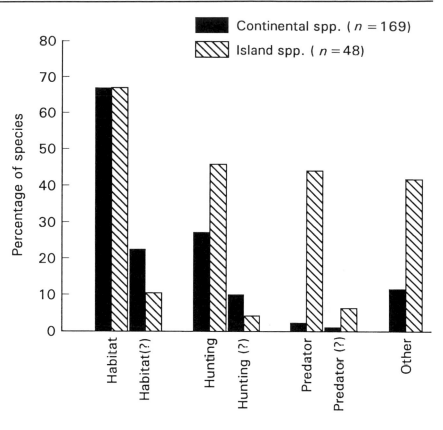

Figure 3.5. A comparison of threats facing continental and island species of endangered birds of the Americas. The "?" implies uncertainty in the identification of the threat. (Modified from Collar NJ, et al. Threatened birds of the Americas. Washington: Smithsonian Institution Press, London: ICBP, 1992.)

3.4 SUMMARY

We chose sets of mammals and birds without regard to habitat, life history, or any attribute that would bias the sample. Yet still the most striking aspect common to the samples is how little is known about many past extinctions. Indeed, we cannot be certain that some species are extinct. Where scientific knowledge is based on one or very few specimens, and no reason is apparent for their decline, then there is the possibility that the extinction was natural, a situation that would be extraordinarily difficult to prove. Common sense tells us, however, that evolution has not stopped and speciation—the loss and gain of species—is still occurring.

Even where causes of extreme rarity or extinction have been attributed to a specific cause, they are for the most part guesses made with greater or lesser degree of enlightenment. In only a few cases can the cause of extinction be deduced with confidence. It is difficult to dissociate the loss of Australian mammals from the spreading European-induced habitat changes and the introduction of alien predators and competitors in the nineteenth and twentieth centuries. Tighter yet is the coincidence between the disappearance of bush wrens and the New Zealand greater short-tailed bat and the arrival of black rats. As convincing are the instances where human hunting caused extinction, as happened to Steller's sea cow. Perhaps those apparently secure instances have contributed more than is warranted to the certainty that man was directly or indirectly responsible for all the others. There is a certain air of common sense to the assumption, and we think it would indeed be the case if enough evidence could be found to analyze each case. In the absence of those opportunities, we must concede that uncertainty or ignorance of cause is the only scientifically credible conclusion for the majority of extinctions that we have described.

4

Case histories

The sample sets in Chapter 3 reveal how little is known about many past extinctions. Categorizing the factors driving historical extinctions does not take us very far in understanding how secure the diagnoses were, yet that is essential before we can treat declines. So here we take case histories of 17 species in trouble and one extinct species to tease out what was known versus what was surmised about the declines. We describe how their causes were diagnosed and discuss the relationship between diagnosis and recovery treatments.

The case histories sample the range of problems that arise when we set out to save species. We start with the dodo's, *Raphus cucullatus*, tale because it illustrates the ease with which assertions are repeated until they become ensconced as facts. For the dodo and the other species we have commented on what we perceive as strengths and weaknesses of diagnoses and treatments. Without first-hand experience of the problem and a feel for the species' ecology, we have not, however, offered solutions. The first five case histories illustrate successful diagnoses. We then give the story of the black robin and the black-footed ferret to illustrate successful but contrasting approaches to rescue of species on the verge of extinction. The next five accounts are, for the reasons we give, less successful in their diagnoses. The last six species represent perhaps more typical problems where we are handicapped by gaps in our knowledge to secure a diagnosis or even the extent or rate of the decline.

4.1   THE DODO

The archetypal extinction is the dodo entrenched in mythology as a slow and stupid bird clubbed to death and then to extinction. The myths about the dodo start with its portraits: the accuracy of the earliest and much-reproduced sketches is doubtful. Savery (1596–1639), who drew the dodo from life, still managed to offer two versions, with and without webbed feet. He portrayed a chunky dodo with a scrunched-up stance, but recently discovered sketches show an upright (straight-legged) bird (Harmanszoon 1601 and 1602, in Kitchener 1993). A reexamination of the dodo's leg bones predict that it was fleet of foot (Kitchener 1993), which fits with Iversen's description that it was able to run fast (in Cheke 1987a). In this section we have depended on Cheke's (1987a) scholarly account of the Mascarenes' ecological history (Figure 4.1).

4.1.1   Problem

The Portuguese and Dutch reached the Mascarene Islands in the sixteenth and early seventeenth centuries and set off a rapid series of

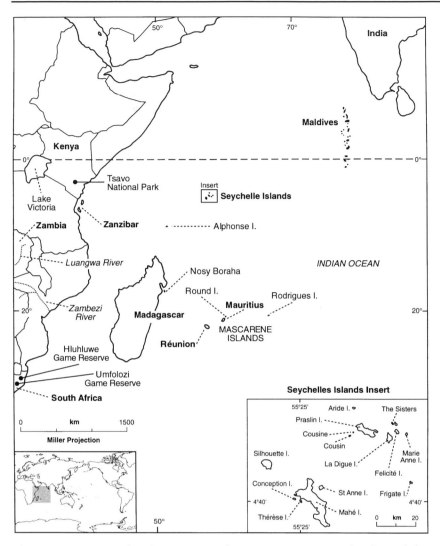

Figure 4.1. Map of Indian Ocean showing Mascarene Islands and the Seychelles.

ecological changes. Historical records frame a complex, even if incomplete, picture. Apart from sailors enthused by prospects of fresh meat, the dodo had to contend with the introduction of pigs, monkeys—crab-eating macaques, *Macaca fascicularis*—cats, and rats. By 1640, the dodo was no longer seen on the Mauritian mainland but lingered on islets until 1662.

4.1.2 Diagnosis of factors driving the extinctions

The conventional picture is of the dodo standing around allowing itself to be clubbed to death. Its tameness was taken as proof of its stupidity—a sad but telling reflection on the observers. Van West-Zanen's picture of sailors killing dodos in 1602 does not convey this image: the birds were making a nimble bid for their lives (in Hachisuka 1953). Plainly the birds were taken for meat, but no one has yet combed the various ships' logs and other accounts to estimate even minimal numbers taken. The forest was still largely intact, and its thick cover would have hindered hunting.

Crab-eating macaques arrived on Mauritius possibly in the late 1500s and quickly became numerous. Some 20 pigs were dropped off in 1606–1607 and achieved pest proportions within decades. Cats were few and are scarcely mentioned until the mid-1800s. Black rats were abundant by the time they were first reported in 1606. Monkeys and pigs would have eaten eggs and chicks, but their greater effect may have been to compete for food with the dodo. The dodo's diet is surmised to have been fruit and seeds of palms and *Pandanus*. Pigs have catholic tastes and readily eat palm seeds (Miller and Mullette 1985). Thousands of pigs and monkeys on Mauritius (1,865 km^2) would have consumed great quantities of fruits, especially palm seeds. The pigs would have cleared the understory, which may have stripped cover as well as food for the dodos. Those examples and the story of another large flightless rail (Section 4.2) point to competition for food and predation of eggs and chicks as pushing the dodo to extinction.

The ecological consequences of the extinctions are unknown. Cheke (1987a) incisively refuted an assertion that the dodo's extinction almost caused extinction of the tambalacoque tree, *Sideroxylon grandiflorum*. This endemic tree has hard seeds, but the claim that without being passed through the dodo's gizzard they would not germinate did not stand scrutiny. Another dodo myth was laid to rest.

Most people believe that the dodo was hunted to extinction because that explanation has been frequently repeated. We suggest otherwise. A different story—pigs competing for food and destroying nests in the lowland forests—can be gleaned from the history of the Mascarenes and a modest amount of ecological insight.

4.2 THE WOODHEN

The woodhen is not striking to look at, but the diagnosis of the factors holding it as a small population marooned in a suboptimal habitat is a telling story. Through ecological research, various hypotheses were examined and rejected until the limiting factor was identified. We relied mostly on Miller and Mullette (1985) and Fullagar (1985) to describe this successful and efficient diagnosis and treatment.

This flightless rail is only found on Lord Howe Island (25 km^2), which lies 570 km off the east Australian coast (Figure 1.1). It is one of

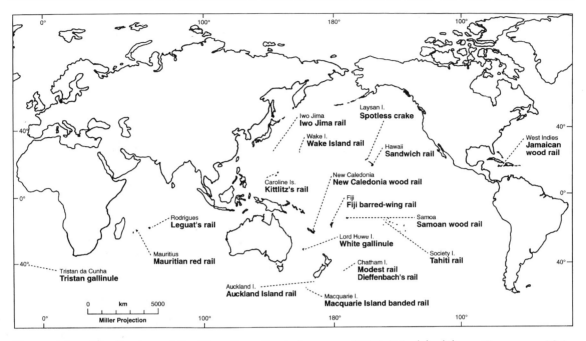

Figure 4.2. Distribution of Pacific rails extinct since A.D. 1600. (Modified from Greenway JC Jr. *Extinct and vanishing birds.* New York: American Committee for International Wildlife Protection Special Publication No. 13, 1958.)

a worldwide family, mainly weak fliers or flightless, and most oceanic islands ended up with their own species. When Polynesians, Micronesians, and Melanesians and their supporting cast of rats, cats, dogs, and pigs reached those oceanic islands (Section 2.4.4), the subfossil records reveal that most rails slid rapidly into extinction. Another 17 species became extinct within historic times (Figure 4.2).

The woodhen of Lord Howe Island came perilously close to joining the list of extinct rails, even though Lord Howe Island was one of the few Pacific islands that Polynesians did not colonize. The problems came with European colonization in 1788, and by 1924, nine of 13 species of bird were extinct (Table 8.2). The woodhen quickly became almost completely restricted to damp forested mountain peaks for almost 120 years.

The brown woodhen is unadorned except for black and chestnut barring on the wings, conspicuous when the bird spreads its wings while sunbathing. It uses its long and stout bill to probe for invertebrates, especially earthworms, in leaf litter and loose soil of the subtropical palm forest floor. Woodhens nest in petrel, *Puffinus* spp., burrows for nests and roost in the burrows or thick patches of ferns.

4.2.1   Problem

In the late 1700s and early 1800s sailing ships regularly visited Lord Howe, and sailors did not stint themselves in taking woodhens as fresh meat. People settled in 1834 and brought pigs, dogs, cats, and goats all before 1851. Before long, Hill (1870) wrote: "[Cats] . . . soon become populous and found an easy prey in the pigeons, parrots, birds like a guinea-fowl and brown hens, decimating the former and driving the latter to the mountains." This was supported by observations from one of the Australian Museum's collecting party in 1887. Etheridge (1889) wrote: "Soon to become extinct on Lord Howe, unless protected, is the woodhen, *Oncydromus sylvestris*, Sclater, a curious and stupid bird. At the present time its range is confined to the extreme southern end of the island, in Erskine Valley, and the ground around the sea-girt base of Mt Gower. It is even now rare and difficult to obtain. . . ."

The woodhens were confined to two mountainous summits surrounded by near vertical cliffs (Figure 4.3). The woodhens had to cope with moss-forest and damp cloudy weather—a different world from the sunny lowland palm forests from which they had been displaced. As early as 1914, Bell (in Recher and Clark 1974) had recognized that the woodhens were in a marginal area.

The woodhen's predicament with only 20–25 individuals was not recognized until 1969 when annual monitoring started. Numbers were stable at between six and ten breeding pairs being capped by space large only enough for ten territories. The species had probably not exceeded 20–25 birds between 1920 and 1980 and perhaps no more than double that over the previous 60 years.

The woodhens' productivity in the early 1970s confined to the summit of Mount Gower was low. Hatching success was unknown as the few nests were deep in petrel burrows and by the time the broods emerged they were mostly a single chick. Miller and Mullette (1985) record circumstantial evidence that two of three clutches were destroyed by currawongs, *Strepera graculina crissalis*—a large, omnivorous magpie-like bird, also an endangered endemic. The reasons for the low productivity were unknown.

The eight to ten territorial pairs on Mount Gower fell to only three to six pairs in 1978–1980 when nine adults (seven females) disappeared compared to four in each of the 2 previous years. Four of the seven presumed deaths were in the unusually wet winter of 1978. The disproportionate loss of females left unpaired males alone in their territories for at least three breeding seasons.

4.2.2   Diagnosis of factors driving the decline

The immediate problem was why the woodhens were confined to the summit of Mount Gower and why productivity was so low. Research

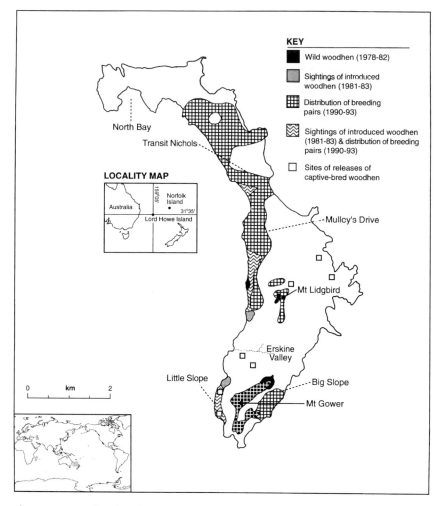

Figure 4.3. The distribution of the woodhen before and after its population recovery. (Modified from Miller B, Mullette KJ. Rehabilitation of an endangered Australian bird: the Lord Howe Island woodhen *Tricholimnas sylvestris* (Sclater). Biol. Conserv. 1985;34:55–95 and Harbin B. Personal communication, 1994.)

started and the first step was to band the woodhens and monitor numbers of breeding pairs and chicks fledging. Territory size for breeding pairs was established from sightings of marked birds over almost 5 years. Such details as how to classify the birds as male or female and adult or juvenile were established.

Research was intensified in 1978, and the key to the eventual success was that the biologist tested and refuted several hypotheses as to why the birds were presently restricted to the mountain tops. He compared weights of invertebrates at lowland sites and Mount Gower and showed that lack of food for the woodhens was not limiting them to their isolated summit. Likewise, shortage of nests or roosting sites was not a problem.

The investigation turned to whether introduced animals were the problem. While carrying out a biological survey of the island in 1969–1971, Recher and Clark (1974) noted high numbers of black rats in Mount Gower's plateau summit and thought that the rats were a factor in the woodhen's decline. Miller trapped around the island and found that rats were significantly more common on Mount Gower than elsewhere (Box 8.8). Rats might have taken woodhen eggs, but Miller found that the rats did not take domestic hen, *Gallus gallus*, eggs left out as bait for them.

Feral cats were scarce as the result of a concerted effort since 1976 to eradicate them. Examination of scats and stomach contents revealed nesting seabirds and black rats figured the most in their diet. Cats were unrecorded from Mount Gower, but they might account for birds straying onto the lowlands.

Tasmanian masked owls, *Tyto novaehollandiae*, had been introduced in the 1920s to control the rats. They failed at their appointed task, but they did spread everywhere, including the southern mountains. As they took domestic hens and a woodhen from the captive breeding pens, they might have killed woodhens, but the owls are nocturnal and the woodhens roost in cover at night.

Pigs were common on the lower slopes of Mounts Gower and Lidgbird and potentially would take eggs, incubating birds, and chicks as well as compete for food. The pigs ate large numbers of earthworms, *Lumbricus*, but comparison of invertebrates at the one site where pigs had extensively foraged with other sites was insignificant. Pigs and woodhens do not overlap even though the boundaries of each approached to within a few map meters, as cliffs and rock faces separated them. This line imposed by the steep terrain splitting the distribution of the pigs and woodhens identified pigs as the factor restricting the spread of the woodhens.

It is worth emphasizing the logic of Miller's approach. He did not succumb to building a case on a few observations alone, which would have pinpointed the currawong as a force to be reckoned with, but he compiled the evidence for and against and then discarded a series of possible causes. The causes of the high mortality of adults and low breeding success were not determined except to suggest that food shortage was not a problem. The birds were seen to shelter from rain and to sunbathe when possible, which suggests that the cool, wet, cloudy summit weather was not to their liking.

4.2.3. Recovery treatments

To open up areas for the woodhens to spread into, some 180 pigs were shot, reducing the population to one boar (1979–1981). The woodhens responded quickly, with a territory established in 1981 at a site where previously the pigs had been.

A second step in 1980 to supplement the low numbers was to start a breeding program. Translocation of individuals to unoccupied areas from the remnant Mount Gower population was considered too risky. Relying on natural spread would be slow and uncertain given the low productivity of the birds on Mount Gower. The risks of captive breeding were high as the woodhens had not been bred or even maintained before in captivity, and the three pairs removed from Mount Gower were, at the time, the only productive ones.

Once the decision for captive breeding was made, the next step was where to breed. The choice was the island itself, to minimize transfer and acclimatization times and the risk of disease. In 1980, the breeding pens were built at sea level within a large enclosure of lowland palm forest surrounded by a catproof fence. Besides the pens, a laboratory with incubators was built. Thirteen chicks were reared in the first season of captivity (1980) and by the time the program ended only 4 years later, 82 woodhens had been raised. Twenty-two progeny from the breeding program were released in four batches at one location between May 1981 and June 1982. Their survival was high, as only two were not resighted during the close monitoring. By February 1983 two pairs were seen with chicks, and both were birds that had been raised artificially, as were most birds. No difference was noted between the survival of the birds raised by their parents (six) and those raised artificially (65). In 1983 the rest of the birds were released.

The first release site (Little Slope) had no woodhens, so any evidence of woodhens could be assumed to be from the released woodhens, and it had a dense breeding colony of petrels whose burrows would double as nests and roosts for the woodhens (Figure 4.3). By timing the woodhen release with the occupation by petrels the biologists hoped the chances of owls taking the woodhens would be lessened. The release site was bounded by cliffs and the ocean, which would initially hold the birds to an area of a practical size to monitor.

The shortage of territorial space on Mount Gower for the woodhens was becoming more apparent. When three pairs were removed for captive breeding their territories were quickly occupied by non-breeding birds. The productivity of captive birds was high, as it was when birds were released on the lowlands, so it was the conditions on Mount Gower and not some inherent problem with the birds that were capping productivity. The brief pulse of captive breeding quickly succeeded. By 1987, and today (1993), the population is stable at about 50–60 breeding pairs which saturate the available habitat (mainly palm forest) on the island (Figure 4.3).

Miller and Mullette (1985) acknowledge that the woodhen had some practical advantages that helped the program's success: "The woodhen has turned out to be an ideal subject for study and management. It is tame, easy to census, adaptable, almost omnivorous, a prolific breeder in captivity and non-migratory." The key, however, was not the captive breeding or the acquiescence of the woodhen, but the correct diagnosis and correction of the problem. Without the virtual eradication of the pigs, the releases would have failed, and it would have then only taken bad luck for the woodhen to have gone extinct.

4.3 THE BERMUDA PETREL (CAHOW)

Sailors called the Bermuda petrel the cahow after its wild and haunting call, which they likened to a devil's cry. People and their pigs ate petrels and their eggs, and within 20 years of its discovery, in 1603, the petrel was gone until its rediscovery this century. The few survivors had a precarious time breeding in a suboptimal habitat. The key to successful diagnosis of which factor was limiting the petrels was behavioral observations, and we describe that diagnosis in this account drawn from Beebe (1935), Verrill (1902), Zimmerman (1975), Collar et al (1992), and, especially, Wingate (1978, 1985), who is closely associated with the petrel's recovery.

4.3.1 Problem

Subfossils show that the cahow nested throughout Bermuda (Figure 4.4), digging its nest burrows in the forest floor. Its decline preceded settlement and was triggered by hunting and the stocking of the island with pigs in 1560. The first account of the cahow was in 1603, and by 1621 petrels were believed extinct until rediscovery between 1906 and 1948. The rediscovery took four decades for misidentifications to be corrected (Murphy and Mowbray 1951).

The cahow's fortunes slowly took a turn for the better when its nesting area was found in 1951. The petrels had been driven to nest on small rocky islets free from pigs, dogs, cats, and people but populated by rats. There was no sign that the cahows were increasing from the 18 pairs counted in 1951. The reason became apparent during observations of nesting. The lack of soil for burrowing forced the cahows to nest in natural burrows and crevices on cliffs where the white-tailed tropic bird, *Phaethon lepturus*, outcompeted them. The fledging loss was 60% and had been as high as 100% in the early years after the cahow's rediscovery in 1951.

With the discovery of nesting competition and then artificial baffles to exclude the competitors, the cahow's troubles were not quite over. Wurster and Wingate (1968) reported decreased fledging success between 1958 and 1967. By the early 1970s, apparently, fledging success improved as DDT—the putative cause—left the ecosystem after a series of bans in the late 1960s. The number of breeding pairs slowly increased from 18 pairs in 1962 to 43 pairs in 1992.

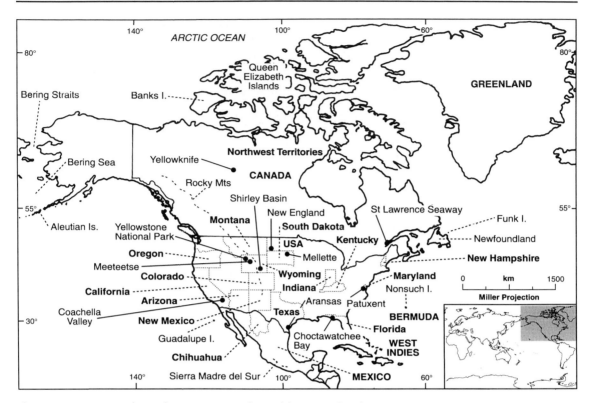

Figure 4.4. Map of North America and neighboring islands.

4.3.2 *Diagnosis of factors driving the decline*

Historical accounts strongly implicate feral pigs and people taking birds and eggs for food as responsible for the virtual extinction. The timing of the pigs' arrival and then people colonizing Bermuda and the petrels' decline coincided. In 1609, one of a shipwrecked party (Strachy, in Verrill 1902) observed that: "... [cahows] gather themselves together and breed in those Ilands which are high, and so farre alone into the Sea, that the Wilde Hogges cannot swimme over them, and there in the ground they have their Burrowes, like Conyes in a Warren."

By the time of settlement in 1612, feral pigs had reduced cahows presumably by destroying their burrows and eating birds and eggs. Newly arrived settlers ate birds and their eggs, especially when starvation hit the settlement in 1615 after rats had destroyed the first crops. Black rats, cats, and dogs arrived with people (the Norway rat, *Rattus norvegicus*, was not on the island until the eighteenth century), but their effects on the cahows are essentially unknown. The role of rats was assumed once the petrels were rediscovered in 1951, based on finding rats on the same islets as the birds, and rat remains and cahow bones were found together (Murphy and Mowbray 1951).

White-tailed tropic birds also use the same crevices as cahows, and as they breed later they simply oust the cahow's chick. Petrels leave their chicks unguarded while they are searching for food (see Section 4.12 for a description of petrel breeding behavior). Observations of white-tailed tropic birds taking over the only ten cahow nests found in 1951 and killing the single chicks led to recognition of nesting site competition as to why cahows had not recovered. This diagnosis was later strengthened by observations of successful fledging when white-tailed tropic birds were either naturally or artificially excluded. It was not until 1961 that Wingate (1978) finally located a mere eight nests in deep crevices beyond the reach of tropic birds or nests with small entrances that excluded the competitors. Those competition-free sites explained the cahow's persistence.

Cahows declined between 1958 and 1967. Fledging success had decreased by 50% (Wurster and Wingate 1968), which coincided with finding 6.44 ppm/wet weight of DDT in four unhatched eggs in 1967. A causal relationship was not established, but such levels exceeded those associated with breeding failures in other birds (Wurster and Wingate 1968). These investigators rejected inbreeding as a likely explanation for loss of chicks since previously the fledging rate had been higher. DDT levels presumably dropped during the late 1960s, when fledging success apparently rose, although levels of DDT were not measured.

4.3.3   Recovery treatments

The cahow was on the verge of extinction when, in 1616, Bermuda's governor recorded that he had issued a proclamation (Butler, in Verrill 1902): ". . . against the spoyle and havock of the cahowes, and other birds, which already wer almost all of them killed and scared awaye very improvidently by fire, diggeinge, stoneinge, and all kinds of murtheringes."

This initial treatment—protection against hunting—was too late. When 300 years later, the cahow was discovered nesting on two tiny rocky islets in 1951, the government of Bermuda moved swiftly to protect the sites as sanctuaries. Subsequently, in 1961, the nine small islands in the Castle Harbour group (10 ha) were designated as a national park.

The next management treatment was poison baiting for rats. The rats were eradicated from Nonsuch Island (one of the Castle Harbour islands) where the petrels nested. No benefit to the petrels is recorded, which reflects a lack of evidence for the effect of rats in the first place.

Most efforts to foster the cahow's recovery were practical measures to reduce competition for nest sites in the cahow's suboptimal habitat. Initially the treatment was removal of competitors, but as the tropic birds—Bermuda's national bird—floated ashore, this approach lacked local support. Exclusion of the tropic birds from the nest crevices was the next step, but a bevy of administrative and human failings delayed the baffles to selectively exclude tropic birds for almost 10 years.

Construction of artificial burrows is another measure to increase nesting of the cahow. Burrows on a low-lying area required protection of a seawall constructed in 1991 after four floods resulted in loss of two breeding pairs (Collar et al 1992). The treatments resulted in an increase from 18 pairs (eight chicks fledged) in 1962 to 43 pairs (23 chicks fledged) in 1992 (Collar et al 1992).

The initial suite of factors that drove the cahow almost to extinction seems to have been first pigs and people taking birds and their eggs and conversion of evergreen forest to tobacco fields. Once reduced to a suboptimal habitat where its particular nest requirements were scarce, the cahow fared poorly. Rat predation was blamed, but that depended on skimpy and coincidental evidence. Once attention was paid to cahow behavior, their nest competitors were identified. Then it took persistence to apply the technical measures to reduce nest competition, and the numbers of cahows started to increase.

4.4   THE LARGE BLUE BUTTERFLY

The story of the large blue butterfly, *Maculina arion* (Figure 4.5), is possibly one of the most illustrative case histories for conservation biology. The answer to the plight of this large and attractive European butterfly was arrived at through ecological study. The decline had been known for decades and was well documented. Various conservation measures had failed simply because the cause had been incorrectly diagnosed.

Our information comes from Elmes and Thomas (1992) and Thomas (1984a,b), who describe the conservation of the large blue and other species of the genus. However, it would be an injustice not to have included some of the intricate detail of the large blue's ecology as unraveled by J.A. Thomas and co-workers. Not only is the detail absorbing, but it is pertinent to management of the butterfly's habitat.

The large blue is a conspicuous butterfly in a country with a long tradition of natural history, and its plight attracted public concern and interest. Its larva is an obligate parasite in ant colonies, *Myrmica* spp., and this has been known since 1916, but it took thoughtful research to realize the parasitism's significance for the butterfly's conservation.

4.4.1   *Problem*

The large blue butterfly's extinction from southern England had been forecast in the 1880s when adverse weather may have driven a decline (Muggleton and Benham 1975). By 1930, the butterfly was no longer

The Large Blue
Maculinea arion

Kathie Hollis '94

Figure 4.5. Butterfly.

found at many sites. From the late 1940s to the 1960s, it was increasingly restricted to southwestern England, and by 1974 the number of colonies had fallen from 30 in the 1950s to only 1 or 2 (Muggleton and Benham 1975). Estimated numbers of adults had fallen from 100,000 in 1950 to 250 in 1972. Finally, two successive droughts during the breeding season extinguished the last known British colony in 1979 (Elmes and Thomas 1992). The drought caused caterpillars to overcrowd and then perish in ant nests.

The habitat of the large blue had changed greatly. The larval plant food is wild thyme, *Thymus drucei*, which is a common and widespread colonizer of bare ground in grasslands. Plowing, forestry, and changes in stock management along with less rabbit grazing after myxomatosis epidemics all led to disappearance of short-turf grasslands.

4.4.2   *Diagnosis of factors driving the decline*

The initial guess was that collecting of butterflies drove the declines. A reserve was established in the 1930s to exclude collectors and, unfortunately, domestic stock. It became overgrown and the butterflies disappeared. Pesticides, weather, and even inbreeding depression were all advanced as causes, but none were tested or substantiated.

Changes in habitat were also assumed to be a major factor, but as the habitat with the larval food plant, wild thyme, still existed it was not clear why the large blue had declined. Its parasitism of ants was known but not investigated until an intensive 6-year study of the last butterfly colony in England. The low survival (4–37%) of the caterpillars in the ant nests was limiting the population. Caterpillars feed on thyme flower buds but gain little weight. In August they wander until they are adopted by an ant after ritual of deception and bribery. The caterpillar offers the ant sugar and through smell and tactile signals cons the ant to carry it back to the ant nest. There the caterpillar's tough cuticle, acquired smell, and sugar bribes protect it as it feeds on ant larvae.

Thomas (1977, in Elmes and Thomas 1992) demonstrated that the caterpillar's survival depends on the right species of ant. If the turf increased by only 2 cm in height, the microclimate cooled, allowing the replacement of *M. sabuleti* by *M. scabrinodes* and caterpillar survival dropped from 15% to less than 2%. Until then changes in the ants had not been noticed when the grasslands had changed even slightly with less grazing.

The ants must not be short of food, as otherwise they become too discerning to accept the caterpillar's rather crude mimicry. The ant colonies have to be within 2 m of the thyme plant (the ant's foraging distance) and have to have at least 400 workers to provide about 230 larvae needed by a caterpillar. Even then, all that is not quite enough. Caterpillar survival was three times greater when the queen ant was absent than when she was present (Thomas and Wardlaw 1990). If a queen is present, nurse ants attack larger female ant larvae to stop them from developing into virgin queens (gynes). The nurse ants are triggered to attack by the gynes' pheromones which rub off onto the *M. arion* caterpillar as it feeds on the gyne larvae. The ants respond by attacking the caterpillar. Once the caterpillar has grown enough, it can engulf the larvae without becoming smeared with pheromones and thus avoids attack by the nurse ants.

There is a practical conservation significance of the relationship between the survival of the *M. arion* caterpillars and the presence of the queen ants. The queen effect is weakest (and survival of *M. arion*

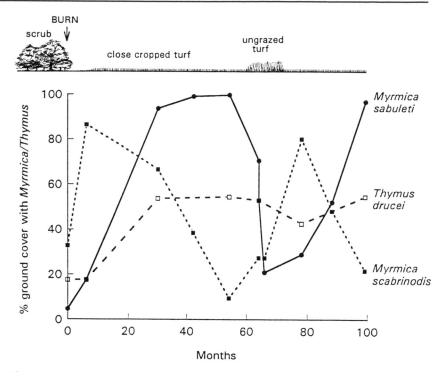

Figure 4.6. Changes in ground cover of thyme and ants as habitat management practices change. (Modified from Thomas JA. The conservation of butterflies in temperate countries: past efforts and lessons for the future. In: Vane-Wright RI, Ackery PR, eds. The biology of butterflies. Symposium of the Royal Entomological Society No. 11. London: Academic Press, 1984a:333–353.)

caterpillars highest) when an ant colony is newly established and is increasing for 3 to 4 years. Burning and grazing scrub create new habitat rapidly colonized by *M. sabuleti* (Figure 4.6). Survival of the large blue caterpillars was 52% in ant nests in newly burnt grassland compared to 27% in long-established turf, but, as Thomas (1980) notes, the results are a hypothesis to be tested to measure the queen effect and caterpillar survival under different habitat treatments.

4.4.3 Recovery treatments

Early attempts at recovery failed because diagnosis was incorrect. Once ecological research revealed that the critical factor was survival of ant colonies of *M. sabuleti*, succession of plant communities was returned to an early stage suitable for the ants by burning and grazing (Elmes and Thomas 1992). Then large blue butterflies from northern Europe were successfully repatriated in southern England. The lesson is clear: treating the butterfly's decline came from understanding its ecology.

4.5 THE SEYCHELLES MAGPIE-ROBIN

Introduced predators had exterminated the Seychelles magpie-robin, *Copsychus sechellarum* (Figure 4.7), from eight islands until only Frigate Island (Figure 4.1) held a population. We chose it as a case history because it illustrates how monitoring and ecological studies revealed why the remnant population remained capped despite removing the cause of the initial decline.

The magpie-robin is actually a thrush, and others of its genus are found in Madagascar and Asia. It is a confidingly tame ground-feeding bird with black plumage highlighted by a blue sheen and a white bar on the wing. It is relatively long-lived (8 years) and usually lays a single egg. The magpie-robins fledge about 10 days before their flight is strong and their weak fluttering flight would leave them vulnerable to predators. Frigate is a low rocky island (202 ha) settled probably in the early 1800s. The forests were quickly cleared, first for coconut

Seychelles Magpie-robin
Copsychus sechellarum

Kathie Hollis '94

Figure 4.7. Magpie robin.

plantations and then for agriculture. Currently, secondary forest, coconut palms, crops, and grasslands cover the island. Our account draws from Watson et al (1992), Gretton et al (1991), and Wilson and Wilson (1978).

4.5.1   Problem

Magpie-robins on Frigate Island have been monitored at irregular intervals since 1960. Sightings of color-banded adults and nest checks revealed an increase from a low of 12 birds in 1965 to 38 to 41 birds in 13 territories by 1979. Then productivity dropped, with most losses during egg laying and incubation. By August 1981 only 20 birds were alive. Plainly something had changed for the magpie-robins despite concerted trapping and poisoning of feral cats. In 1984 the birds still only numbered 25, and, by 1988, 21 birds occupied eight territories. The survival of young per pair had dropped from 0.6 in 1978 to 0.13 in 1988–1989.

4.5.2   Diagnosis of factors driving the decline

The diagnosis for loss of magpie-robins from the Seychelles is based on coincidences in timing and overlap in distribution. In 1866, Newton (1867) commented that dogs, cats, and rats had reduced the birds on five islands. Robins were common on Aride Island in the late 1800s. Cats were introduced in 1918 and may have increased to 200 by 1925. Hunting exterminated the cats by 1932. The robins had survived but then disappeared after another introduction of cats in 1937. The robins may have survived the cats as long as they did because seabirds nest on Aride Island and the cats would have had other prey when the robins were fledging. By 1938, Vesey-Fitzgerald (1940) recorded that magpie-robins were only found on Alphonse and Frigate Islands, neither of which had cats. Alphonse is a coral island (11.3 km^2) about 400 km southwest of the Seychelles, and the magpie-robins there had been introduced late in the nineteenth century. They thrived until cats were introduced in the 1950s, and one robin survived to 1959 when it, too, disappeared.

The case against the cats was strengthened when the low number of magpie-robins in the mid-1960s on Frigate Island coincided with the earlier establishment of cats. After cats were reduced (86 cats were killed in 1960), the magpie-robins slowly increased. The next decline in the robins in 1981–1984 coincided with an increase in cats killed, but magpie-robin numbers did not recover to levels of the late 1970s.

Newton (1867) mentioned introduced Indian mynah, *Acridotheres tristis*, as a possible threat to eggs and nestlings, and Watson et al

(1992) commented on the magpie-robins' aggressive defense of their nests against mynahs and skinks, *Mabuya* spp. Aride Island and Frigate Island were the only islands not apparently (up until at least 1976) invaded by black rats, and Norway rats had not reached the Seychelles.

The apparent stability in the numbers in the late 1970s raised the question of what was capping the population. A field study of the robin's distribution and behavior was to investigate whether food or feeding habitat was limiting. The methods were simple and effective. Sightings of magpie-robins were marked on a map gridded with 1-ha squares where habitat had been mapped. The results were that the magpie-robins were associated with habitats that had extensive bare earth and leaf litter: those areas were 43% of the island in the late 1970s. Most (93%) feeding bouts were on the ground and on bare earth or litter.

The magpie-robin's need to forage for invertebrates in soil or leaf litter may explain why the population did not bound back in 1980. A switch in the island's economy toward tourism led to people abandoning their vegetable plots. As those plots reverted to herbaceous growth, they were not attractive to the birds and five territories were lost.

The lack of food in the territories as the bare ground was revegetated forced the adults to spend more time foraging and less time at their nest. Support for this comes from the resumption of nesting in two territories after supplemental feeding. The unguarded nests are more vulnerable to predation. This was compounded by a shortage of nest sites. Magpie-robins prefer rotted out tree holes (and not, presumably, palm trees) which may be remote from feeding areas.

4.5.3 Recovery treatments

The first treatment—the reduction of cats in the 1960s and then in the 1980s—twice reversed the robin's decline. The second treatment was supplemental feeding and to provide suitable foraging sites through habitat management, and this is still under way. The third treatment was an attempt to translocate birds to another island. Five mated pairs were taken in 1978–1979 to Aride Island, but in 1980 only one survivor was found. The reasons for the failure might have been the stress of transfer or possible use of insecticides, or the magpie-robins may have become entangled in the sticky fruits of a tree that no longer grows on Frigate Island.

The magpie-robin's case illustrates the importance of looking beyond the conspicuous factor, in this case an introduced predator. It took monitoring and ecological field work to reveal the consequences of changes to an already modified habitat. Those changes caused a food shortage which capped the single-island population.

4.6 THE BLACK-FOOTED FERRET

The story of the black-footed ferret is not just the high-profiled rescue of North America's only wild ferrets. It is also the tale of a decline to virtual extinction that took place this century. The decline was predicted and the driving cause known but it still happened (Box 4.1). Ferrets may have numbered some 800,000 in the 1920s, yet the last ones in the wild were taken into captivity only 60 years later. The rescue was a last-ditch effort with federal, state, and private interest groups all involved, mostly cooperatively.

We chose the ferret's story because it is an example of what happens when we fail to act on a decline until it is almost too late and when the

---

Box 4.1. Chronology of events in the decline and recovery of black-footed ferrets

---

1964—Rediscovery at Mellette County, South Dakota; studies for next 11 years.

1965—U.S. Fish and Wildlife Service list ferret as endangered.

1968—Reproductive studies of related species begin.

1971—Four females and two males captured from Mellette to begin captive breeding in Patuxent; four females die from vaccine-induced CDV.

1972–1973—Two females and one male taken from Mellette to Patuxent.

1974—FWS establishes recovery team; Mellette colony dies out.

1976–1977—Captive pair at Patuxent breed producing two litters; no young live.

1978—Recovery plan approved (no extant ferrets known, including none at Mellette).

1979—Last ferret at Patuxent died.

1981—Ferret colony discovered at Meeteetse, Wyoming.

1982–1984—Ferrets radio-collared to sample demography, home range, and movements.

1983—Captive-breeding plan again (1982) considered premature.

1984—Meeteetse colony reached peak of 129 ferrets.

1985—Meeteetse colony rapidly declined to 58; sylvatic plague outbreak and massive efforts to control it by dusting burrows for fleas. Six ferrets taken for captive breeding; all died of CDV. Six more ferrets captured and quarantined.

(Continued)

Box 4.1.   (Continued)

1986–1987—All of the Meeteetse colony of ferrets captured.

1987—Two litters produced in captivity (seven young survived).

1990—A total of 188 ferrets in captivity; 80 kits born and 58 weaned.

1991—Forty-nine 4-month-old ferrets released Shirley Basin, southwestern Wyoming.

1992—A total of 349 ferrets held in seven institutions. Nine released ferrets seen and two litters. A further 90 ferrets released.

1993- -Nineteen ferrets seen and four litters. Forty-eight ferrets released.

1994—Only six ferrets found but no evidence of CDV.

(Modified from Cole BP. Recovery planning for endangered and threatened species. In: Seal US, Thorne ET, Bogan MA, Anderson SH, eds. Conservation biology and the black-footed ferret. New Haven: Yale University Press, 1989:201–209; Thorne ET, Oakleaf B. Species rescue for captive breeding: black-footed ferret as an example. Symp. Zool. Soc. London 1991;62:241–261; Barton M. Black-footed ferret rides again. BBC Wildlife 1992;10(9):63; They're going wild again in Wyoming. BBC Wildlife 1992;10(1):56; Miller BD, Biggins D, Hanebury L, Vargas A. Reintroduction of the black-footed ferret (*Mustela nigripes*). In: Olney PS, Mace GM, Feistner ATC, eds. Creative conservation: interactive management of wild and captive animals. London: Chapman and Hall, 1994:455–464.)

opportunity to come to grips with its ecology is almost lost. Doubts linger about whether enough is known about the ferret's decline and ecology to be confident about establishing it back in a prairie ecosystem which has itself changed.

The black-footed ferret is a strikingly masked and lithe specialist predator of prairie dogs (ground squirrels, *Cynomys* spp.) in North America. Its long slender body allows it to slip down burrows to find its prey while they sleep—a tactic that lessens the ferret's risk of injury when killing prey the same size as itself. The ferret ranged from the Canadian prairies, west to the intermontane basins of the Rockies and south to Texas. Their dependence on the prairie dog is total: They live in the colonies and take prairie dogs for food and use the burrows for shelter and to duck out of reach of predators.

The recovery efforts have spawned a mass of publications. The 1985 workshop proceedings detail various models and considerations for captive breeding (Seal et al 1989). Thorne and Williams (1988) and Thorne and Oakleaf (1991) give a balanced view of the events as they unfurled through the planned recovery. Miller and Anderson (1993) describe the ferret's behavior. It is, however, the Wyoming Fish and Game reports that reveal the extraordinary amount of planning and research that has been behind the breeding and release of the ferrets (Oakleaf et al 1992 and 1993). The U.S. Fish and Wildlife Service's (1988) Recovery Plan summarizes other information.

4.6.1 Problem

In 1918 Nelson (1918) predicted the fate of the black-footed ferret, which had impressed him

as one of the strangest [of the North American fauna] and most like a stranded exotic . . . [which] exist as parasites in the prairie-dog colonies, making their homes in deserted burrows and feeding on the hapless colonists . . . With the occupation of the country and the inevitable extinction of the prairie-dog over nearly or quite all of its range, the black-footed ferret is practically certain to disappear with its host species.

Contemporary data suggest that a ferret needs a prairie dog colony of at least 12.5 ha to exist for a year and at least 50 ha to raise a litter. In the 1920s, an estimated 40.5 million ha of prairie dog colonies may have supported 800,000 black-footed ferrets. Ranchers and farmers saw the prairie dogs as a scourge and eradicated them with poison and plow. The onslaught which began in the early 1900s eradicated 90–95% of the historically occupied prairie dog range. And it is not over. Prairie dogs, despite the magnitude of their decline, are still being poisoned. Even in the 1990s federal government agencies authorize and subsidize the annual clearing of 80,000 ha of colonies despite evidence that the colonies have a minimal effect on cattle grazing, and poisoning costs more than any cattle weight gain (Miller et al 1994b).

By the 1960s, ferrets occurred in only ten U.S. counties compared to 130 in the 1800s. A small pocket of survivors was found in 1964 in Mellette County, South Dakota, and some were taken into captivity (Figure 4.4). Ferrets were seen in 17 different prairie dog colonies in Mellette County between 1964 and 1973 that were increasing after control had virtually stopped since 1964. The ferrets died out both in the wild (for reasons unknown) and in captivity by 1979. Another colony at Meeteetse, Wyoming (Figure 4.4), was discovered in 1981 after a local sheepdog killed a ferret raiding his food. After canine distemper virus reduced the colony, the last ferrets in the wild were taken into captivity in 1987.

4.6.2   Diagnosis of factors driving the decline

The diagnosis is based on our understanding of the black-footed ferret's dependence on prairie dogs. The ferret's disappearance coincided with the destruction of most prairie dog colonies. The ferrets became isolated in small populations as the prairie dog colonies were reduced in size and number. Other than the certainty that their population dynamics were linked, little is known about the predator–prey relationships between ferrets and prairie dogs. The ferrets occasionally take other prey, but their dependence on the prairie dogs is also for burrows of the right diameter that they can slip into away from predators such as coyotes, *Canis latrans*.

Prairie dogs are vulnerable to epidemics of sylvatic plague, *Yersinia pestis*. Plague at Meeteetse destroyed 22% of the prairie dog colony in 1985, but outbreaks can be more extensive: The epidemology of sylvatic plague as the colonies of the prairie dogs became reduced and fragmented is unknown. An added complication would be any susceptibility of ferrets to sylvatic plague. Other predators of prairie dogs are resistant, but at least one ferret succumbed after eating an infected prairie dog. Possibly ferrets have had not had enough exposure to the plague, which may have only reached North America this century (Williams et al, 1994). The ferrets are also highly susceptible to canine distemper virus, which is endemic in carnivores on the prairies.

The ferrets taken into captivity were from a small population presumed to have been isolated for an estimated 50 years (numbers of breeders probably varied between 15 and 35). A limited electrophoretic survey of 12 ferrets estimated a heterozygosity of $H = 0.008$, which was described as being low (O'Brien et al 1989). The measure begs the question of, low relative to what? "Relative to other species of the family" is only partly revealing as their heterozygosity was low, but then it is also within the range of 18% of 169 mammal species tested (Caughley 1994). O'Brien et al (1989) concluded that the ferret's genetic variation was as low as that of the cheetah, which implied a problem for the ferrets. The cheetah's low heterozygosity had been believed to underlie various reproductive shortfalls and susceptibility to disease. That was a risky conclusion, to judge from a reevaluation of data on the cheetah (Caughley 1994; Caro and Laurenson 1994). And low heterozygosity based on protein electrophoresis alone is not an adequate basis to presume that the genetic variation is low (Section 6.3.2).

4.6.3   Recovery treatments

Black-footed ferrets have been legally protected since the mid-1960s and listed as endangered in 1973. The fortuitous finding of a ferret colony in 1981 raised the possibility that others existed, but adequate surveys did not cover the historic range. Efforts went instead to rescuing the ferrets after an outbreak of canine distemper, and the few survivors were used to found a captive population in 1985. The captive breeding built on experiences from the first attempt in 1971 when the black-footed ferret's close relative, the Siberian ferret, *Mustela eversmanni*, was a surrogate to investigate breeding physiology. Black-footed ferrets increased in captivity, and by 1992, there were 349 ferrets in seven captive colonies.

Choice of a release site depended on rigorous criteria and ecological research, which included studies of badgers, *Taxidae taxus*, as possible competitors for prairie dogs, reservoirs of canine distemper, and predators of ferrets. The extent of the prairie dog towns was mapped, and serosurveys of carnivores and prairie dogs for sylvatic plague and canine distemper revealed both diseases (Williams et al 1992). An outbreak of canine distemper had been present in coyotes and badgers in 1989–1990 at the release site but was presumed over by 1991 (Oakleaf and Luce 1992).

Planning for releases took into account the effects both numerically and genetically on the captive population as well as for the released population. Avoidance of canine distemper is a priority, and released ferrets are vaccinated with the killed vaccine even though it only offers short-term protection. Releases totalled 187 juveniles in the falls of 1991, 1992, and 1993 at Shirley Basin, Wyoming (Figure 4.4). The ferrets bred during the first winter, and survival was low as predicted for juveniles. Predation, dispersal, and disappearance reduced numbers, and starving or injured ferrets were rescued. Radio tracking is being replaced by searchlight surveys, and monitoring is used to modify release techniques. The second release in 1992 was to compare the survival of ferrets raised under two contrasting regimens in the same year. Ferrets raised in enclosures with prairie dogs survived better than those in cages. By 1994 the aim was realized to liberate adult females with their litters which had experience of being raised in enclosures with prairie dogs living in burrows. The year 1994 was also when ferrets were released in Montana and South Dakota (B. Oakleaf, personal communication, 1994).

Development of release techniques was paralleled by experiments testing the ability of captive Siberian ferrets to learn to avoid predators and catch prey (Miller et al 1994a). Survival of neutered ferrets released into a prairie dog colony was highest for five wild-caught ferrets (from China) and least for cage-reared ferrets. Siberian ferrets captive-raised but with experience of catching food in enclosures and exposed to a surrogate predator (domestic dog) fared better than cage ferrets but not as well as the wild-raised ones.

Considerable effort led to successful captive raising and release. The detailed plans acknowledge the need to prepare sites for repatriations, but the availability of suitable habitat may become a bottleneck as it takes years to identify possible sites, let alone to negotiate suitable management plans and agreements for the sites. It took 7 years to increase the average area of prairie dog colonies from 37.4 to 54.8 ha in Montana. The Russian steppe marmot, *Marmota bobak*, was successfully repatriated in Russia; its numbers are increasing and it is recolonizing large areas (Bibikov 1991), which may hold some lessons for enhancement of prairie dog colonies.

The ecology of the prairies has changed. Cattle have usurped the role of bison, *Bison bison*, grazing in keeping the grasses short, which favors "prairie dog" colony expansion (Reading et al 1989). Coyotes, a predator, are more common now that wolves, *Canis lupus*, have disappeared from prairies. Those changes underlie problems encountered with the swift fox, *Vulpes velox*, repatriation on the prairies (Section 9.5.2), which may have parallels for the black-footed ferret.

Public conflict arises from the dependence of the ferrets on prairie dogs, which are still seen as threatening profitable ranching. Organized or incidental poisoning and shooting to control prairie dogs and coyotes continues on rangelands. The statewide inventories of prairie dog colonies rank them by size, and in 1989 the guidelines were that complexes less than 400 ha could be cleared (Biggins and Crete 1989). This continued control can only be a threat to the long-term survival of prairie dogs, ferrets, and other members of the prairie ecosystem. Spatial organization of the prairie dog complexes (groups of colonies) may be as important as size for their perpetuation and their colonization by ferrets.

We draw attention to May's (1991) comment that he doubted the long-term success of reintroducing black-footed ferrets until the epidemiology of canine distemper is understood. And we would add to that the need to understand what sylvatic plague does to the patchiness and success of prairie dog colonies and its effect on the ferrets themselves.

It is too soon to unequivocally rate the black-footed ferret's repatriation as successful, but it has had an auspicious start thanks to its detailed planning. Diagnosis of the ferret's decline as caused by prairie dog eradication seems clear-cut as a proximate factor and disease as at least one ultimate factor. What is less obvious is the long-term interactions between ferrets and prairie dogs and disease epidemics. Perhaps with the captive populations looking more secure, emphasis will shift to ecological studies of prairie dogs and ferrets.

4.7   THE BLACK ROBIN

The black robin is endemic to the Chatham Islands, which lie some 450 km off New Zealand's east coast (Figure 2.1). The species slid to only one breeding pair after introduced predators and habitat changes

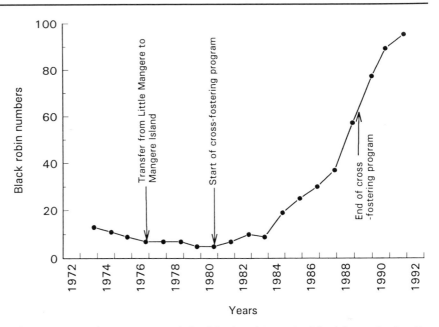

Figure 4.8. The recovery of the black robin. (Modified from Butler D, Merton D. The black robin: saving the world's most endangered bird. Auckland: Oxford University Press, 1992.)

extinguished one and then another island population. We selected the robin's recovery as an example of how ecological and behavioral information combined with intensive and painstaking field work boosted recovery from one breeding pair to two populations which totaled 140 robins in 1992 (Figure 4.8).

This melanistic member of the Australasian flycatcher family is distinguished by having the shortest and roundest wings, presumably the consequence of a predator-free environment (Fleming 1950, in Baker 1991). Our account is drawn from Butler and Merton (1992), and they leave no doubt about the value of expert field ecology combined with meticulous planning.

4.7.1 Problem

The Moriori had, by the fourteenth century, colonized the Chatham Islands (Figure 2.1). Hunting birds for food possibly caused extinction of the larger species in the 18 taxa of birds that disappeared. Habitat stayed largely unchanged until Europeans arrived in 1843. Then fire and domestic stock destroyed forests and birds were taken for food and museums. Cats introduced to control rabbits on Mangere Island (113ha) apparently exterminated the rabbits and 12 species of birds, including the black robin. Black robins were recorded still on Little Mangere Island (22ha) in 1892. The island's remoteness meant that subsequent information was sporadic, but by 1938 the island was clearly the black robin's last refuge.

Counts during visits in 1938, 1961, and 1968 suggested that black robins were stable at 20–35 pairs. Once the birds were individually color-banded in 1972 a different picture was evident. The robins habitually followed observers and that inflated counts; only about 26 robins remained. Annual counts began and tracked a decline in adults. By 1976–1977 when a transfer to Mangere Island was proposed, only seven robins remained, including just two adult females. Both were transferred with their mates and another male to Mangere Island in 1976. One of the females was Old Blue (band number 11384). She was born in 1971 and did not raise chicks until she was 8 years old. From such an inauspicious start, Old Blue made a remarkable contribution to saving her species—she was the only female to leave descendants.

4.7.2   *Diagnosis of factors driving the decline*

The first diagnosis was that black robins were in danger because they were a single and small population. Diagnosis of what reduced the population depends on coincidence. Cats probably exterminated black robins on Mangere Island, although any effect of clearing native woodland for sheep raising is unknown. The only remaining robins were on Little Mangere Island where adult deaths during breeding and low productivity indicted food shortage in a poor-quality habitat. The burrows of nesting sooty shearwaters, *Puffinus griseus* (60,000 pairs), prevented seedling establishment and woodland regeneration. Furthermore, their burrowing and scuffling hindered accumulation of leaf litter, which reduced insect availability. The shearwaters were piling into Little Mangere Island as habitat changes and introduced predators reduced the attractiveness of other islands for nesting.

4.7.3   *Recovery treatments*

Recovery efforts were to supplement the small population by boosting productivity and establishing a second population. It is worth noting how that decision was reached. Captive breeding was rejected because raising small insectivorous birds is difficult. Little Mangere Island was too small and too disturbed by shearwaters to offer potential for supplementing the robins. Removing the shearwaters was impractical and politically unwelcome because of traditional rights to harvest them. Exclosures to keep out shearwaters did increase leaf litter and tree recruitment on the island, but the number of exclosures was too few. Raising live food to supplement natural food was difficult and ran the risks that undesirable competitors would benefit and disease might be introduced: those reasons discounted artificial feeding. The outcome was that the New Zealand Wildlife Service would transfer robins from Little Mangere to another island (Box 4.2).

Early on, the vulnerability of a single locality had been recognized, and transfer of either adults or eggs to be fostered by tits was mooted in 1971. The closely related South Island robin, *Petroica australis australis*, was the surrogate with which to gain experience in translocating birds. Journeys of more than a few hours stressed them and they lost weight. Candidate sites were the only two islands without predators: Southeast Island needed a boat journey of 2 hours from Little Mangere Island and had a dense population of Chatham Island tomtits, *Petroica macrocephala chathamensis*—and robins and tomtits were believed to compete. Mangere Island was closer, had a lower density of tomtits but only enough suitable bush (5 ha) for a few robins. Sheep had been removed in 1961 and tree planting started. The decision was taken, and all the Little Mangere robins were transferred to Mangere in 1976–1977.

The key to the black robin's recovery was cross-fostering. This had not been attempted in the wild, and its potential value arose from a chance observation in 1979. The robins laid again when they lost their first clutch. Recovery efforts then shifted to use cross-fostering as a boost for productivity. Nothing is that simple: the first clutches of robin eggs were placed in the nests of Chatham Island warblers, *Gerogyne albofrontata*, on Mangere but the chicks died. An alternative foster parent would have been Chatham Island tomtits, but they had been removed to avoid any competition with the robins (tomtits were repatriated to Mangere in 1987). Biologists became adept at shuttling robin eggs and chicks between Mangere and Southeast Island where the Chatham Island tomtits successfully raised robins. Another problem then appeared. Fostered robins had imprinted on the tomtit foster parents and would not breed. The solution was to transfer cross-fostered robins back to Mangere as fledglings.

Southeast Island was chosen for the second population despite doubts that black robins had ever been on Southeast Island. The first release of four robins in 1983 established a breeding population.

Bad luck in the form of extremes in the weather threatened the recovery. El Niño caused the 1982–1983 summer to be dry, and robins laid fewer eggs than expected probably because insect food was scarce. Ten of the 38 robins disappeared following a severe storm in July 1985, although avian pox could not be discounted. The high egg fertility (90%) and hatching (80%) suggest that inbreeding was not a problem despite the robins having numbered only 10 to 30 pairs for about 100 years and having been reduced to one breeding pair.

The urgency of finding so few surviving robins in the late 1970s triggered immediate efforts to deal with small population size. The recovery work depended on sustained field work, which took many people—for 20 years up to eight people were involved for 2 to 3 months a year. The planning was meticulous and built on detailed monitoring

---

Box 4.2. Chronology of the Chatham Island black robin's recovery

---

1871—Black robins collected and described.

1938—First sightings since 1892 of only population (Little Mangere Island).

1950—The black robin described as an endemic species.

1968—Comparison of counts with 1961 suggest stable population.

1969—Recognition of need to translocate robins to other islands.

1971—Practice of transfer techniques using South Island robins.

1972—Banding and counts begin.

1973—Population estimate of 17–19 birds; tree planting on Mangere Island.

1974—Monitoring breeding success (11 adults, 6–7 chicks fledged).

1975—Monitoring population (nine adults, two juveniles).

1976—Planting trees on Mangere Island; removal of tits from Mangere Island and Little Mangere Island; successful transfer of two female and three male black robins to Mangere Island.

1977—Both pairs had bred on Mangere Island; last two males moved to Mangere Island.

1978—Two female and five male robins left; a land slip had buried 20% of habitat.

1979—Observation of black robin double clutching; two pairs and a juvenile robin left.

1980—Removal of robin eggs and cross-fostering with warblers.

1981—Trials with warbler's egg, then cross-fostering with tits on Southeast Island.

1982—Monitoring and cross-fostering.

1983—Two robin pairs transferred to, and cross-fostering using tits on, Southeast Island.

1983–1984—Four pairs and nine young fledged.

1984–1986—Cross-fostering continued on both islands.

1988–1989—Comparison of management levels; 38 pairs and 42 young fledged.

1989–1990—Intensive management replaced by monitoring.

1991–1992—Thirty-nine pairs and 32–37 fledged.

(Modified from Butler D, Merton D. The black robin: saving the world's most endangered bird. Auckland: Oxford University Press, 1992.)

and careful behavioral observations. Boosting population size was paralleled by forest restoration to ensure the robins would not be limited by food.

4.8 THE AFRICAN ELEPHANT

The significance of this case history is that it records a serious decline going on right now. It emphasizes the point that conservation problems of continental dimensions are not episodes restricted to past history. And it illustrates a decline that continued for millenia but whose diagnosis was slow in coming.

The African elephant, *Loxodonta africana*, occupies a discontinuous range (7 million km²) south of the Sahara. It was once more widespread but had been lost from north Africa by the early Middle Ages (Carrington 1958), from most of South Africa in the eighteenth and nineteenth centuries (Bryden 1903), and largely from west Africa in the nineteenth and early twentieth centuries (Douglas-Hamilton 1987a,b). By 1979 the minimum estimate was that there were 1.3 million elephants, which was halved by 1987 (Douglas-Hamilton 1987a,b).

4.8.1 Problem

Recognition that the African elephant was in trouble, at least regionally, is ancient. Quoting Parker and Amin (1983): "Pliny said it in Roman times, Fetherick claimed it in 1869, Scheinfurth in 1872, Holder in 1986, Bryden in 1903, Kunz in 1916. . . ." This longstanding picture of decline was interrupted in the late 1950s and 1960s by reports almost exclusively from national parks and reserves. From those enclaves, most studies reported increases in elephants accompanied by a reduction in forest cover caused by the elephants themselves (Buechner and Dawkins 1961; Glover 1963; Steel 1968; Watson and Bell 1969; Laws 1970; Field 1971; Caughley and Goddard 1975; Laws et al 1975; Caughley 1976b). That was "the elephant problem" back then. An acrimonious debate raged over whether elephants should or should not be culled in national parks, and the flavor of that can be sampled in Owen-Smith (1983). The long-term trend was briefly forgotten in favor of the problem in the parks and reserves. That temporary intermission disappeared in the mid-1970s when a wave of poaching reduced elephants in parks except in southern Africa.

Aerial surveys in the mid-1970s confirmed what had been suspected on the ground, that elephants were declining in many parks north of the Zambezi River, which forms the border between Zambia and

Zimbabwe (Figure 4.9). Thereafter the decline was precipitate, continuing through the 1970s and 1980s.

4.8.2 Diagnosis of factors driving the decline

The cause of the decline has always appeared obvious: the human demand for ivory. This age-old theory was immeasurably boosted when in 1969 the international price for ivory commenced a steep rise that continued through the decade in a previously unparalleled manner (Parker 1979). The problem was seen as a quantal expansion of an earlier trend.

An alternative explanation comes from analyzing ivory export records from Africa and customs records of importing countries (Caughley et al 1990). The increase in ivory production from 1970

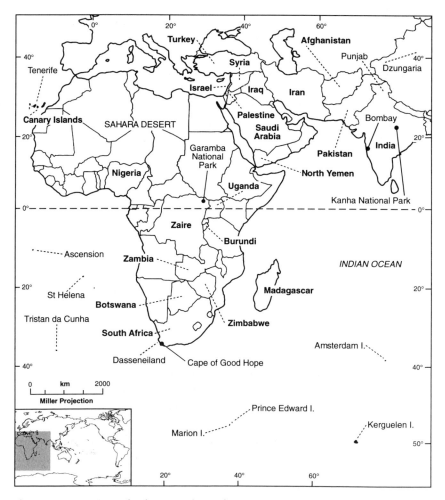

Figure 4.9. Map of Africa and southwest Asia.

reflected a seamless extension of a long-term exponential trend in production effort that had been in place from at least 1950, that trend being perturbed only minimally by fluctuations in the ivory price. Exponential trends can cause unpleasant surprises, nothing much appearing to happen for a long time and then the state of the system suddenly changes. They suggested that the trend in ivory production was inexplicable unless elephants had been declining progressively over the previous 40 years.

The volume of ivory exported parallels the decline in elephants, which is not the same as saying that the export caused, totally or partly, the decline. The graph tells us nothing about why elephants were killed. Ivory came from elephants dying naturally (Box 10.2), or being killed legally to protect crops, during culls, and so on. We have to look further to determine if the production of ivory caused or coincided with the decline.

Parker and Graham (1989a,b) examined and accepted or discarded factors that coincided with the elephant decline. The most telling evidence that the ivory trade was not the only factor is that other large mammals declined over the same time, except people, who were increasing. Between 1931 and 1990 the annual rate of increase of people in East Africa was 3%. The ranges of 30 species of large mammals had contracted by almost half between 1925 and 1975 and most of those species were not harvested for valuable or valued products. Instead they were on the losing end of competition for space between people and wildlife as habitats are converted to agriculture and pasture. In particular in East Africa, human and elephant densities were inversely related. People and their activities excluded elephants from the higher-rainfall areas. As people moved into an area, elephants moved or were moved out.

4.8.3   Recovery treatments

Reserves and restrictions (on harvesting ivory) sum and simplify what has been done for elephants. The debate on slowing then stopping the ivory trade has, in the 1980s and 1990s, hijacked elephant conservation. In 1989 the African elephant was shifted from Convention on International Trade in Endangered Species of Wild Fauna and Flora (CITES) Appendix II (controlled trade) to Appendix I (no trade). That affects mostly the importing countries. But the trade continues on a substantial scale, although production effort may have decreased.

Leader-Williams and Albon (1988) and Leader-Williams (1990) argued that it is entirely possible to protect elephants from unauthorized killing so long as the money available for that conservation effort is concentrated on specific areas rather than being spread across a country. They graphed the money allocated for conservation against rate of change in elephant numbers for 14 African countries, showing that zero change had an intercept at $215 (U.S.) per square kilometer in 1980. We are not aware that this clear demonstration has had any effect upon the way funds are allocated for the conservation of elephants.

Reducing the ivory trade is a stopgap measure. The growing number of people and their requirements have, in effect, sharply reduced carrying capacity for elephants. The solutions to that are not simple, but they start with listening to the people who live with elephants and not just to those who wish to conserve them (Dublin 1994).

4.9 THE PUERTO RICAN PARROT

The Puerto Rican parrot, *Amazona vittata*, has become the flagship for conservation in the Caribbean after intensive efforts to foster its recovery. Numbers had declined over decades until they reached a low of 13 parrots in 1975. The diagnoses of its problems have been shaky and confounded by opportunistic treatment of factors whether clearly diagnosed or not. The parrot has faced several threats which have changed over time. Nest sites may be in short supply, but food shortage and a disrupted social order cannot be discounted. We have relied on Snyder et al's (1987) monograph for this account.

The parrot's green plumage is relieved by a bright red forehead and blue primaries, and it is the only endemic *Amazona* parrot on Puerto Rico. Parrots pair but do not breed until they are 4 years old. The role of nonbreeders and whether they support the breeding pairs is uncertain. Breeding pairs are faithful to their nest cavities formed by heart rot in the palo Colorado, *Cyrilla racemiflora*, tree. Annual breeding coincides with the dry season and fruiting of the sierra palm, *Prestoeo montana*, which may be the triggered to breed. First-year mortality is 30% compared to 9% adult loss based on the disappearance of known birds.

4.9.1 Problems

The parrot was abundant when Europeans reached the island in the 1650s and remained numerous until the nineteenth century. In that time they were taken as food, as pets, and were killed when they raided crops. Between 1850 and 1900 the human population increased to 1 million and 99% of the forests were cleared (Brash 1987). The parrots had became restricted to high rugged land or mangrove swamp, and by the 1930s they were only found in the Sierra de Luquillo (Figure 4.10), a patch of high montane rainforest (1,600 ha). It intercepts the trade winds which annually drench the area with about 400 cm of rain—

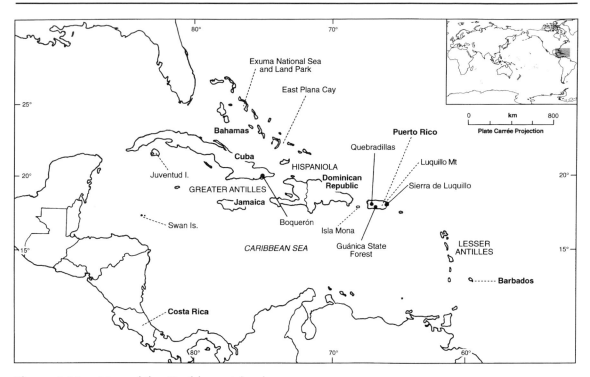

Figure 4.10. Map of the Caribbean Islands.

triple what the parrots on the lowlands would have contended with. The high rainfall and elevation might mean Sierra de Luquillo is suboptimal habitat for the parrots.

The parrots in the Sierra de Luquillo possibly numbered 2000 in 1937, 200 in 1953–1954, and 130 in 1963. The decline accelerated in the mid-1960s and only 13 parrots were left by 1975.

The original range contraction must have been associated largely with the forest loss. The continued decline in the Sierra de Luquillo has been attributed to a variety of factors, none of them quantified. In the late 1940s many large palo Colorado trees were selectively logged, removing actual or potential nest trees. Felling trees to collect honey from bee nests removed further potential nest sites. Nestlings taken for the pet trade and an abortive reintroduction into reforested areas in the mid-1950s together with illegal shooting coincided with the decline between the mid-1950s and 1960s. Tenuous explanations included illegal hunting, guerrilla war activities, large-scale military experiments using herbicides, and high-intensity radiation on forests. The effects of any or all these on the parrots are simply unknown. The parrots face competition for tree cavities with the pearly eyed thrasher, *Margarops fuscatus*, and introduced honey bees, *Apis* spp. Warble flies, *Philornis pici*, parasitize nestlings, and red-tailed hawks, *Buteo jamaicensis*, take nestlings and adults.

4.9.2   *Diagnosis of factors driving the decline*

No clear program of diagnosing the causes of the parrot's decline in the Sierra de Luquillo was undertaken independently from or as a precursor to designing a treatment for recovery. The research was at the same time as and in parallel to the arbitrary treatments. The understandable justification for this was that the parrots were disappearing so fast that those involved felt that they could not wait for the research results.

Few details are available from the first studies in the 1950s when feeding and nesting habits were investigated at a few nests. The first diagnosis was black rat predation, and it was based on four of 11 failed nests having signs of black rats, but the researchers did not distinguish predation and scavenging.

Research started again in 1969 and was intensive (about 20,000 hours of observations in the 1970s). It focused on causes of nesting failure where deficiencies in the nest cavities, competition with thrashers, and red-tailed hawk predation were suspected. Given interventions at and guarding of nest sites, reasons for nest failure are difficult to establish. Three of the five desertions in the 1970s were abandoned eggs and one chick drowned in a wet hole, which suggests a lack of suitable dry cavities.

The population has had only a handful (two to eight) of breeding pairs for almost 40 years despite numbers of parrots changing between 13 and 200. Pairs form and defend territories but only about half lay eggs. The reasons for this have not been investigated. Any deficiencies in availability of food and its annual variation are unknown, although parrots are more productive with the flush of fruiting that follows a hurricane. Hints though these may be, they are no more than conjecture without further investigation.

Parrots may have a complex social system, with the breeding unit being a single pair with helpers. This would explain why each nesting area has only one breeding pair and why when one partner is lost it is so quickly replaced. If this is the case, then it parallels the red-cockaded woodpecker (Walters 1991). We wonder if light would be shed from knowledge of the breeding systems of other *Amazona* parrots, although that may be difficult as many of them are declining.

4.9.3   *Recovery treatments*

The recovery steps began in 1946 with recommendations to set aside reserves in the Sierra de Luquillo for the parrots. Legal hunting was stopped in 1946, and in the 1970s increased patrols by the Forestry Service decreased illegal hunting.

In 1956 and 1957 and then again since 1968 all known nests are protected by rat poisoning and rat guards. Observations of rats and their scavenging after hawk predation at nest sites in the 1970s suggested that those measures are ineffective and perhaps not even necessary.

In the 1970s natural nest sites were repaired and rehabilitated. Nest boxes were designed to be unattractive to the pearly eyed thrashers, and, when offered in conjunction with a nearby box attractive to the thrashers, the problem of competition was solved. The parrot nests were modified to counter the threat from red-tailed hawks. In 1953–1956 only 11% to 26% of nests raised chicks, but nesting success rose to 69% as a result of interventions at nest sites and provision of artificial cavities in the 1970s. For comparison, for the related Hispaniolan parrot, *Amazona ventralis*, 79% to 100% of nesting pairs raised chicks. There was an immediate response with the number of parrots rising from 13 in 1975 before edging up to 25 in 1979 and 47 in 1989. After Hurricane Hugo in 1989 stripped food and cover from the forest, half the parrots disappeared. By 1992 wild birds numbered 39 or 40 (Figure 4.11).

Shortage of nesting sites is now unlikely since artificial boxes and, after a hurricane had destroyed them in 1989, modified natural cavities are freely available. In 1992, parrots used 3 of 47 enhanced cavities.

The low numbers encountered in 1968 and that causes of the decline were unknown prompted the start of captive breeding. A year was spent studying the Hispaniolan parrot's behavior as a surrogate for the Puerto Rican parrot. Originally the captive breeding program was planned to be on the mainland United States. An outbreak of Asiatic Newcastle disease in Puerto Rico necessitated quarantine of the two parrots taken from the wild. One died and rather than risk any more quarantine casualties, an aviary was built in the Sierra de Luquillo.

Between 1973 and 1979 the aviary was stocked with 15 wild birds taken either as eggs or chicks that would not have survived for some reason. After 6 years the first chick was successfully fostered into a wild nest. This captive population had increased to 58 birds by 1992 and had augmented the wild population by 14 parrots. Nonetheless, despite this early success the captive population had only four breeding pairs.

The Hispaniolan parrots bred freely while the Puerto Rican parrots were ill-assorted pairings who did not readily breed, and then only half of them laid fertile eggs compared to 92% fertility for Hispaniolan parrots. Reasons for the low fertility are obscure. Relatedness between pairs laying fertile eggs and pairs laying infertile eggs was insignificant as measured from DNA (deoxyribonucleic acid) probes (Brock and White 1992). Artificial insemination was unsuccessful because the male parrots were intractable even after a month's training. Whole and sequential clutch replacements did increase the productivity. Survival to 1 year of captive-bred chicks placed with wild pairs was similar in fostered and wild birds.

Trials of different techniques were preceding direct releases of fledglings and older birds, but releases were suspended in 1986 to build

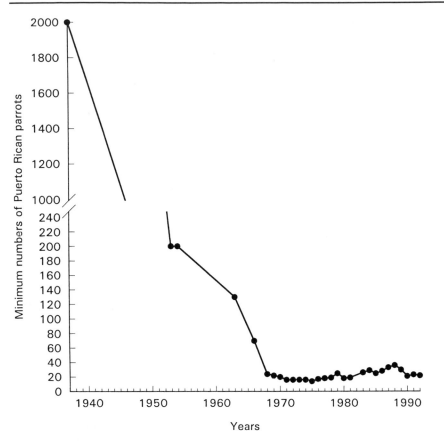

Figure 4.11. The decline and recovery of the Puerto Rican parrot. (Modified from Snyder NFR, Wiley JW, Kepler CB. The parrots of Luquillo: natural history and conservation of the Puerto Rican parrot. Los Angeles: Western Foundation of Vertebrate Zoology, 1987 and Collar NJ, et al. Threatened birds of the Americas. Washington: Smithsonian Institution Press, London: ICBP, 1992.)

up the captive population after recommendations from a population viability analysis. The simulations led to the idea of establishing other captive populations to reduce the risks from catastrophe and to use expertise available on the U.S. mainland. A second aviary on Puerto Rico was built, but the idea of sending parrots to a U.S. mainland zoo was overturned. Splitting the small population was risky given difficulties in getting the parrots to pair and either the risk of disease transmission between Puerto Rico and the mainland or the exposure of captive parrots to diseases (Wilson et al 1994).

The Puerto Rican parrot declined over a long period and finally ended up as a remnant population in what may be a suboptimal habitat. The suboptimal and modified habitat had a shortage of suitable nest cavities and this diagnosis is supported by the increase in parrots after nest sites were augmented. The low rate of occupancy of supplemented nest sites suggests other problems are unsolved. The population may too small for an effective breeding system, or food may be short. Until we know why the proportion of wild pairs and captive pairs breeding remains so low, the population remains at risk.

4.10   THE PINK PIGEON

This distinctive-looking pigeon, *Nesoenas mayeri*, with its pink body and brown tail, wings, and mantle was the most endangered pigeon in the world. It is only found on Mauritius (Figure 4.1) where another forest-dwelling pigeon (the blue pigeon, *Alectroenas nitidissima*) became extinct by the 1850s. The pink pigeon's initial decline coincided with the clearing of lowland forests, and it had become restricted to a patch of upland forest (Figure 4.12). Forest clearing continued, and by the mid-1980s only 20 birds remained. Its story illustrates, like the Puerto Rican parrot, difficulties when diagnoses fail to separate the effects of natural environmental variations, habitat changes, and introduced animals.

Mauritius has a flora and fauna comparable in scientific interest and conservation concern to Hawaii or New Zealand. The three islands (Mauritius, Réunion, and Rodrigues) have historically lost between 52% and 83% of their endemic birds (Cheke 1987b). Diamond (1987) covers ecology in scholarly detail, and we have relied on the chapters by Cheke and by Jones. Information on the captive breeding of the pink pigeons and their release is from Jones et al (1992).

4.10.1   Problem

Clearing lowland forests picked up momentum after the 1730s, and by the 1800s pigeons were already rare and restricted to the southwestern uplands. After the 1830s those upland forests were rapidly felled, and only some 4% of the original forest was left by 1880 (see Figure 4.12). Large areas of secondary forest (cut over but not cleared) supported many native birds, but by the 1950s they were replaced by tea and pine plantations. In 1940 the pink pigeon had available about 16,000 ha of native vegetation and pole forests. Within 40 years that area had contracted by almost 80% to 3,600 ha. The pigeons only use just over half this area (see Figure 4.12). Although native forest still covers the area northwest of a deep gorge, the birds stopped being sighted in the area after 1978.

There were 40–60 pigeons in the 1950s, but they dwindled to about 20 birds in the mid-1970s (Figure 4.13). The most recent count was 22 pigeons in 1992. About the only salient fact known about the pigeon's

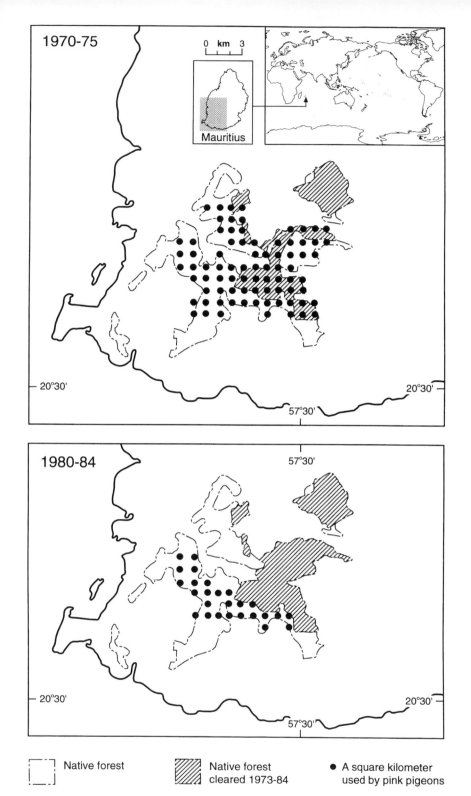

Figure 4.12. The contraction in native forest and distribution of the pink pigeon, Mauritius. (Modified from Jones CG. The larger land-birds of Mauritius. In: Diamond AW, ed. Studies of Mascarene Island birds. Cambridge: Cambridge University Press, 1987.)

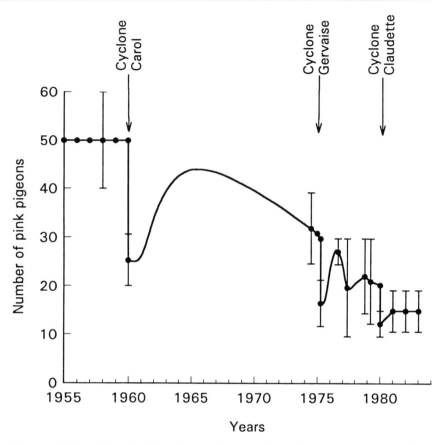

Figure 4.13. The decline in numbers of the pink pigeon, Mauritius. (Modified from Jones CG. The larger land-birds of Mauritius. In: Diamond AW, ed. Studies of Mascarene Island birds. Cambridge: Cambridge University Press, 1987.)

population dynamics is that fledging success is low. About 70% of the pairs nest, the usual clutch size is two and on average one chick fledges. The overall fledging success for 116 nesting attempts between 1976 and 1981 was 13 fledglings (11%). It is unclear if it was eggs or chicks that disappeared. Nesting is from December to June, which is the cyclone season, when nests are vulnerable to be swept to the ground by the winds (unless protected by being in a tree cavity).

4.10.2 Diagnosis of factors driving the decline

The diagnosis of forest clearing as causing the initial decline depends on coincidence between large-scale clearing and contraction of the pigeon's distribution to upland forest. Natural history studies in the 1970s and 1980s did not obtain a clear picture of what was limiting the pigeon, but they did describe feeding and nesting. Pink pigeons feed in trees and on the ground, searching for leaves, buds, flowers, fruit, and seeds from native and introduced plants. They would have faced seasonal food shortages toward the end of winter until the summer rains started or when cyclones ripped fruit and leaves from the trees.

Cyclones and seasonal food shortages would have always been part of the pigeon's ecology. Yet the recoveries in numbers after cyclones since the 1960s have only been partial (Figure 4.13), possibly because of the removal for captive breeding of nine eggs, two fledglings, and 11 adults from the only 24 to 30 birds in 1975–1976.

Whether forest clearing has led to food as a limiting factor is unclear and another possibility is nest site shortages. The pigeon is not a colonial species, but pigeons in the 1980s now nest in one small grove (2.4 ha) of an introduced evergreen tree. The shift may have been a response to a decrease in trees with open holes as the original forests were felled.

The long list of plants and animals introduced to Mauritius includes four pigeons and doves, but competition with the pink pigeons is unlikely as habitat and feeding preferences differ. The red-whiskered bulbul, *Pycnonotus jocosus*, from India is common and also feeds on *Nuxia* flowers at the end of the dry season when food is in short supply. Jones (1987) points out the destructive feeding of the crab-eating macaques, which strip fruit before it is ripe, as well as leaves and flowers. Nonetheless these observations leave us wanting as to whether the pigeons do face a shortage of food.

Mongooses were classified by two authors as either incidental or serious predators—a divergence that conveys the extent of facts available. Mongooses arrived in the mid-1850s, after the pigeon's initial decline. Feral cats are also common, which led Jones (1987) to write that "Cat predation is likely to be significant but as yet undetected source of mortality among adult pigeons feeding on the ground."

By the early 1700s monkeys were rated at plague proportions, and in 1741 they were recognized for taking bird eggs. Colonel Meinertzhagen wrote in 1912 (in Hachisuka 1953) that "Monkeys are doubtless responsible for their [pink pigeon's] extirpation." But the pink pigeon's decline and spread of black rats and monkeys coincided with forest clearing, which confounds diagnosis for the initial decline.

The only study of monkey predation was on the lowlands and blamed black rats more than monkeys for taking birds' eggs. Evidence against the monkeys was missing eggs and partially destroyed nests. Monkeys took domestic hens' eggs placed in old weaverbird, *Ploceus cucullatus*, nests (McFarling 1982, in Jones 1987). The case against black rats for taking pigeon eggs is more secure as causing nesting failure (Jones et al 1992). The telltale signs left by the rats were recognized from using dove eggs as bait in boxes that excluded monkeys compared to eggs fed to captive monkeys. Rats were responsible for 7 of 13 missing pigeon clutches in two breeding seasons compared to one of eight clutches after poison baiting was started in 1991 (other causes were not reported).

4.10.3 Recovery treatments

Pink pigeons were first protected in 1910 and then in 1977 when native birds were permanently protected. Nature reserves established in 1951 were extended in 1971 to cover most of the forest now used by the pigeons (Cheke 1987b). The recovery efforts were first to investigate natural history and identify causes of the decline. Jones (1987) acknowledges lack of evidence for a seasonal reduction in food as limiting, but then recommends supplementary feeding. He also denigrated protecting nest sites on the grounds that the habitat could not support more pigeons—this follows logically from his position that the pigeons are food limited.

Rat predation and possible food shortage cannot be separated because both supplementary feeding and rat poisoning were implemented at the same time. The combined treatments coincided with an increase in fledging success. In 1992 fledging success was 71% compared to 11% between 1976 and 1981.

The studies led to recommendations for captive breeding to supplement the wild population. Given what even one cyclone can do to the birds and their food supply, this was a safe step to take. The risks were that the pink pigeon was almost unknown in captivity (Jones et al 1983) and the source population was, by then, small. Colonies were started at the government aviaries on Mauritius and the Jersey Wildlife Preservation Trust (Figure 8.8).

Initially, the breeding record was poor: half the 1,447 eggs laid in captivity (1977–1986) were fertile and only 200 fledglings lived. Domestic pigeons, *Columbia livia*, used as foster parents transmitted a herpes virus and the pink pigeon chicks died (Friend and Thomas 1991). Success came with double clutching using Javan doves, *Streptelia risoria*, as foster parents, but hatching success decreased as the number of multiple clutches increased (Lind 1989).

Efforts to improve the poor breeding record (Bell and Hartley 1987) helped, but then lack of space temporarily halted the program in 1982. By 1990, 13 institutions had 95 pink pigeons which successfully bred at four zoos, producing 35 chicks, of which 29 were from Jersey (Olney and Ellis 1991).

The first release was 22 pigeons in the botanic gardens (26 ha) on the north end of Mauritius in 1984–1985 to establish a semiwild population supported by supplemental feeding (Todd 1984). It did not increase, partly because Indian mynahs (which are rare in the native forests) took at least five of nine clutches and by 1986 people had killed or injured 15 pigeons (Jones et al 1989).

Jones (1987) recommended that the captive-bred pigeons not be released into the range of the wild pigeons, since the reasons for the decline are not well understood. Instead, between 1987 and 1992, 51 pigeons were released into the forest abandoned by the pigeons in 1978. Some birds have bred and at least five young have fledged. The birds were banded, radio-tagged, and initially held on site in an aviary which is maintained as a feeding station.

The natural history studies offered possible hypotheses to be tested, but then diagnosis of factors limiting recovery slipped into disarray. Seasonal food shortage and predation during breeding are the favored contenders, and a viability analysis run in 1991 recommended increasing productivity through supplementary feeding, controlling nest predators, and releasing captive-bred birds (Seal and Bruford 1991, in Jones et al 1992). In the absence of a definitive diagnosis, however, the future is not assured.

4.11   THE HAWAIIAN GOOSE (NENE)

The Hawaiian goose is often known from its Polynesian name—nene, an onomatopoetic rendering of its call. It is a celebrated cause but a doubtful success story. Captive breeding and releases did little to help restore the population as the causes of its decline were still present. The recovery efforts had bypassed the critical step of diagnosis and this is still being remedied. To trace this story we have drawn extensively from Kear and Berger (1980), Black et al (1991), Banko (1992), and Black and Banko (1994).

4.11.1   Problem

In the 1800s, as many as 25,000 nene occurred on the Hawaiian Islands. Their pathway to virtual extinction had started by the 1820s, and by 1890 the geese were gone from Maui and scarce on Hawaii (Figure 4.14). In 1951 about 30 geese were left. The nene's range had contracted, and most of the range lost was below 1,500 m, but the significance of this was slow to be realized in the recovery efforts.

Nene used to migrate to lower elevations (less than 330 m) during the breeding season (November to April) to feed on sprouting vegetation in moist grasslands and open forest. As the weather became hotter and drier they would return to the lava flow uplands and feed on berries in vegetation patches. Changes in land uses prevented the seasonal migration, and restriction during breeding to a suboptimal habitat likely reduced the availability of high-quality food.

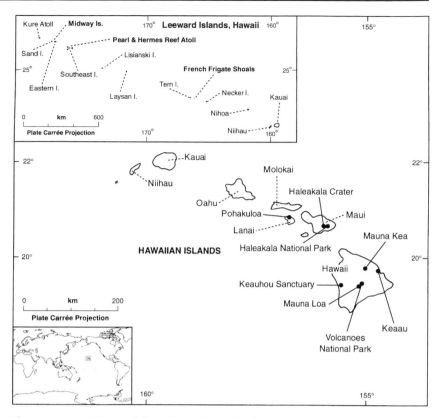

Figure 4.14.   Map of the Hawaiian Islands.

Starting from 1960, 2,127 captive-bred nene were released on Hawaii and Maui, but numbers did not increase beyond the increment of the releases and declined after 1981 when releases were temporarily reduced. Only two feral flocks have increased above the total numbers released. By 1990, wild geese numbered 230 on Hawaii. Breeding success is low (only 58% of wild pairs on Hawaii bred from 1978 to 1981), and only 7% of those 140 breeding attempts fledged a gosling. The geese released on Maui in the 1960s declined from a peak population of 225 to 325 in 1977 to 184 in 1990.

The geese at release sites are vulnerable to mongooses, cats, and dogs, which take eggs, fledglings, or molting adults. Unlike their close relatives, molting nene cannot take refuge from potential predators by swimming away.

Only about half the released captive-bred birds survived to 2 years, and they were less likely to breed and then had smaller clutches than wild-raised geese (unpublished references in Marshall and Black 1992). The emphasis on a rapid buildup of the captive breeding stock meant that goslings were raised in large groups without adult birds (Marshall and Black 1992). Such birds, "uneducated" by experienced parents, were at a grave disadvantage in foraging and avoiding predators.

Traffic is a recent problem for the nene in Volcanoes National Park (Figure 4.14), which has to annually contend with almost half a million cars. The geese cross roads to feeding areas, and traffic killed 5% of the park population in 1992 (Ling 1992).

4.11.2   Diagnosis of factors driving the decline

The nene's decline coincided with the arrival of cattle, *Bos taurus*, horses, *Equus caballus*, sheep, goats, and pigs between 1778 and 1803, but before the introduction of mongooses in 1883. Causes were reported to be hunting, especially after the introduction of shotguns, lowland habitat loss from domestic and feral stock grazing, and pigs eating nene eggs and goslings.

Diagnosis of factors limiting the nene's recovery did not really start until the mid-1980s despite a 1952 recommendation for intensive field studies in the restoration project that started captive breeding. The biologist who designed that project wrote: "The only effort which is likely to produce lasting and concrete results in conserving the nene is that directed toward discovering the factors responsible for the nene's population decline" (Smith 1952 in Kear and Berger 1980).

Diagnosis of predation was made in Volcanoes National Park. Nest success was higher (83% of 30 nests) in enclosures (predators excluded and trapping in the vicinity) compared to unprotected nests in the wild (44% of 70 nests). The experimental design was, however, confounded by supplementary feeding of geese in enclosures. Using artificial clutches of chicken eggs and live traps baited with eggs identified mongooses as the dominant predator.

Also in Volcanoes National Park, hatching success was higher for nests <1 km from pasture than for nests >5 km away as the parents had to forage less, which gave them more time to defend their nest. Incubating females in poor shape were less attentive, which increased their vulnerability to predators. A hint to the significance of food availability and predation comes from two feral flocks. They are the only populations to increase, and they live on irrigated lowland grass pastures where mongooses are absent (Kauai) or controlled (Keaau, Hawaii).

4.11.3   Recovery treatments

The first step to reversing the decline was to seasonally restrict hunting, and then followed captive breeding, which is the dominant theme in the nene's recovery. It started in 1949 with two pairs from a private collection sent to rearing pens at the Hawaii breeding center. Problems with diet, rearing techniques, and infertility made for slow progress in building up the flock. A pair was sent to the Wildfowl Trust at

Slimbridge (England). But both laid eggs and so a gander was hastily sent over to join them. Between 1952 and 1979, the Wildfowl Trust had raised 888 nene, some of which were loaned or sold to other collections. When it came time to repatriate them to the Hawaiian Islands, fears that they had contracted avian pox in captivity limited their release to Maui, which at that time did not have any geese.

Agreements between landowners and the state of Hawaii established first the Keauhou Sanctuary in 1958 and then, in 1974, three other sanctuaries. Other protected areas are Volcanoes and Haleakala National Parks (Figure 4.14). The releases on Hawaii stared from the local breeding center in 1960, and for the next 30 years some 1800 geese were placed in state sanctuaries and Volcanoes National Park. There was little harmony among the different views as to how the geese should be released (Zimmerman 1975). In Volcanoes National Park the geese are reared by captive parents and are then free to fly from the pens. Predators are controlled by trapping, and the geese can reach grass pastures. The state sanctuaries are at higher and drier elevations, and annual survival of released birds has not exceeded 14%. The geese are fed and are protected by some predator control, but most were reared in incubators without parental guidance.

The releases on Maui have been more successful, which probably reflects food availability and predator control. Between 1962 and 1981, 494 nene from Slimbridge and Hawaii were released in Haleakala Crater, Maui (Figure 4.14). Current estimates are some 37% of the number released, and the population was probably stable between 1981 and 1990. It is a wet volcanic area (2,500m elevation), where the geese can feed on grass growing year-round in horse pastures. Mongooses and feral cats are trapped.

Predation, poor nutrition during the breeding season, and inadequate rearing practices were confounded in their diagnosis. A population viability analysis promoted predator control as the key to self-sustaining populations, but its parameters were based on artificial feeding as well as predator control. Habitat management to offset the suboptimal habitat is proposed in the recovery plan. Time will tell how secure the nene populations are and the extent to which they will need management in perpetuity.

The nene survived Polynesian colonization, but seven other species of geese—all flightless—disappeared during that time. The nene declined during the myriad changes that followed European colonization: hunting, changed lowland habitats, and introduced predators left it marooned in a suboptimal habitat. Food was limiting productivity, but this was slow to be diagnosed either by ecological observations or efficient experimental design. Instead captive breeding and predator control were the focus. Captive breeding brought its own problems. It has barely supplemented population size as releases have balanced losses, but it has established a second island population.

4.12   THE DARK-RUMPED PETREL

The dark-rumped petrel is a medium-sized ocean-living bird that dips fish, crustaceans, and squid from the productive upwellings of the equatorial undercurrent in the Pacific Ocean. One subspecies returns to nest only on Galápagos Islands (Figure 4.15). The other subspecies nests on the Hawaiian Islands, and its colonies folded after Polynesian and European colonization. The Galápagos Islands were settled about 300 years ago, and the petrel's decline may not have started until the 1930s when a suite of introduced animals became common. The petrel's

Figure 4.15.   Map of South America and neighboring islands.

story is not an account of a clear-cut diagnosis, which is why we chose it as a case history. It does illustrate the need to follow a sequence of actions for diagnosis and treatment. We have drawn on Harris (1970), King and Lincer (1973), Simons (1985), Coulter (1984), Tomkins (1985), and Cruz and Cruz (1987).

The petrel's ecology was little known until recently when nesting behavior and vulnerability to introduced predators were described. Petrels annually nest in colonies usually in forested or thickly vegetated sites with friable soil where they dig nest burrows. After the single egg hatches, the parents guard their chick only for 2 or 3 days before they both leave it in the burrow. They return infrequently with food until it fledges some 107 days later. Dark-rumped petrels probably do not breed until they are 7 years old and they may live as long as 20 to 30 years.

4.12.1　Problem

The Polynesians' arrival on the Hawaiian Islands probably led the petrel's nesting areas to contract to mountainous areas and disappear from Oahu (Figure 4.14). Even after Europeans arrived, petrels were still locally common and nesting in large colonies in the mountains in the early 1900s (Perkins 1903). Thereafter the colonies declined and disappeared until a petrel was rediscovered in 1948 on Hawaii. Of the two known colonies, the larger is on Maui with 430 breeding pairs in 1985. Fledging success was low (39%) in 1979 but adult survival was high (80–93%). The colony on Hawaii was small and decreasing in the early 1980s.

People had reported that petrels were declining on the Galápagos Islands, but it was not until the 1970s that the decline's rapidity was recognized (Table 4.1). Populations on Isla Santa Maria and Isla Santa Cruz were estimated to have declined by 33% between 1977 and 1981. The threats differ for each island (Table 4.1), and the salient points are the low fledging success, adult deaths, the variety of predators involved, the effects of introduced pigs, goats, and cattle, and clearing of vegetation.

4.12.2　Diagnosis of factors driving the decline

The petrel's initial decline on Hawaii coincided with hunting and the introduction of rats and pigs. Pigs, while rooting though the lowland forests, probably destroyed burrows and fed on petrels. That might be the explanation for the petrel colonies being restricted to the walls of an extinct volcano on Maui. We draw a parallel with the pigs and the restriction of Lord Howe Island woodhens to a mountain top (Section 4.2). The decline in the early 1900s coincided with hunting (Perkins

Table 4.1. A summary of the status of dark-rumped petrels and their introduced predators on the islands of Galápagos

Island (area)	Recent estimate		Previous estimate		Introduced predators				
	Year	No. pairs	Year	No. pairs	Rats	Cats	Dogs	Pigs	Goats
Isla Santa Maria (173 km²)	1981	1,000	—	—	●	●	●	●	●
Santa Cruz (986 km²)	1985	1,000	1980	9,000	●	?	●	●	—
Santiago (585 km²)	1984	500	1979	11,000	●	?	—	●	●
San Cristóbal (558 km²)	1980	6,750	—	—	●	●	●	?	?
Isabela (4,588 km²)	Unknown		—	—	?	?	?	?	?

Modified from Harris MP. The biology of an endangered species, the dark-rumped petrel (*Pterodroma phaeopygia*) in the Galápagos Islands. Condor 1970;72:76–84; Coulter MC. Seabird conservation in the Galápagos Islands, Ecuador. Int. Council Bird Preserv. Tech. Publ. 1984;2:237–244; Tomkins RJ. Breeding success and mortality of dark-rumped petrels in the Galápagos, and control of their predators. Int. Council Bird Preserv. Tech. Publ. 1985;3:159–175; Cruz JB, Cruz F. Conservation of the dark-rumped petrel *Pterodroma phaeopygia* in the Galápagos Islands, Ecuador. Biol. Conserv. 1987;42:303–311.

1903) and the disappearance of 30 taxa of forest birds (Section 3.1) between 1892 and 1930, which is when the black rat spread (Atkinson 1977).

A study of the relic petrel population on Maui set out to identify factors affecting its survival in the late 1970s. Previously, black rats were named simply because petrels nested above the assumed ranges of feral cats and mongooses and within the reach of rats. Improved trapping techniques revealed both mongooses and feral cats. Both predators were implicated in 60% of the petrel's breeding failures.

The colony on Maui is on the sparsely vegetated walls of a volcanic crater at 2,550 to 3,000m. The petrels use crevices in debris at the base of rock outcrops and lava tubes, or dig burrows under rocks. The lack of vegetation and friable soil contrasts with the humid scrub-forested nesting areas on the lowlands. The nesting colony is above the range of the mosquito (*Culicidae*) vector of avian malaria, which together with avian pox possibly played a role in the decline of other Hawaiian birds (Warner 1968). Nonetheless Simons (1985) checked but found no signs of blood parasites in a few adult petrels.

On Galápagos Islands, fledging success was only 6% in 1967 in the thickly vegetated uplands of Isla Santa Cruz. Monitoring petrel burrows in 1978 and 1979 strongly implicated black rats as the leading killer of eggs and chicks but also that adults were taken by dogs, cats, and pigs. Identification was based on the characteristic signs left by different predators. Petrels on the different islands vary as to whether they are nesting in rock piles or burrows, and this was mentioned as a factor determining their vulnerability to predators. The trampling of burrows by goats and pigs also occurs.

The petrel's decline occurred after black rats reached Santiago in the late 1600s. In the 1800s rats were introduced to Isla Santa Maria and spread to Isla San Cristóbal. They arrived on Isla Santa Cruz after 1934 and reached exceptionally high densities. Norway rats did not arrive until 1984, when they were found on Isla Santa Cruz and on Isla San Cristóbal (Key and Heredia 1994).

On Isla Santa Maria the diagnosis of rats as causing the decline was supported by an apparent increase in fledging success after rat control during the petrel's breeding season. However, the normal range of annual variation in the petrel's breeding success is unknown.

4.12.3 Recovery treatment

On Maui, live trapping to remove rats, cats, and mongooses started in 1979 and the loss of eggs and chicks decreased (34% in 1980 and 6.4% in 1981). Once the seriousness of the declines on the Galápagos Islands was realized in the early 1980s, seasonal poisoning of rats started on Isla Santa Maria in 1983. The program also aimed to control cats and dogs, although that might have unwanted consequences if the cats and dogs were killing many rats.

Habitat improvements to nesting areas could reduce predation on the adult petrels. Recommendations were to build artificial rock piles and lay wire mesh over the soil to mitigate damage to burrows from trampling and stop large predators from digging out adults. That would leave those predators to prey on the rats. These habitat improvements seemed not to have been implemented by the time their description was published (Tomkins 1985).

The published information on the decline and recovery of the dark-rumped petrels is fragmentary and leaves the impression that problems threatening the dark-rumped petrel are not fully resolved. The studies and treatments have focused on fledgling success. Diagnosis of introduced predators was supported by increases in fledging success after mongooses and feral cats (Hawaii) or black rats (Galápagos Islands) were killed. The efficiency of control is undetermined and effects of environmental variation such as El Niño on food availability are indistinguishable from predator control. To rest the case for the declines against the introduced predators may foreclose incentives to investigate improving habitat for nesting colonies. The effect of bolstering fledging success on population size is unclear, but as in a long-lived species, variations in adult survival may largely determine trend in population size; protecting breeding adults is, at least, as important.

4.13   THE RED KITE

The story of the red kite, *Milvus milvus*, reveals what happens when only one obvious factor is diagnosed and treated. Persecution has figured in the decline of many large raptors throughout the world and the red kite is no exception. Their numbers declined and their range contracted. In 1990, 11,000 to 13,000 birds were scattered as small populations in 21 European and Asian countries. Britain's only remnant population was in central Wales, and it had as few as a dozen kites in the early 1900s, which barely increased beyond 10 to 20 breeding pairs for 60 years. Despite protection and some 30 years of monitoring and intermittent studies, the Welsh kite's problems were slow to be diagnosed. Wales is on the edge of the species' range and possibly a suboptimal habitat, which could have been a hint to other factors limiting the recovery. We use Lovegrove's (1990) account as a basis for describing the red kite's decline and recovery.

4.13.1   Problem

Several hundred years ago in England and Wales the red kites were protected as useful scavengers in the towns, but that ended when garbage disposal became organized. Parish ledgers registered bounties paid, which dwindled as 200 years of kite trapping and nest raids took their toll. By the early 1800s, kites had disappeared from large areas and, as they declined in their remaining strongholds, egg collecting was often the final blow. By the time protective legislation was enacted in 1880, the British kites were reduced to four to five pairs in Wales.

4.13.2   Diagnosis of factors driving the decline

The Welsh red kites were annually monitored: records of bird numbers, nests, young fledged, and nest robberies extend from 1901 until the present. The remnant population barely grew until 1960 despite protection during nesting from egg collectors. In the 1940s and 1950s the number of breeding pairs still fluctuated between 10 and 15, but once nest protection improved after 1960, pairs slowly increased to number 69 by 1989 (Figure 4.16).

Nest robberies continued despite the improved protection at nest sites. In 1985, 10 of 45 known nests were robbed. Between 1951 and 1980, human disturbance and robbery accounted for 21% of all clutch failures compared to 8% for natural predation (Davis and Newton 1981).

A second problem for the kites is poisoning of adults. Kites feed on carrion, and farmers poison sheep carcasses to reduce foxes and crows,

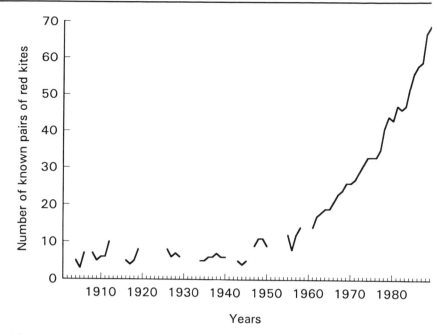

Figure 4.16. Recovery in the numbers of red kites breeding in Wales. (Modified from Lovegrove R. The kite's tail: the story of the red kite in Wales. Sandy, Bedfordshire, UK: Royal Society for the Protection of Birds, 1990.)

Corvus. Despite publicity campaigns against poison baits, poison killed at least 18 kites between 1980 and 1989, with 11 of those poisonings being in 1989.

A third problem is low productivity. Annually 90% of the territorial pairs build nests; the mean clutch size is 2.2, but a pair only fledged an average of 0.5 fledgling/nest, which is less than half the rate observed for red kites in continental Europe (Davis and Newton 1981). Nonlaying, hatching failure and brood mortality were the reasons for failure, which in other raptors are associated with food shortage. Remote video monitoring revealed that, while incubation and hatching were normal, chick development was not. Studies are under way (1992) to estimate if food shortage is the reason why chicks die.

Circumstantial evidence supports the idea that food shortages lead to the low breeding success. In 1954, myxomatosis struck rabbits and the main food for the kites virtually disappeared. The following year, half the usual number of nesting pairs laid eggs and only one young was fledged. Neither eggs nor young per pair related to the availability of sheep carcasses during the winter, but the kites did raise significantly more young in years when vole, *Microtus*, numbers were high compared to years when they were low (Davis and Newton 1981).

In the refuge areas in Wales, habitat changes only accelerated after the 1940s. There was no evidence that forestation with conifers or the increase in sheep stocking rates correlated with the kite's low productivity.

The evidence is then suggestive that the kites face food shortages during nesting. How that relates to the kites being on the western edge of their range and how their refuge may be deficient in food is undescribed.

In 1961 and 1962, two thirds of the nests failed to hatch any eggs. This coincided with the widespread use of the insecticide dieldrin in sheep-dips. Egg hatching increased after withdrawal of dieldrin in 1965. Levels of organochlorines and eggshell thinning were only sampled from 35 eggs taken between 1964 and 1978 and were less than levels associated with breeding failures in other species (Davis and Newton 1981).

Fears of inbreeding depression (Section 6.3.7) had long been raised, but there is no evidence for reduced fitness such as low productivity, for which there are alternative explanations (food shortage). DNA analysis using allele frequencies at a single locus revealed that variation of the Welsh kites was less than two European populations that had not been squeezed through numerical bottlenecks (May et al 1993). The Welsh kites nest in two areas separated by 10 km, and the two sub-populations differed in allelic frequency and heterozygosity (Section 6.3). The southern kites had heterozygosity of 24% compared to 66%, but their productivity was significantly higher probably because the habitat is more varied and productive.

4.13.3   *Recovery treatments*

A series of acts from 1880 to 1981 made the taking of kites and their eggs illegal in the breeding season and prohibited using traps and poison bait, but the legislation was largely ineffective. Some guardians were also egg collectors themselves or in the pay of the same. Local farmers were paid a bonus for successful kite nests on their property, and essentially this protection has continued but became more effective in 1960 when full-time staff took over the monitoring and nest protection. The protection of those remaining few pairs for an uninterrupted 100 years is possibly the longest-running protection effort for any species.

The first attempt to increase genetic variation was in 1923 with a proposed introduction of kites from Europe. Two of four fledglings from France turned out to be black kites, *Milvus* spp., and all four went to the London Zoo. In 1927, 74 kite eggs from continental Europe were given to wild buzzards, *Buteo* spp., to incubate but the fate of the 32 kites that fledged was unknown. Four Spanish red kites released in 1956 and 1957 disappeared.

In the late 1980s translocations were with the intention to supplement the population rather than restore genetic variation (inbreeding depression had not been demonstrated). Eggs removed from deserted nests or nests vulnerable to robbery were artificially incubated and the chicks returned to the nests. Success (to fledging) was about 30%. Captive breeding using rehabilitated red kites unsuitable for release started in 1989, but by 1992 no chicks had been raised.

The next step was translocations to establish new populations. Between 1989 and 1992, 122 kites mostly from Sweden were released from large enclosures at a site in Scotland and another in south England. The birds became established and started breeding in 1992 (Evans et al 1994).

For the red kite the immediate and obvious problem was robbing eggs from nests, and although it continues it is at a lower level after full-time protection in 1960, and this respite was enough for the kite population to increase slowly. The diagnosis of whether food is limited and whether that is because the population is marooned on the edge of the historic range is still unconfirmed. This leaves the treatment, if any, uncertain. Instead efforts have now gone into establishing other populations in areas where food shortage is assumed not be a problem. The three populations are a chance to test food abundance against rate of change in the populations.

4.14   THE NUMBAT

The numbat is a unique termite-eating marsupial now restricted to a few areas in Western Australia. After disappearing from the arid eastern portions of its range during the late 1800s and early 1900s, only remnant populations were left, which abruptly collapsed to just a few small populations by 1979 (Figure 4.17). We profile the numbat as an example of a species restricted to habitat remnants in reserves where it is vulnerable to introduced predators. Efforts have focused on managing free-ranging populations by reducing foxes, and any shortage of food or shelter is undiagnosed. Treatment included translocating individuals from the largest population to new sites. Fears that this source population could withstand only limited removals led to a captive breeding program begun in 1985 but that, up until 1993, had not contributed to wild populations. Our account draws on a meticulous study of the numbat's natural history (Calaby 1960) and Friend (1990).

Numbats are of medium size and strikingly marked by black and white streaks running across their back and a thin black eye stripe: a bottlebrush tail completes their attractive appearance. They forage almost exclusively on termites, mostly *Coptotermes acinaciformis*, licking them from galleries which they expose by digging. The termites kill the woodland's dominant tree, wandoo eucalyptus, *Eucalyptus redunca*, and fallen trees are indispensable shelters for numbats. Numbats are

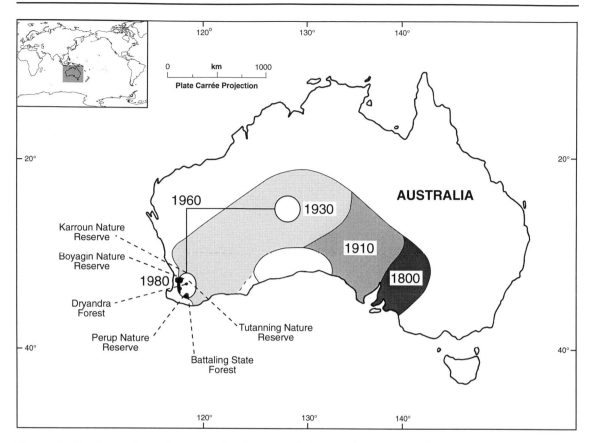

Figure 4.17. Historic and current distribution of the numbat in Australia. (Modified from Friend JA. The numbat *Myrmecobius fasciatus* (Myrmecobiidae): history of decline and potential for recovery. Proc. Ecol. Soc. Austral. 1990;16:369–377.)

solitary, but males and females overlap in their territories and an average density would be about one pair per 50 ha. Females breed when a year old, and most females annually bear four young. Natural predators are raptors and reptiles, including the carpet python, *Python irregularis*.

4.14.1   Problem

The numbat's decline is only known from the contraction in its range from southern and central Australia to isolated woodland areas in the southwestern part of Western Australia's wheatbelt by the 1950s (Figure 4.17). At that time the numbat was abundant, but in the 1970s those remnant populations collapsed to four small populations.

The only index to the changes in the numbers comes from Dryandra Forest (Figure 4.17), which held one of the last populations. The index is crude, since it is numbats sighted from a vehicle driven slowly along forestry tracks for at least 200 km. The forest understory is relatively open, but as numbats take cover in fallen logs an unknown proportion is missed. In 1955–1956 the average was 3 numbats/100 km compared to 1979 and 1980 when the average had fallen to 0.7 numbat/100 km.

4.14.2   Diagnosis of factors driving the decline

Explaining the numbat's decline has to encompass the initial decline and range contraction and the collapse of the remnant populations. The initial decline followed well after European colonization and mostly this century. Loss of escape cover as land practices changed probably left numbats vulnerable to foxes and cats. Feral cats had been present in the central deserts in the late 1800s and foxes arrived in the mid- 1900s. The contraction of the numbat's range appears to partially precede the initial spread of the fox. A factor that changed at about the time of the numbat's disappearance was the change from traditional small burns to widespread summer fires after the aboriginal peoples left their traditional lands. Widespread burning would have left fewer patches of escape cover. The spread of rabbits and their role in supporting foxes as an alternative prey could have enabled foxes to eradicate even low densities of endemic mammals such as numbats. The spread of rabbits mostly preceded the foxes' arrival (Figure 4.18). Comparing Figures 4.17 and 4.18 suggests that relating the contraction in the numbat's range to the foxes' spread is not simple. The other land use change in the numbat's more western range was forest clearing (for example, see Figure 8.5).

The distribution of numbats contracted to southwest Australia's forests and woodlands where rainfall is highest (more termites) and fallen logs provide cover. A third factor may have been poisonous shrubs (*Gastrolobium* and *Oxylobium*). Rabbits cannot survive eating the shrubs, and foxes, either through scarcity of rabbits or secondary poisoning from eating rabbit, were few. In the 1950s the numbats were still reasonably abundant in the southwestern woodlands. Calaby (1960) suggested that the future security of the numbats lay in woodland reserves.

Clearing of the forests continued (Figure 8.5), and the numbats dwindled until only a few populations survived in forest reserves by the late 1970s. Then even in the reserves, numbers of numbats suddenly collapsed. Three factors may have acted singularly or in concert. Rainfall was less than average in the 1970s, and drier years would presumably decrease termite availability. However, rainfall was less in the 1980s when the numbats increased. A change in the frequency and intensity of fires would have either decreased termites or hollow-log shelters. In the 1970s wide-scale rabbit control with the poison 1080 stopped, and without deaths from secondary poisoning the number of

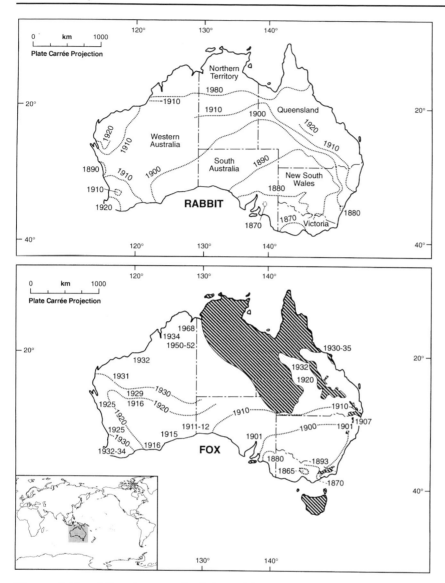

Figure 4.18. Historic spread of the rabbit and the fox over mainland Australia. (Modified from Thompson HV, King CM. The European rabbit. The history and biology of a successful colonizer. Oxford: Oxford University Press, 1994 and Jarman P. The red fox—an exotic, large predator. In: Kitching RL, ed. The ecology of exotic animals and plants; some case histories. Brisbane: Wiley, 1986:44–61.)

foxes increased (King et al 1981, in Friend 1990). This is assumed to have occurred where the numbat numbers collapsed.

The diagnosis of fox predation depends on a comparison of numbat sightings in two blocks (2,000 and 2,500 ha) separated by only a 1-km buffer. More numbats were seen on the block baited with the poison 1080 compared to the unbaited block, but the difference was insignificant. Apart from the lack of replication, the treatment and control block were close enough that foxes or numbats could move between blocks. Fox predation may interact with forest fires which burnt the understory and fallen logs, but any effect of reduced cover was not tested.

4.14.3   Recovery treatments

The two recovery treatments are controlling introduced predators and repatriation using wild-caught numbats. In Dryandra Forest the first-line treatment since 1982 is poison baiting every month to control foxes. In Perup Nature Reserve (Figure 4.17) the numbats recovered from as low as 50 to 300 with little intervention other than sporadic poison baiting in the 1970s that became more regular in the 1980s.

The second treatment was to establish other populations using wild-caught numbats from Dryandra Forest. By 1993 numbats had been released in nature reserves at Boyagin, Karroun, and Tutanning, and Battaling State Forest (Figure 4.17). In Boyagin numbats had disappeared after 1970 and were repatriated in 1985. A fox killed Rocky, the first numbat released, so a program of monthly poison baiting for foxes started. The 42 numbats released between 1985 and 1987 were radio-collared, but subsequently the numbats were monitored from their sign (scats and diggings for termites). In 1992 the population was estimated to be 80–100, extrapolated from the numbers of territories determined by radiotelemetry.

A total of 97 numbats were released between 1986 and 1992 in the Karroun Nature Reserve (Figure 4.17), which is a large semiarid open woodland. The area is aerially poison baited for dingoes and foxes twice a year with some additional ground baiting. Initially the numbat survival was high but then decreased. Possibly reduction of foxes led to feral cats increasing. Three cat-killed, radio-collared numbats were found compared to none killed by foxes. Two other releases were into nature reserves, both of which are poison-baited for foxes to foster the recovery of other endangered marsupials (brush-tailed bettong, *Bettongia penicillata*, and western native cat, *Dasyurus geoffroii*).

Captive breeding was started early (1984), and Friend and Whitford (1993) describe the successful development of an artificial diet, the early breeding success, and then the subsequent failures of breeding. By 1993 only two captive-raised numbats were released, and numbats taken from the wild about equal the number raised over the 10 years.

The increase in numbats during fox control supports the diagnosis of fox predation as a limiting factor. The experimental design was weak, and the continued poison baiting for foxes was not used to repeat and improve the experiment. It is uncertain whether the future of numbats is tied to control of foxes (or possibly feral cats on more arid eastern sites) and whether foxes have been overemphasized as a limiting factor. The causes of the initial decline are less certain, other than at the level of the obvious loss of woodlands. The populations of numbats are all in nature or forest reserves, which raises the question of how to manage the habitat. Future fire management of the woodland reserves to maintain a supply of termites and fallen trees for shelter needs to be investigated.

4.15 THE CHINCHILLA

Chinchillas, *Chinchilla* (Figure 4.19), are appealing-looking rodents from the Andes, South America. The Incas appreciated them for their thick, soft gray fur. Then Europeans discovered the attractiveness of chinchilla pelts. In less than 50 years, despite protective legislation, this colonial mammal was hunted almost to extinction in the wild. It was domesticated and is one of the best-known furs on the world market. Yet the wild chinchilla, after its brush with extinction, has remained in obscurity. Although overharvesting seems obvious enough as a diagnosis, it is only secured by coincidence in timing for the decline and harvesting. The chinchilla's plight illustrates the problems we face when we do not have enough information to know what factors are limiting recovery, let alone what to do about them. We compiled this account mostly from Jiménez et al (1992), Grau (1984), Thornback and Jenkins (1982), and Allen (1972).

The long-tailed chinchilla, *C. laniger*, was widespread in the Chilean foothills and coastal mountains but is found now only in remote areas of the Coastal Cordillera (Las Chinchillas National Reserve) 300 km north of Santiago (Figure 4.15). Long-tailed chinchillas are bred by the tens of thousands in captivity for their pelts, as pets, and as laboratory animals. The short-tailed chinchilla, *C. brevicaudata*, is the larger species (600–850 g body weight) and preferred high elevations (3,000–4,500 m)

Chinchilla,
Chinchilla laniger

Figure 4.19. Chinchilla.

in the Andes where the borders of Bolivia, Chile, and Argentina meet (Figure 4.15). Subsequently, there are only scattered reports for northern Chile and none from Peru for the last 50 years.

4.15.1 Problem

Chinchillas were hunted for their meat and fur before European colonization. Once discovered by Europeans in 1829, however, trade in pelts started quickly. Specially trained dogs and grisons (*Galictis* spp., a large weasel) were used to hunt chinchillas in western Peru for the most prized fur in the mid-1800s. At least 7 million pelts were exported from Chile between 1840 and 1916 (Miller 1980 in Miller et al 1983). As populations buckled under this onslaught, the geographic range contracted, and the Chilean trade dropped from 400,000 pelts in 1899 to 3200 in 1915. Pressure on the few surviving colonies must have been severe as the price jumped from $13 (U.S.) in 1890 to $200 (U.S.) in 1930 (Rice 1988). Trade statistics usually failed to separate the two species.

The chinchilla's problems in the 1920s continued with demand for illegal hunting and stock for ranching. Only after 1980 were strong measures taken to reduce illegal hunting. CITES reports that 2,144 *C. laniger* parts were exported from South America to the United States between 1982 and 1984 (Jorgenson and Jorgenson 1991). It is unclear whether this illegal hunting or another factor prevented recovery. A candidate factor is overgrazing by goats and cutting for firewood, which changes the scrub and bunch grass habitat. One of the chinchilla's preferred foods, algarrobilla, has itself become listed as endangered in Chile from gathering its tannin-rich pods and burning.

The long-tailed chinchilla's ecology is not totally unknown (Jiménez et al 1992; Rodriquez 1980 in Miller et al 1983; Durán et al 1987). Observations in Las Chinchillas National Reserve revealed that they feed and shelter at the base of spiny-leaved bromeliads on semiarid and rugged north-facing slopes and foxes, *Pseudalopex culpaeus* and *P. griseus*, kill them. Even though we know that chinchillas breed at 6 months and have a litter of one or two young twice a year, we do not seem to have enough on population trend to determine if long-tailed chinchillas are secure in the reserve or elsewhere.

4.15.2 Diagnosis of factors driving the decline

Harvesting as a factor is surmised from the volume of pelts traded. The contraction in the ranges of both species coincided with pelt exports. Capture for breeding stock for ranching may have accelerated the decline, and illegal hunting continued to at least 1980. Habitat destruction apparently started as chinchillas were declining from harvesting.

4.15.3 Recovery treatments

The initial protection was an agreement between South American countries in 1910 to ban hunting and trade. National legislation followed in Argentina, Bolivia, and Peru by 1922. Legislation against hunting is still in place, and although CITES lists chinchillas on Appendix I (Chapter 12), as late as 1981 wild-caught fur was marketed in Japan.

The sad fact is that despite its high profile as a valuable furbearer, chinchillas in the wild have been largely ignored. The long-tailed chinchilla is only known in any number in a Chilean national reserve inaugurated for it in 1983, and long-term security of its colonies is unknown. Whether the short-tailed chinchilla even still exists seems uncertain. We do not know enough about their status and ecology to either diagnose why their numbers never recovered or what we could do about it. Neglect of chinchillas is not unique in the conservation of South American rodents (Gudynas 1989) or, come to that, most of the world's rodents.

4.16 THE HISPID HARE

The hispid hare, *Caprolagus hispidus* (Figure 4.20), is, in fact, a rabbit (and was sometimes called the Assam rabbit; Gee 1964). It has already

Hispid Hare *Caprolagus hispidus*

Figure 4.20. Hispid hare.

once disappeared from the view of science. And now, despite rediscovery in 1971, it faces habitat loss on the northern Indian continent. It is a small, unglamorous animal confined to a shrinking habitat under pressure from people trying to make their living. We chose it to represent species for which the assumption of their decline is based on contraction of habitat, as nothing is known of their numbers and little enough about ecology. Yet without that information, its survival in reserves may be doubtful.

The dark brown hare is about 2.0 to 2.5kg in body weight with coarse, bristly hair, short ears, and tail. The pygmy hog, *Sus salvanius*, shares the hare's grasslands, and both species have parallel stories of decline, rediscovery, and conservation efforts. We have used Bell (1986), Oliver (1977, 1981, 1984, 1989), and Bell et al (1990).

4.16.1   Problem

Hispid hares historically occupied tall grasslands and scrub along the southern Himalayan foothills of Nepal west through Bengal into Assam (Figure 4.21). Whether it was common or not is unclear. By 1929 the hare was "not so common as formally owing to the decrease in grass jungle in the district" (Inglis et al 1929 in Ali et al 1960). In Bangladesh it was considered extinct since the 1920s until unconfirmed reports of single specimens in 1974 and 1980. In Nepal a hare seen in 1982 was the first reliable report since Lt. Col. Faunthorpe shot one in 1926. On the Indian side of the border in the same area, hares were common in the 1920s but became scarce: one record was a hare collected in 1956 after one of the Rajah of Kasmanda's tame elephants, *Elephas maximus*, trod on it. Another hare was sighted in 1960. In Assam a hare was collected in 1956, and none were recorded until 1971. By 1981 the only hares were in the Manas Wildlife Sanctuary (Bhutan) and were threatened by flooding from proposed dams.

Surveys in sanctuaries and forest reserves with likely habitat identified a few hispid hares. The 12 populations known in 1990 were scattered across the historic ranges (Figure 4.21). They are isolated in reserves with conflicting land uses (Section 10.3), and neither size nor trend in the populations is known. Their distribution is probably becoming more fragmented, with the populations becoming isolated and then disappearing: between 1971 and 1981 two of four populations in northern Assam vanished (Oliver 1981). It is not clear whether all possible locations have been checked and whether more populations will be found.

Hispid hares live in dense tall grasslands along rivers where flooding or periodic burning maintain them. They use tall grasslands growing in forest clearings or as an understory in forests and abandoned villages. In Nepal the remaining tall grasslands are only found in reserves. Cutting tall grasses for thatch and their burning to force production of

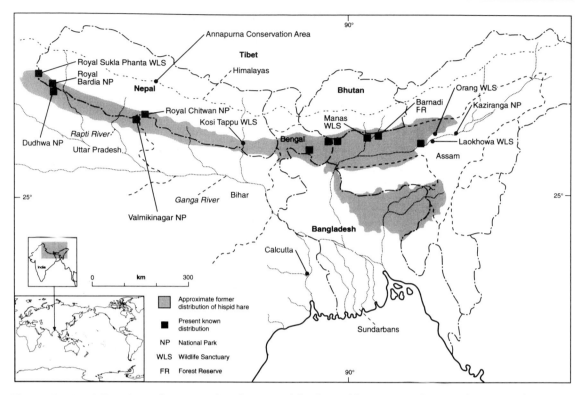

Figure 4.21. Historic and current distribution of the hispid hare in northern India. (Modified from Bell DJ, Oliver WLR, Ghose RK. The hispid hare *Caprolagus hispidis*. In: Chapman JA, Flux JEC, eds. Rabbits, hares and pikas; status survey and conservation action plan. Gland, Switzerland: IUCN/SSC Lagomorph Specialist Group, International Union for the Conservation of Nature and Natural Resources, 1990.)

green shoots for stock grazing are frequent, even in wildlife sanctuaries. In Assam increasing numbers of people encroach on state forestry reserves, and agriculture has replaced grasslands in northwest India (Uttar Pradesh and northwestern Bihar). Extensive flood control systems have modified riverine grasslands and forests and further reduced hare habitat.

4.16.2 Diagnosis of factors driving the decline

Diagnosis of habitat loss driving the hispid hare's decline rests on coincidence between lack of hare sightings and habitat loss. Hares have disappeared from some reserves after encroachment by people needing thatch and grazing for their domestic stock. A clue as to why human use of the grasslands and hares may be incompatible is that, where the grasses are burnt, hares lived in strips of unburnt marshy grasses where they were vulnerable to predation (Bell 1986). Human disturbance and hunting are implicated in the hare's decline but without substantiation.

4.16.3   Recovery treatments

The hispid hare is listed in the Indian Wildlife Protection Act (1972), but it needs more effective habitat protection. The hare's survival depends on management of the tall grass habitat. Recommendations to that end have not apparently yet been reconciled with people's needs (Section 10.3).

Efforts to keep the hispid hare in captivity have been unsuccessful (Oliver 1984), but the action plan recommends captive breeding. It seems to us that more on the hare's habitat requirements and how they could be met through management of reserves might be more useful than whisking the hares into captivity, a point also made by Oliver (1979). It is not clear whether the hare's ecological requirements are well known enough to manage them in reserves beyond trying to limit people also using the reserves. In short, we are left wanting for what has happened beyond the generalized diagnosis of habitat loss.

4.17   THE JAMAICAN HUTIA

The Jamaican hutia, *Geocapromys brownii* (Figure 4.22), illustrates the difficulties in diagnosing a species' decline without knowing its status

Jamaican Hutia
Geocapromys b. brownii

Kathie Hollis '94

Figure 4.22.   Hutia.

or ecology. The extant hutias are a remnant of the extraordinarily distinctive Caribbean mammals that predated the arrival of Amerindians (about 7000 B.P.). Most species became extinct in the last 4,000 years, and all that is left of the 12 insectivore, 16 ground sloth, 3 primate, and 52 rodent species are 13 rodents and 2 insectivores (Woods 1989). The giant hutias, *Heptaxodontidae*, weighed up to 150 kg and were the first to go. More modest in size are the hutias, *Capromyidae*, and most species disappeared after people arrived. The problems that face the surviving species are habitat loss, hunting, and introduced predators (Box 4.3).

Capromyids (hutias) look like a cross a between a large rat and a large guinea pig, *Cavia*, being stoutly built and weighing 4 to 7 kg. They are vegetarian, taking fruit, leaves, and bark, but their specific ecologi-

---

Box 4.3. The extinct and surviving hutias (*Capromys, Geocapromys, Hexolobodon, Plagiodontia, Isolobodon,* and *Hyperplagiodontia*) in the Caribbean

Species	Status	Islands	Possible causes
C. acevedoi	Extinct	Cuba	?
C. angelcabrerai	Endangered	A few cays, Cuba	Hunting
C. antiquus	Extinct	Cuba	?
C. barbouri	Extinct	Cuba	?
C. pilorides	Endangered	Cuba	Hunting? (middens)
C. nana	Endangered	Cuba	Agriculture, hunting
C. auritus	Endangered	One cay, Cuba	Hunting
C. sanfelipensis	Endangered	One cay, Cuba	Hunting, habitat, rats
C. garridoi	Endangered	A few cays, Cuba	Hunting
C. melanurus	Rare	Cuba	Deforestation
C. prehensilis	Unlisted	Cuba	
C. meridionalis	Indeterminate	One island, Cuba	Hunting, dogs, cats
C. gundlachi	Indeterminate	One island, Cuba	Hunting, dogs, cats
G. brownii	Indeterminate	Jamaica	Hunting, habitat
G. ingrahami	Rare	Bahamas	Storms, cats, dogs
G. columbianus	Extinct	Cuba	Hunting? by 1850
G. pleistocenicus	Extinct	Cuba	?
H. poolei	Extinct	Haiti	?

(Continued)

Box 4.3.   (Continued)

Species	Status	Islands	Possible causes
H. phenax	Extinct	Haiti	? by 1600s
P. ipnaeum	Extinct	Dominican Republic	Hunting? (middens)
P. spelaeum	Extinct	Haiti	Hunting? by 1750
P. araeum	Extinct	Dominican Republic	?
P. caletensis	Extinct	Dominican Republic	?
P. velozi	Extinct	Dominican Republic	?
P. aedium	Rare	Dominican Republic	Hunting, habitat
P. hylaeum	Unlisted	Dominican Republic, Haiti	
I. portoricensis	Extinct	Puerto Rico	Hunting? by 1700
I. levir	Extinct	Haiti	Hunting? by 1550
I. montanus	Extinct	Haiti	Rats? by 1600
Hy. stenocoronalis	Extinct	Dominican Republic	?

(Modified from Alvarez VB, González AC. The critical condition of hutias in Cuba. Oryx 1991;25:206–208; Woods CA. Endemic rodents of the West Indies: the end of a splendid isolation. In: Lidicker WZ Jr, ed. Rodents—a world survey of species of conservation concern. International Union of the Conservation of Nature/Species Survival Commission Occasional Paper No. 4, 1989:11–19; Groombridge B. Global biodiversity: status of the Earth's living resources. London: Chapman and Hall, 1992.)

cal requirements are little known. Their fecundity is low, as litter size is usually two and the gestation is long, about 123 days.

The Jamaican hutia, *Geocapromys brownii*, is the country's only endemic terrestrial mammal (Figure 4.10). It was assumed to be on the brink of extinction because its habitat had been largely cleared and it was still sought for the table. Similar problems plague hutias on Cuba (Alvarez and González 1991). Five Cuban species became extinct probably in historic times, and of the ten extant hutia only one species is relatively common (Box 4.3).

4.17.1   Problem

Subfossil bones in caves and kitchen middens of the Arawaks and other Amerindians reveal that the hutia had been widely distributed on Jamaica. Its decline was assumed to be due to native vegetation being changed for cultivation and to the impressions of naturalists and hunters. In the mid-1970s, the Jamaican hutia was supposedly known only from three local areas where its numbers had declined (Clough 1976). The disappearance of the Swan Island hutia, *Geocapromys thoracatus*, in 1960 from Swan Island (Figure 4.10) added to concern that the Jamaican hutia was in trouble. Subsequently, however, a survey in 1982 documented hutia at 16 sites. The hutia is nocturnal, and generations of hunting have made it shy and elusive, so it was little wonder that there was an impression of its rarity. The Jamaican hutia was not in immediate danger of disappearing.

4.17.2 *Diagnosis of factors driving the decline*

Most hutias have disappeared with no definitive epitaphs. Some species occur in early Indian middens, but it is only conjecture to say that hunting caused their extinction. The loss of Swan Island hutia was recent (1960), and although something is known about what happened, the diagnosis is still uncertain. The hutia was common on this small and uninhabited rugged limestone island (100 ha). Lowe (1911), a naturalist who visited the island, hoped that "[Little Swan Island] . . . in the future represent the last stronghold of this peculiar and old-time race of animals, for here they are left absolutely unmolested; and no enemies, human or otherwise, seem likely to disturb them."

But his hope was in vain. First, in 1955 an exceptional hurricane hit the island and then goats were introduced. Clough (1976) surmised that the goats had no effect on the forage before they disappeared by 1960 and that the rains of the 1955 hurricane had drowned hutias. Before their naturally slow rate of reproduction could allow recovery, the hutias, especially the juveniles, were faced with introduced cats. Accounts of the timing of this are unclear: Cats were released before 1960 or in 1960, or a succession of cats was released on Little Swan Island during the previous 15 years. The effect was not in doubt: no hutias were found in 1960 and 1974 while cats and Norway rats were still there.

On Jamaica the hutia may have owed its survival to being able to hide in the fissures and holes of the limestone plateau that dominates the Jamaican landscape. Oliver (1982) described hunting hutias and noted that they used those bolt holes, but whether the hutias could escape from the introduced mongooses is unknown. Hutias lived in xeric scrub forest through lowland rainforest up to mist and elfin forest at 1,000 m above sea level and disturbed habitats if there is cover. Oliver (1982) surmised that as hutias still occurred both in areas where they were hunted and in areas where there were mongooses, neither of these were critical factors compared to deforestation. The specifics of how deforestation operates—loss of cover and critical seasonal food availability—are unknown.

4.17.3   Recovery treatments

The Jamaican hutias have been protected from hunting since 1922, but in the 1970s most hunters were apparently unaware or unimpressed that hutia hunting was illegal. Areas designated as forest reserves are also listed under the 1937 Forest Act as game sanctuaries, but again compliance is lacking.

The perception in the early 1970s that the Jamaican hutia was facing extinction led to a captive breeding program in 1972. By 1984, 95 hutias had been raised and, despite losing half the stock to viral pneumonia in 1979, the captive population had exceeded space available in zoos. To reduce the captive numbers and to learn more about hutia ecology, they were repatriated to Jamaica. The 41 hutias arrived for acclimatization and quarantine at a Jamaican zoo before release in an isolated block of wet limestone forest. Despite care taken in their release, the effort failed, but without monitoring we do not know why (Oliver, personal communication, 1994).

The Jamaican hutias were treated as a small population with captive breeding and repatriation being used to supplement their numbers. Their apparent state as a small population was assumed because forests (their habitat) were being cleared. But the brief ecological study shed little light on what effect forest clearing had, and we are left wondering about the hutias's status, ecology, and limiting factors.

4.18   THE PUERTO RICAN CRESTED TOAD

We have taken the Puerto Rican crested toad, *Peltophryne lemur*, as an example of a common conservation problem, namely few data on status, extent of decline, and ecology. Yet despite those gaps, captive breeding and releases are under way. Many of the world's amphibians are declining (Baringa 1990), but a global cause of their declines is not on the firmest of footings and, given all the ecosystems that they inhabit, is unlikely.

Subfossils of Puerto Rican crested toad are widely distributed on the limestone semiarid lowlands. After its discovery in 1868 it was known from scattered localities throughout the island and was locally common. Specimens were collected in 1929 and 1931 and then not until 1967. Sources for this account were Johnson (1990, 1994), Miller (1985), Paine and Duval (1985), Paine et al (1989), and Pregill (1981).

4.18.1   Problem

The Puerto Rican crested toad's ecology is tied to hot and dry weather broken by torrential rains during the hurricane season when it breeds.

It copes with a semiarid habitat by estivating in underground holes in limestone rocks with its head plugging the entrance. When the toads emerge at night to breed after a downpour of at least 18 cm of rain, their numbers annually vary, which hampers estimating trends in numbers of breeding toads. Only two breeding populations have been found in recent decades, and one appears to have subsequently disappeared.

The toad's rediscovery in 1967 was on the northeastern coastal plain (Quebradillas), where toads were found breeding in large, concrete cattle drinking ponds (Figure 4.10). No more than 25 toads were seen, but it calls late at night and earlier visits in the evening would underestimate the numbers of crested toad. Nonetheless, after 1989, no more were seen.

Fortuitously, in 1984, Miguel Canals (manager, Guánica State Forest, Figure 4.10) recognized crested toads in a flooded car park on the southern coast. Some 3,000 individuals were counted after 33 cm of rain fell in 24 hours during a hurricane. Another hurricane in 1985, however, broached a breeding pond and toads were washed out to sea. Fewer toads than were seen in 1984 have bred at intervals, such as the 300 toads that emerged in 1990. Either the toads had declined or not enough rain fell to trigger migration to the breeding pond.

4.18.2   *Diagnosis of factors driving the decline*

Uncertainty about the decline compared to environmental changes confounds diagnosis. After the 1850s, sites were drained and breeding ponds filled in, and agriculture replaced forests. The rate of loss of the original forest was 99% by 1899, but the collapse of the sugarcane industry in the 1940s allowed large areas (32% by 1978) to revert to secondary forest (Brash 1987). How those changes affect the Puerto Rican crested toad are uncertain until its habitat requirements are described. Hotel development along the shoreline has obliterated its limestone karst habitat. Spraying insecticides and acaricides may have taken their toll.

Introduced mongooses eat adult toads, but again the effects are unknown. Another complication may be competition for spawning sites with the giant marine toad, *Bufo marinus*, introduced in the 1920s to control insect pests in the sugarcane. This is speculative, but an observation of newly emerged toadlets suggests another interaction between the two species. *B. marinus* toadlets emerge from the rapidly evaporating ponds and huddle under rocks a few days before the crested toadlets, which then may be hard put to find shelter.

Once the toad was apparently reduced to two small populations in modified environments, the potential for bad luck loomed larger. Why the northern population breeding in the cattle troughs disappeared is unrecorded. The southern population in the Guánica State Forest has suffered through environmental vicissitudes such as drainage of the parking lot whenever it flooded and then destruction of the second breeding pond after Hurricane Hugo pummeled the area in 1985.

4.18.3   Recovery treatments

Prompt action by the forest manager at the Guánica State Forest probably saved the remaining population of the crested toad. The ditch that drained the parking lot where the toads bred was dammed, cars stopped parking in the lot and people were informed about toads.

The northern and southern populations were used in the 1980s to start captive populations. Raising and breeding the toads were not difficult (Paine et al 1989). Some 3,000 toadlets have been raised and released, but signs of success are few (Johnson 1990). Fieldwork to describe the ecology and chemistry of the existing breeding pond has helped in locating and constructing new breeding ponds. In 1993, 12,000 tadpoles were released in a specially constructed pond. Head-starting (Section 9.5.1) as a treatment to supplement population size or establish new populations will take years to evaluate.

Field surveys to find out the toad's status, natural history, and limiting factors are not sufficiently under way, although Miller (1985) wrote of the need for "complete understanding of natural population dynamics before reintroductions are made" after the first release of captive-bred toads (1983). Captive breeding has taken precedence over studying the toad's ecology. This is understandable when there is only one natural and one artificial breeding pond. In the long term, however, the captive breeding efforts are barely palliative unless supporte4d by ecological studies and fieldwork to rectify shortage of breeding ponds and low recruitment.

4.19   SUMMARY

Our aim was to scrutinize the diagnosis of factors driving declines, and case histories are revealing of that (Table 4.2). The case histories were ten birds, six mammals, one amphibian, and one insect. Eight birds were, or had become, a single population on an island. The mammals had been widespread on a large island or mainland, and the insect became extinct at the edge of its range. The amphibian (the Puerto Rican toad) had been widespread on a single large island.

When it comes to the diagnoses, the only generalization possible is that there is no substitute for systematic fieldwork to weigh up and then accept or reject hypotheses for a decline. The woodhen's story

Table 4.2. Summary of diagnoses of causes of decline for case histories

	Habitat[a]			Introduced animals[a]	Hunting[a]
	Suboptimal	**Modified**	**Intact**		
Puerto Rican parrot	Observational	Observational	N/A	Treated	Observational
Nene	Associated	Observational	N/A	Treated	Previous
Cahow	Associated	N/A	N/A	Observational	Previous
Black robin	N/A	Observational	N/A	Associated	N/A
Dark-rumped petrel	?	Observational	?	Treated	Previous
Red kite	Observational	Observational	N/A	N/A	Treated
Woodhen	Treated	N/A	N/A	Experimental	Previous
Magpie-robin	N/A	Experimental	N/A	Treated	N/A
Pink pigeon	?	Observational	N/A	Treated	N/A
Numbat	N/A	Observational	N/A	Experimental	N/A
Black-footed ferret	N/A	Associated	N/A	N/A	N/A
Chinchilla	N/A	Observational	N/A	N/A	Previous
Hispid hare	N/A	Observational	N/A	N/A	Observational
Jamaican hutia	N/A	Observational	N/A	?	Observational
African elephant	N/A	Associated	N/A	N/A	Treated
Crested toad	N/A	Observational	N/A	?	N/A
Large blue butterfly	N/A	Experimental	N/A	N/A	N/A

[a]Observational = circumstantial evidence such as observation; associated = coincidence of space or time with decline; treated = treatment effected a recovery but no experimental design; experimental = designed experiment; previous = early decline.

exemplifies success with this approach, but then so do those for the cahow, magpie-robin, and large blue butterfly.

Most species had been declining over long periods. The effects of that initial decline played a part in diagnosing current problems. Seven species ended up, to varying degrees, in a refuge where they persisted because the original cause of their decline was excluded. Again, the woodhen is the clearest example, but cahows, black robins, numbats, hispid hares, red kites, and dark-rumped petrels (in Hawaii) persisted, sometimes only just, in refuges. The refuge may be on the periphery of the species range, and it may be suboptimal habitat lacking specific resources (food, shelter, nest sites). Realization of this should aid diagnosis, but in our examples this was not always obvious. The magpie-robin is an exception and one that underwrites the value of monitoring after the initial treatment.

We note, without wanting to dwell on mistakes, the tendency to launch into introduced predator control with correspondingly less emphasis on modified or suboptimal habitat. The effect of this is that factors appeared to be nested—once one factor was treated, another factor became obvious.

Theory played a minor role in diagnosis and treatments for these 17 species, although 11 species had very small populations. The small population paradigm did, however, help set direction notably for the black-footed ferret captive breeding and release. The declining population paradigm with its stock in pragmatic management was to the forefront. We could see where more use of existing theories might be fruitful in both diagnosis and treatment. Theoretical population dynamics predicts that relatively long-lived species can withstand fluctuations in recruitment more so than decreases in adult survival, especially adult females. Yet this was not reflected in either diagnosis or treatments. Emphasis tended to be on indices of recruitment such as fledging success, which are more easily monitored. Behavior, particularly social behavior, and breeding systems are another neglected aspect.

Loss of prey and hunting were relatively easy to diagnose for the black-footed ferret and elephant, respectively, but that left habitat modification or loss in abeyance. Yet habitat changes are key to planning subsequent treatments, as was shown for the Seychelles magpie-robin and black robin. Those species, after a rescue by controlling introduced predators, were capped by limiting a resource (insects) until habitat management ensured its availability.

For the chinchilla, hispid hare, Jamaican hutia, and Puerto Rican toad, there was little that we could conclude about which factors had driven their declines and which had or were currently limiting their populations. In fact for the chinchilla, hutia, and toad we cannot with certainty even state their status, a necessary step to determining their problems.

CHAPTER 5

Population dynamics

This chapter deals with how the lower-order parameters of a population—age-specific rates of fecundity and mortality, and the age distribution—combine to determine its upper-order parameters—birth rate, death rate, generation length, and rate of increase. It provides the minimum theory of population dynamics needed to deal with conservation problems. A modest acquaintance with that theoretical background is needed to understand how the demographic and environmental stochasticity affects a population (Chapter 6) and to appreciate the strengths and weaknesses of population viability analysis (Chapter 7).

This chapter introduces metapopulations. Most species, except during their infancy or extinction, occur as a number of local populations or colonies. Their individual population dynamics and immigration or emigration between them determine their fate. We describe some models that have been assembled to describe the behavior of a metapopulation's component populations, and we discuss implications for conservation biology.

5.1   THE LIFE TABLE

Table 5.1 provides a life table for a stationary population (i.e., rate of increase $r = 0$) of Himalayan thar, *Hemitragus jemlahicus*, a mountain goat introduced 90 years ago to the Southern Alps of New Zealand's South Island. As an aside, the thar's history in New Zealand exemplifies a population that thrived from small beginnings. The founders were 13 thar that came from a captive population started by 29 individuals between 1894 and 1909 (Caughley 1967a). The introductions in 1904 and 1909 had increased by 1970 to 30,000 to 40,000 thar (Tustin 1988). For some reason the name *tahr* used elsewhere in the English-speaking world became *thar* in New Zealand.

The data (Caughley 1966) come from a sample of 623 females aged 1 year or older shot in the Godley Valley (Figure 2.1). Because animals in mountain country must often be shot at considerable distances, and this can lead to an unthinking selection of large targets over small, great care was taken to sample at random with respect to age. Subsequent testing detected no bias in that respect. The animals were shot during the season of births and so their ages, as determined from the number of growth rings on their horns, closely approximated whole numbers of years.

Independent evidence was available to argue that the population from which this sample came had a rate of increase very close to zero and that it held to zero for several years. Hence, the age distribution was of the form termed *stationary*: the standing age distribution records age-specific survival.

Table 5.1.   Life table for a stationary population of female Himalayan thar in New Zealand

Age (yr) (x)	Sampled frequency (F_x)	Smoothed frequency (f_x)	Survivorship (l_x)	Mortality (d_x)	Mortality (q_x)	Survival rate (p_x)
0	—	205	1.000	0.533	0.533	0.467
1	94	96	0.467	0.006	0.013	0.987
2	97	94	0.461	0.028	0.061	0.939
3	107	89	0.433	0.046	0.106	0.894
4	68	79	0.387	0.056	0.145	0.855
5	70	68	0.331	0.062	0.187	0.813
6	47	55	0.269	0.060	0.223	0.777
7	37	43	0.209	0.054	0.258	0.742
8	35	32	0.155	0.046	0.297	0.703
9	24	22	0.109	0.036	0.330	0.670
10	16	15	0.073	0.026	0.356	0.644
11	11	10	0.047	0.018	0.382	0.618
12	6	6	0.029			
>12	11					

Modified from Caughley G. Mortality patterns in mammals. Ecology 1966;47:906–918.

The column labeled F_x in Table 5.1 gives the raw age frequencies of the shot sample. They are a little ragged, mainly because of sampling variation, and must be smoothed so that each age frequency is lower than that preceding it. The fitted regression equation

$$\log_{10} y = 1.9673 + 0.0246x - 0.01036x^2$$

achieves that admirably, and the zero-class frequency was then estimated from fecundity rates (Section 5.2). The survivorship column is filled in as $l_x = f_x/205$, which gives the probability at birth of surviving to age x. The next column, labeled "mortality," is the probability at birth of dying in the age interval $x, x + 1$ and is calculated as $d_x = l_x - l_{x+1}$. Mortality rate, the probability at age x of dying before age $x + 1$, is $q_x = d_x/l_x$, and its complement, the probability at age x of surviving to age $x + 1$, is $p_x = 1 - q_x$.

That exercise was straightforward because the population was known to have a stationary age distribution appropriate to a zero rate of increase. But suppose the population were increasing or decreasing. Table 5.2 gives such a case. The data (Caughley 1970) come from 82 female thar aged 1 year or older shot in the Rangitata Valley of the New Zealand Southern Alps (Figure 2.1). This population was increasing at about $r = 0.12$, so these data must be treated differently. Three operations are needed to convert the raw frequencies F_x to f_x. First the f_0 frequency is filled in as before from the number of female births that

Table 5.2. Construction of a life table from a stable age distribution of female Himalayan thar in New Zealand ($r = 0.12$)

Age (yr) (x)	Sampled frequency (F_x)	Corrected frequency (f_x)	Survivorship (l_x)	Mortality (d_x)	Mortality (q_x)	Survival rate (p_x)
0	—	43	1.000	0.374	0.374	0.626
1	25	27	0.626	0.018	0.029	0.971
2	18	26	0.608	0.027	0.044	0.956
3	18	25	0.581	0.035	0.060	0.940
4	19	23	0.546	0.043	0.079	0.921
5	11	22	0.503	0.052	0.103	0.897
6	12	19	0.451	0.059	0.131	0.869
7	8	17	0.392	0.065	0.166	0.834
8	2	14	0.327	0.062	0.190	0.810
9	3	11	0.265	0.058	0.219	0.781
10	4	9	0.207	0.050	0.242	0.758
>10	5					

Modified from Caughley G. Eruption of ungulate populations, with special emphasis on Himalayan thar in New Zealand. Ecology 1970;51:53–72.

would have been produced that year by this sample of females. Second, each F_x frequency is multiplied by e^{rx}, where $r = 0.12$, to allow for the effect of the rate of increase on age distribution. Third, the corrected frequencies are smoothed to enforce a continuous decline with age. Thereafter the life table is run up as before by calculating l_x from f_x and filling in the other columns. Note the difference between the p_x column of this life table compared with that of the stationary population in Table 5.1.

Those are but two methods of constructing a life table. Other methods are appropriate to different kinds of data. For example, a life table can be constructed from the age distribution of animals dying naturally, their age frequencies forming a multiple of the d_x series. Alternatively the q_x series can be estimated directly by mark-recapture. Caughley (1977) gives these and other methods.

5.2 THE FECUNDITY TABLE

The age-specific reproductive performance of a population is presented as a fecundity table. By convention it is given as the mean number of female live births produced at the season of births by a female aged x, a variable labeled m_x. Alternatively it could be given for the male

Table 5.3. Age-specific fecundity rates of Himalayan thar from two populations, one stationary and the other increasing at $r = 0.12$

Age (x)	Fecundity rates of stationary population (m_x)	Fecundity rates of increasing population (m_x)
0	0.000	0.000
1	0.005	0.000
2	0.135	0.333
3	0.440	0.389
4	0.420	0.500
5	0.465	0.500
6	0.425	0.458
7	0.460	0.437
8	0.486	0.500
9	0.500	0.500
10	0.500	0.250
>10	0.411	0.500

Modified from Caughley G. Eruption of ungulate populations, with special emphasis on Himalayan thar in New Zealand. Ecology 1970;51:53–72.

segment of the population but for the difficulty of establishing paternity. Constructing a fecundity table shares none of the difficulties faced in constructing a life table. All it requires is a sample of females whose ages are known and the information on which of them bred that season. The calculations are independent of rate of increase. Table 5.3 gives fecundity tables for the two populations of Himalayan thar used to illustrate life-table analysis. Because thar produce only one offspring at a time, their m_x values can be interpreted as the probability of a female aged x giving birth to a daughter at the annual birth pulse. Note how the two populations differ in m_x, particularly at 2 years of age. The values of m_x given here are ragged, a reflection of the moderate size of the samples used to estimate them.

5.3 AGE DISTRIBUTION

Age distributions are best presented as each age frequency divided by that of the first age class. The age distribution will be stable if the population has been increasing or decreasing at a constant rate for a couple of generations. The stable age distribution S_x is related to survivorship and rate of increase by $S_x = l_x e^{-rx}$, where r is the exponen-

Table 5.4. The stable age distributions resulting from a single survivorship schedule l_x interacting with different rates of increase r

Age (x)	Survivorship (l_x)	Stable age distribution S_x				
		$r = -0.4$	$r = -0.2$	$r = 0.0$	$r = 0.2$	$r = 0.4$
0	1.000	1.000	1.000	1.000	1.000	1.000
1	0.700	1.044	0.855	0.700	0.573	0.469
2	0.600	1.335	0.895	0.600	0.402	0.270
3	0.400	1.328	0.729	0.400	0.220	0.120
4	0.100	0.469	0.223	0.100	0.045	0.020

tial rate of increase. The stationary age distribution appropriate to a zero rate of increase is, by the same formula, $S_x = l_x e^{-rx} = l_x e^{-0x} = l_x$. Thus, the stationary age distribution is simply the special case of the stable age distribution that obtains when $r = 0$.

Table 5.4 shows how rate of increase interacting with survivorship determines the standing age distribution of a population. It indicates the fallacy of the commonly made assumption that because age distribution did not change over the course of a study, the rate of increase was zero over the same period.

5.4 RELATIONSHIPS AMONG PARAMETERS

Life tables and fecundity tables are difficult to interpret on their own because they hide the dynamic behavior of a population behind a mass of detail. They are best summarized by the upper-order parameters: birth rate, death rate, generation length, and rate of increase. These can sometimes be calculated without having to estimate the lower-order parameters of l_x and m_x. The equations are usually written for populations with no seasonal trend in the production of births because most theory of population dynamics comes from human demography. Instead we give them here for seasonally breeding (birth pulse) populations because that is the most common breeding system in nature.

5.4.1 Birth rate and death rate

The crude birth rate, B, is the number born during a birth pulse divided by the number other than newborn in the population at that time. Its estimation is usually confined to the female segment of the population. This is the birth rate most commonly calculated from field data. It is related to the lower-order statistics by

$$B = \frac{1}{\Sigma S_x - 1}$$

where S_x, the stable age distribution, is $l_x e^{-rx}$. The finite birth rate, e^b, is defined as the finite rate at which a birth pulse population with a stable age distribution would initially increase if deaths ceased. It is measured as the size of the population immediately after the annual birth pulse divided by its size immediately before it:

$$e^b = \frac{\Sigma S_x}{\Sigma S_x - 1}$$

The exponential birth rate is then simply $b = \ln e^b$. The relationship between B and b (Caughley 1967b) is

$$B = \frac{e^b}{\Sigma S_x}$$

The crude death rate D is the proportion of the population that dies over a year:

$$D = \frac{\Sigma d_x e^{-rx}}{\Sigma l_x e^{-rx}}$$

where l_x and d_x are, respectively, age-specific survivorship and mortality taken from the life table. The finite death rate e^{-d}—that finite rate at which the population would decline if births ceased—is $1 - D$. The exponential death rate is $d = \ln(1 - D)$.

5.4.2 Rate of increase

A population's rate of increase can be estimated over a year as $\lambda = N_{t+1}/N_t$, the ratio of numbers (or an index of numbers) in one year to numbers in the preceding year. It is called the finite rate of increase. Its meaning is easy to grasp, but rate of increase presented in that way is difficult to use. It has the annoying property of taking values between zero and infinity when the population is increasing but only between 0 and 1 when it is decreasing. A more useful measure is the exponential rate of increase r, which is related to λ by $\lambda = e^r$, e being the base of natural logarithms ($e = 2.71828$). The meaning of r is not as easy to grasp, but the statistic is considerably more useful. When the population doubles each year, $r = 0.69$; when it halves each year $r = -0.69$, thereby maintaining the symmetry of equivalent rates of increase and decrease and making the point that a rate of decrease is simply a rate of increase of the same magnitude with sign reversed. Rate of increase r

can be estimated as ln λ or, if several, years' data are available, as slope of the linear regression of ln N on time t.

The relationship between a population's life table, fecundity table, age distribution, and the rate of increase is given by the fundamental equation of population dynamics (Lotka 1907a,b):

$$\Sigma l_x e^{-rx} m_x = 1$$

where the summation is across all ages x. This equation indicates that if we had a life table l_x and a fecundity table m_x, we could estimate an unknown rate of increase r. It would have to be estimated by iteration (substituting successive trial values of r in the equation until the left-hand side sums to 1) because the equation does not lend itself to a direct solution. Never estimate r in that way. A much more accurate and quicker result can be obtained by measuring population size (or an index of population size) over 2 or more years.

The finite rate of increase λ is related to finite rates of birth e^b and death e^{-d} by

$$\lambda = e^b e^{-d} = e^{(b-d)}$$

and their exponential rates by

$$r = b - d$$

The intrinsic rate of increase r_m is defined as the rate at which a population with a stable age distribution would increase if no resource were limiting (in practice the rate of increase from minimal density where there is no competition for resources). It is calculated in exactly the same way as r, the only difference being the circumstances under which it is measured. Intrinsic rate of increase r_m is the r of the logistic equation and of most other population models. Whether the realized rate of increase r or the intrinsic rate of increase r_m is being referred to must often be deduced from context.

5.4.3 Generation length

The rate at which things happen is an important consideration in conservation biology. But how do we measure time in a way that is appropriate? The year is a useful unit because it spans the range of seasonal variability, but it does not allow for the difference between species in the speed of physiological processes. Sometimes we need measurements in physiological time which would allow us to compare the dynamics of, say, elephants and mice. Time measured in heart beats would do that. So does time measured in generations.

Generation length is a slippery concept. There are many definitions and they provide quite different estimates. The one we use here discounts generation length $\bar{T}$ for the population's rate of increase. It is the mean age of mothers producing an offspring during a season of births. If the age of the female breeders is represented as a histogram with frequency f against age x for the whole population or a sample of it, the mean of that or any other histogram is given by

$$\bar{x} = \frac{\Sigma\,fx}{\Sigma\,f}$$

where Σf is the number of breeders in the population or sample. The exact equivalent of that equation in terms of the lower-order parameters of a stable age distribution is, replacing $\bar{x}$ with $\bar{T}$.

$$T = \frac{\Sigma l_x e^{-rx} m_x x}{\Sigma l_x e^{-rx} m_x}$$

$$= \Sigma l_x e^{-rx} m_x x$$

$$= \Sigma S_x m_x x$$

5.5   STOCHASTIC EFFECTS

Table 5.5 gives a summary of life-table and fecundity data from the previously discussed population of thar that was increasing at $r = 0.12$.

Table 5.5.   Summarized life table and fecundity table for an increasing population ($r = 0.12$) of Himalayan thar in New Zealand

Age (x)	Females (n_{xf})	Males (n_{xm})	Stable age distribution (S_x)	Survivorship rate (l_x)	Survival (p_x)	Fecundity rate (m_x)
0	5	5	1.000	1.000	0.626	0.000
1	3	3	0.561	0.626	0.971	0.000
2	3	3	0.488	0.608	0.956	0.333
3	3	3	0.418	0.581	0.940	0.389
4	3	3	0.352	0.546	0.921	0.500
5	2	2	0.290	0.503	0.897	0.500
6	2	2	0.233	0.451	0.869	0.458
7	2	2	0.182	0.392	0.834	0.437
8	2	2	0.136	0.327	0.810	0.500
9	1	1	0.098	0.265	0.781	0.500
10	1	1	0.069	0.207	0.758	0.250
11	0	0	0.047	0.157	0.731	0.500
12	0	0	0.030	0.112	0.000	0.500

Data are from Tables 5.2 and 5.3.

Its survival rate over the first year of life is $p_x = 0.626$. That would translate to about 63% of females born still surviving on their first birthday if the population were large. But what if only five females were born as given in that table under the symbol n_{xf}?

Table 5.6 gives the probabilities for this case and for cohorts of lesser size. The coefficient for z survivors from an initial cohort of size n is simply the number of ways of selecting a subgroup of size a:

$$\frac{n!}{z!(n-z)!}$$

Note that the combinations of p and q, and the powers to which they are raised, follow an orderly sequence row by row. You should have little trouble writing the case for $n = 6$.

The only problem might come with the numerical coefficients, but these are also predictable. They form Pascal's triangle, each coefficient being the sum of the two flanking it in the row above:

$$\begin{array}{ccccccccccc} & & & & 1 & & 1 & & & & \\ & & & 1 & & 2 & & 1 & & & \\ & & 1 & & 3 & & 3 & & 1 & & \\ & 1 & & 4 & & 6 & & 4 & & 1 & \\ 1 & & 5 & & 10 & & 10 & & 5 & & 1 \end{array}$$

For cohort size $n = 5$ and $p = 0.626$ we have

Number surviving:	0	1	2	3	4	5
Probability of surviving:	q^5	$5pq^4$	$10p^2q^3$	$10p^3q^2$	$5p^4q$	p^5
Probability (for $p = 0.626$):	0.005	0.046	0.175	0.334	0.319	0.121
Percent occurrence:	0.5	4.6	17.5	33.4	31.9	12.1

Thus the most likely result, survival to an age of 1 year for three out of five newborn, occurs in only about 33% of cases. Survival of four is

Table 5.6. Probability of $0, 1, \ldots, n$ animals surviving over a year, where p is probability of survival and $q = 1 - p$ is probability of death

Number at risk (*n*)	Probability that the number below will survive					
	0	**1**	**2**	**3**	**4**	**5**
1	q	p				
2	q^2	$2pq$	q^2			
3	q^3	$3pq^2$	$3p^2q$	q^3		
4	q^4	$4pq^3$	$6p^2q^2$	$4p^3q$	p^4	
5	q^5	$5pq^4$	$10p^2q^3$	$10p^3q^2$	$5p^4q$	p^5

almost equally likely (32%), and other outcomes, including survival of none (0.5%), are entirely possible. The spread of those values is summarized by their standard deviation, which is

$$s = \sqrt{npq}$$

A simple numerical solving will reveal that s reaches a maximum when $p = 0.5$ and that it decreases as p deviates from 0.5 in either direction.

A more important measure of spread in this context is the coefficient of variation, $CV = s/n$, because it depicts the spread of outcomes relative to the number of animals. For the case considered here of $p = 0.626$ and $n = 5$ the standard deviation is 1.082 and $CV = 0.216$. With the same value of p but a cohort of $n = 500$ the standard deviation is 10.820 and $CV = 0.022$. Thus in relative terms the outcomes for the larger cohort is more predictable by a factor of 10.

To summarize, a population's rate of increase is determined by the probabilities of the life table and the fecundity table interacting with the age distribution. The lower the population size the more likely that the actual rate of increase will be displaced by "sampling error" from expectation. Look at Table 5.5 again. There are a total of 26 probabilities contained in the p_x (or l_x) and m_x columns. Each probability in combination with the number of animals in the age class (n_x) can provide a spread of outcomes, their relative magnitude increasing with decreasing population size. It is no wonder then that the rate of increase of small populations under the influence of this demographic stochasticity is erratic from year to year even when environmental conditions are constant.

The variation in r caused by these sampling effects is measured as V_d, the variance in r attributable to demographic stochasticity. That variance is simply the square of the standard deviation of r values that can be produced by a population of given size with fixed schedules of fecundity and survival. It is related to population size N by

$$V_d = \frac{V_{d1}}{N}$$

where V_{d1} is the stochastic variation in r contributed by an average animal. We can use the example in Table 5.5 to calculate a rough value of V_{d1} by simulation for that population of thar. Starting with the age distribution given as n_{xf} for females and n_{xm} for males, we put each cohort through 1 year by drawing the appropriate number surviving from a distribution of mean $n_x p_x$ and variance $n_x p_x (1 - p_x)$. The number born to each age class is drawn for both males and females from a distribution with mean $n_x m_x$ and variance $n_x m_x (1 - m_x)$. Then the process is repeated for the next year. The easiest way to make the random draws with a computer is to take each individual and declare it a survivor if its p_x exceeds a random number drawn from a rectangular

---

Box 5.1. QuickBASIC subroutine to generate random draws from a binomial distribution with specified *p* and *n*

BINOMIAL
Subroutine to calculate stochastic number of survivors (Y) out of N animals where probability of survival is P. Values of N and P are imported from the main program.

```
Y = 0
FOR I = 1 TO N
R = RND    RND calls up a random number between 0 and 1
IF P >= R THEN Y = Y + 1
NEXT I
RETURN    Return to main program bearing a stochastic
          value of Y
```

---

distribution bounded by 0 and 1. The same system may be used to estimate whether it bred. Box 5.1 gives a subroutine in QuickBASIC to allow random draws from a binomial distribution.

The population was thus "grown" each year until it reached or exceeded a size of 500, its rate of increase for each year being stored together with the size of the population at the beginning of the year. Thereafter it was restarted with the initial age distribution given in Table 5.5. A large number of runs provided the variance of *r* for each population size, and we used the 300 estimations of mean *r* for population sizes between $N = 101$ and $N = 400$ to calculate V_{d1} from the average of the 300 values of V_d/N. It came out as $V_{d1} = 0.25$. From this we can estimate V_d for any population size as $V_d = 0.25/N$.

The value of V_{d1} given here should not be taken as a constant for thar. A change to one probability of the life table or fecundity table will change V_{d1}. For example, were we to raise by 50% the first p_x of Table 5.5 from 0.626 to 0.939, V_{d1} would decline to 0.18 from the previous 0.25, a reflection of the variance of outcomes for that age being reduced by the shift of that p_x further away from 0.5.

5.6 METAPOPULATION STRUCTURE AND DYNAMICS

Heterogeneity of habitat (patchiness) and behavior restrict species into local populations (demes) whose members mostly breed within their population. Local populations may disappear over time, but usually their habitat patches are reoccupied. This pattern of local extinctions and colonizations is relevant to averting species extinctions in the context of habitat fragmentation (Section 8.1.6), captive breeding (Section 9.5.1), and reserve systems (Section 10.2.1).

The dispersion of animals as local populations has long been known: Andrewartha and Birch wrote in 1954, that "A natural population occupying any considerable area will be made up of a number of . . . local populations or colonies. In different localities the trend may be going in different directions at the same time." Although, then, the idea of local populations was not new when Levins (1968) coined the term "metapopulation," he established a mathematical relationship between their extinction and recolonization. Levins (1968) specified that a metapopulation is "any real population [that] is a population of local populations which are established by colonists, survive for a while, send out migrants, and eventually disappear."

Levins (1969, 1970) modeled the proportion of habitat patches occupied by the subpopulations of a metapopulation as a logistic equation which in its familiar form depicts the change in numbers within a single population as

$$\frac{dN}{dt} = r_m N\left(1 - \frac{N}{K}\right)$$

Levins's model depicts the change in the proportion of patches occupied as

$$\frac{dp}{dt} = (m-e)p\left[1 - \frac{p}{1-e/m}\right]$$

in which p is the proportion of occupied patches and m and e are respectively the rate at which patches are colonized and emptied when most of the patches are empty. Thus m and e are equivalent to maximum birth rate and minimum death rate of a population growing logistically. Hence $m - e$ is equivalent to r_m, and $1 - e/m$ is the analogue of K in the population model, both of which represent equilibria. The equation is usually written in a form masking its logistic nature:

$$\frac{dp}{dt} = mp(1-p) - ep$$

It is essentially identical to Wright's (1940) model depicting change in the frequency of a nondominant gene within a metapopulation.

The behavior of Levins's model is discussed by Hanski (1991), by Hanski and Gilpin (1991), and by Burgman et al (1993). It is an "occupation" model in which patches are either occupied or not. Paraphrasing Burgman et al (1993), the assumptions of this model are as follows:

(a) Subpopulations have no dynamics. The model assumes that each subpopulation is either extinct or it is extant at carrying capacity *K*. A recolonized patch goes from empty to a population size saturated in a single time step.

(b) There are an infinite number of patches. Because *p* is a continuous variable, it implies that patch number is infinite. Further, all subpopulations are assumed to have the same probability of extinction, implying that the patches are the same size and the subpopulations occupying them have equal numbers of individuals.

(c) Extinctions are independent of each other. There is zero correlation between extinction events in neighboring patches, implying that patches experience independent environmental variation.

(d) There is equal probability of dispersal, dispersal and recolonization are assumed equally likely from one patch to any other.

It is obvious that Levins's model greatly simplifies what actually happens in nature, a point made particularly by Harrison (1991). She contrasted models based on wild populations to show that not all populations have the "classic" metapopulation structure characterized by local extinctions and recolonizations. Instead there is a continuum of population structures which at one end has patchy populations in localized and ephemeral habitats such as fallen fruit. Those populations are characterized by high rates of individual dispersal between patches which lessens the chance of local extinctions. At the other end of the continuum is the mainland-island population structure (MacArthur and Wilson 1963, 1967) where a mainland population continually supplies immigrants to restock islands as local extinctions claim their populations. The large size of the mainland population protects it from the random processes that obliterate small populations on islands (Chapter 6). Many examples have come from the study of birds (Diamond 1975; Pimm 1991).

In contrast to the mainland-island relationship, the difference between a "source" population from which individuals disperse to replenish "sink" populations is not size but habitat quality. The sink populations occupy poor-quality habitat patches that do not support them for long, especially in the face of droughts or other environmental variations. The source population on a high-quality habitat not only exists longer but fares well enough for individuals to disperse to recolonize the other patches. Most of the published examples of this population structure come from butterflies (Thomas 1994, Harrison 1991).

The simplicity of Levins's model is not necessarily a fault if it is used only to explore qualitative outcomes. Further, if the simple model captures the essence of processes more complex than portrayed, it might even furnish an adequate summary of the outcome of those more complex processes. For example, Verboom et al (1991) showed that the Levins model, applied in stochasticized form to the ecology of the European badger, *Meles meles*, yielded a close approximation in outcome to a more detailed model incorporating subpopulation dynamics.

A number of authors have elaborated the Levins model to add sex and age structuring and the dynamics of subpopulations, stochasticization, correlation of extinction probabilities among sub-populations, and varying the probability of colonization according to the distance between patches of habitat. Burgman et al (1993) review these methods and contributed others. Hanski and Thomas (1994) found congruence between their elaboration of Levins's model of patch occupancy and the behavior of three species of butterfly.

An ingenious application of metapopulation theory is Lande's (1987) modeling of an undivided population of a territorial species as a set of subpopulations occupying patches of the size of individual territories. Each "subpopulation" comprised a single mated pair. Unoccupied patches were those unsuitable for breeding. We agree with him that this kind of model will be useful for predicting the effect of habitat fragmentation.

May (1991) followed a similar path in utilizing the Levins model to predict the effect of fragmenting the habitat. He took the equilibrium of the Levins model,

$$p^* = 1 - \frac{e}{m}$$

and modified it to allow for the number of patches being reduced (as in habitat clearing) by a proportion $1 - f$, leaving only a fraction f available for occupation by the species. The equilibrium proportion of the re-maining patches occupied by the metapopulation is thereby reduced to

$$p^* = 1 - \frac{e}{mf}$$

Clearly p^* gets smaller as f gets smaller. The metapopulation dies out when f falls below the critical value

$$f_c = \frac{e}{m}$$

This makes the important point that it is not necessary to destroy all the patches for the metapopulation to go extinct. May (1991) drew a paral-lel between eradicating a metapopulation by destroying only a propor-tion of the patches with eradicating a disease by vaccinating only a proportion of the hosts (patches). The mathematics is essentially iden-tical (Anderson and May 1985).

The genetic heterogeneity of a metapopulation's constituent popu-lations will differ and will be less than for the metapopulation as a whole. The equations emphasize the importance of dispersal in deter-mining how many patches will be occupied and (indirectly) the extent to which the populations will be genetically uniform. With selective neutrality the genetic divergence between local populations is a function of the number, not the proportion, of migrants. An average exchange of one migrant per generation (provided it successfully reproduces) will maintain the same alleles in the subpopulations, but a much higher exchange is needed to maintain similar allelic frequencies. Heterozygous advantage reduces the divergence in allelic frequency between local populations (Allendorf 1983). In this context the metapopulation has become a conceptual model for captive breeding (Section 9.5.1) and the design of reserves (Section 10.2.1).

5.7 SUMMARY

Rates of birth and death and generation length can be calculated directly or as summaries of the interaction between age-specific rates of fecundity and mortality and age distribution as displayed in life tables and fecundity tables. Rates of increase are preferably estimated directly from changes in population size. Scrutinizing the probabilities in a life table makes plain why the rate of increase in small populations can be so variable. We supply the way to estimate that variance in the rate of change caused by demographic stochasticity.

Most species exist as a metapopulation as habitat heterogeneity, and breeding systems divide it into a series of populations which interact by dispersal between them. A variety of models have been derived to predict the probabilities for the fate of those individual populations. A metapopulation has more genetic variation than a single population, and the risk of extinction is lessened as all populations are unlikely to be exposed to the same event simultaneously.

6

Risks faced by small populations

The specter of small populations haunts conservation biology with good reason. Very small populations such as might be found in a nature reserve, on a tiny island, or in a zoo are at risk simply because of their size. Knowledge about the fate of small populations is mostly theoretical because we rarely witness their end. Simberloff (1986) made this point when he found that there was not enough known about the final stages of extinctions to survey ultimate causes.

Systematically collected observations of small populations show that once they are down to a few breeding pairs, the probability of extinction is high but variable. Models of persistence time have standard deviations comparable in size to the means (Goodman 1987; Pimm 1991). The variability makes the point that species can exist as a handful of breeding pairs for decades, and they can persist from a small starting population, even a single pair. Table 6.1 gives a scatter of examples. The converse of extinction—speciation—stems from small populations that persisted.

Examples of persistence do not detract from the reality of threats that face small populations. The probabilistic nature of extinction is such that some small populations survive, but most lose out when faced with the three categories of risk characteristic of small numbers. The risks rise from sampling probabilities—the effects of random draws from small pools of individuals. We outline the consequences of random draws for demography, environment, and genetics. Another threat facing some small populations is not probabilities but that numbers may be too low to maintain a viable social structure. Lastly, we review what is meant by a small population—the concept of the minimum viable population size.

6.1   DEMOGRAPHIC STOCHASTICITY

A tossed coin has a 50:50 chance of coming down heads. We say that the probability of its landing heads is $p = 0.5$. The probability of its landing tails is $q = 1 - p$, which in this case is also 0.5. The probability of the coin landing heads is not affected by the results of previous tosses: The events are independent of each other—a fact not often appreciated by gamblers.

Five hundred tosses are likely to yield close to 250 heads and 250 tails. Five tosses behave differently. First, they cannot yield anything near a 50:50 result. The closest approach to parity would be three heads and two tails or two heads and three tails. Second, almost any result is possible, including a run of five heads. In our sample of 500 tosses the probability of 500 heads in a row is vanishingly small. The point we are laboriously making is that the result of a large number of events is highly predictable. It conforms to the law of averages. The result

Table 6.1. Examples of small populations where the number of breeders was less than ten pairs and the population has persisted at least one generation

Species	Location	Minimum number of pairs (or individuals)	Release Date	Subsequent Status		Source and comments
				Numbers	Date	
Introduced animals						
Feral cat[a]	Kerguelen Island	1	1956	10,000	1983	After control 1973–1977
	Marion Island	(5)	1948	3,045	1977	Box 9.7
Mouflon[a]	Haute Island (off Kerguelen Island)	1	1957	250–550	1978	Eruptions and crashes
Reindeer[a]	Grande Terre (off Kerguelen Island)	2	1956	2,000	1982	
Translocations of endemic species						
Black robin	Chatham Island	2	1976	120	1992	Section 4.7
Laysan rail	Eastern Island	2	1891	Numerous	1940	Table 9.2
Hutia	Bahamas	(11)	1969	1,200	1985	Section 4.17
Remnant populations						
Elephant[b]	South Africa	(11)	1931	118	1987	Fenced reserve
Indian rhino[c]	India	(12)	1908	1,080	1989	Protected reserve
Black robin	Chatham Island	(10–30)	1938	10–38	1972	Section 4.7
Woodhen	Lord Howe Island	(20–25)	1920	10–25	1980	Section 4.2
Pink pigeon	Mauritius	(10–20)	1970s	22	1992	Section 4.10
Red kite	Wales	4–5	1880	15 pairs	1960	Section 4.13

[a] Data are from Chapuis JL, Boussès P, Barnaud G. Alien mammals, impact and management in the French subantarctic islands. Biol. Conserv. 1994;67:97–104.
[b] Data are from Hall-Martin A. Elephant survivors. Oryx 1980;15:355–362; and Cumming DHM, DuToit RF, Stuart SN. African elephants and rhinos. Status survey and conservation action plan. Gland, Switzerland: IUCN, 1990.
[c] Data are from Ramono WS, Santiapillai C, MacKinnon K. Conservation and management of Javan rhino (*Rhinoceros sondaicus*) in Indonesia. In: Ryder OA, ed. Rhinoceros biology and conservation. Proceedings of an International Conference, San Diego, CA, 1991. San Diego: Zoological Society of San Diego, 1993:265–273.

of a small number of events is highly unpredictable and not worth wagering on.

As with coins, so it is with animals. Whether a population of a few animals increases or decreases over a year is less dependent on the age-specific probabilities of survival and reproduction than on raw chance, which is to say that they follow the binomial distribution. The individual fortunes of each animal swamp the effect of the probabilities that determine the outcome for a large population. A small population can die out entirely by chance even when its members are healthy and the environment favorable. In a single unlucky year a mature female might break a leg, another produces a litter only of males, and that is that, or, as Ecclesiastes 9:11 succinctly tells us: "I returned and saw under the sun, that the race is not to the swift, nor the battle to the strong, neither yet bread to the wise, nor yet riches to men of understanding, nor yet favor to men of skill; but time and chance happeneth to them all."

6.1.1 Persistence time under demographic stochasticity

The variance in r caused by demographic stochasticity in a constant environment depends on the size of the population (Section 5.5). A halving of population size leads to a doubling of the variance in r. A small population is therefore subject to double jeopardy: It is likely to suffer erratic swings in size under the influence of demographic stochasticity alone, and its small numbers provide no buffer against a decrease ending in extinction.

Suppose a population is limited in size by a shortage of nesting sites. A seabird such as the Bermuda petrel (Section 4.3) that nests in holes in cliffs will serve as an example. Its numbers are thus capped at a level, say K. In such circumstances, under demographic stochasticity alone, the mean time to extinction increases almost exponentially when graphed against K (MacArthur and Wilson 1967). The slope of that curve increases with r_m (Lande 1993). Thus the risk imparted by demographic stochasticity to such a population declines sharply according to its ability to recover quickly from low numbers and according to the maximum level to which it can increase.

Pielou (1977) gives an exact solution, in terms of instantaneous birth rate, b, and death rate, d, for the probability that a population increasing or decreasing at a constant exponential rate will die out over a specified interval:

$$p_E(t) = \left(\frac{de^{(b-d)t} - d}{be^{(b-d)t} - d} \right)^{N_0}$$

where t is the time interval and N_0 the starting population size. The expected rate of increase r is the difference between b and d (Section 5.4.2), whereas the probability of extinction over a specified time interval is a function of the ratio b/d. Although her model of constant r is special, and not particularly realistic, the influence of the b/d ratio on persistence under demographic stochasticity is probably general.

6.1.2 Effective population size (demographic)

Section 5.4.2 explained that a population's rate of increase is determined by its age-specific rates of mortality and fecundity interacting with the population's age distribution. If the age distribution has converged to its stable form, its rate of increase will be constant. A stable sex ratio is a further requirement for a constant rate of increase. Age distributions and sex ratios can deviate markedly from their stable forms, rate of increase being boosted above the level suggested by the life table and fecundity table or reduced below it. For example, when the St. Matthew Island reindeer, *Rangifer tarandus*, herd crashed after eating out its food supply, it ended the slide with over 30 females but only one male (Klein 1968). The population was thus poised to recover much faster than if the sex ratio had been 50:50. (The fact that it did not is attributable to a sad case of fumble-finger when a few specimens were shot to gauge condition. The male was included in the sample.)

Shepherd (1987) showed that kangaroo emerging from a drought in Australia had the age distribution trimmed of younger and older animals. These effects boost the size of the next crop of newborn: the populations are acting dynamically as if they were larger. Alternatively, a population coming out of an atypically mild winter will contain an excess of young animals below breeding age. The birth rate measured as newborn per head of population will thus be lower, the population acting reproductively as if it were smaller.

Demographic stochasticity almost closed the story of the woodhen when seven of the nine woodhens that died in a 2-year span were female. Three pairs were left and three unpaired males (Section 4.2). In the absence of treatment (removal of predators), the outcome of the imbalanced sex ratio would have been uncertain. The black robin had an even closer call—when only seven robins were left, only two of which were females and only one of them left living descendants (Section 4.7). The Fiordland population of New Zealand's endangered and enigmatic kakapo was less lucky. The last 18 parrots in this, one of the last two populations, were all males (Merton 1989; D. Merton personal communication, 1994). They lasted 10 years until the final three disappeared in 1987.

The degree to which population growth is enhanced or held back by the instability of age distribution and sex ratio can be measured as the size of an "ideal" population producing the same net increment per year as the population of concern. It is called the *effective population size* (*demographic*) and symbolized N_{ed}. In this context an ideal population has a 50:50 sex ratio and a stable age distribution.

A disparate sex ratio may have a large effect upon net change in numbers over a year and hence upon N_{ed} at the beginning of that year (Caughley, unpublished data). The net change in numbers will be

$$\Delta N = NP_f pb - N(1 - p)$$

in which, at the beginning of the year which starts immediately after the annual birth pulse, a population of size N contains a proportion P_f

of females; p is the probability of surviving the subsequent year averaged over individuals of all ages, and b is the number of live births produced per female at the birth pulse terminating that year. The first term ($NP_f pb$) gives the number born at the birth pulse, and the second ($N[1 - p]$) gives the number dying over the 12 months preceding it. It indicates that net change in numbers in a population of any given size is a linear function of the proportion of females in the population. The regression of ΔN on P_f has a slope of Npb and an intercept (the value of ΔN when $P_f = 0$) of $-N(1 - p)$.

The equation for net annual increment may also be written

$$\Delta N = N(P_f pb + p - 1)$$

and in that form the net annual increment ΔN_I to an ideal population of the same size is

$$\Delta N_I = N(\tfrac{1}{2}pb + p - 1)$$

in which P_f, the proportion of females, is set at $\tfrac{1}{2}$ by definition. The size of an ideal population that would produce the same net annual increment as the population under study is

$$N_{ed} = \frac{N(P_f pb + p - 1)}{\tfrac{1}{2}pb + p - 1}$$

Similar reasoning leads to an estimate of N_{ed} in terms of age distribution (Caughley, unpublished data). We follow the demographic convention of considering only the female segment of the population because of the difficulty of establishing paternity of offspring. Defining the age distribution F_x as N_x/N_0, where N_x is the number of females in age-class x and N_0 is thus the number of female newborn, we arrive at

$$\Delta N = \Sigma N_x(p_x m_x + p_x - 1)$$

where p_x is survival rate of females between ages x and $x + 1$ and m_x is the number of daughters born to females of age x. For a population with a stable age distribution (i.e., the "ideal" population) $F_x = f_x = l_x e^{-rx}$ (Section 5.3) where l_x is survivorship and r is the rate of increase. The effective population size in terms of age distribution is then

$$N_{ed} = \frac{N\Sigma F_x(p_x m_x + p_x - 1)}{\Sigma l_x e^{-rx}(p_x m_x + p_x - 1)}$$

The value of N_{ed} may be greater than or less than the census population size N, both in terms of sex ratio and age distribution. That is in contrast to the effective population size (genetic), written N_{eg}, which is the size of a genetically ideal population having the same loss of heterozygosity as the population of concern (Section 6.3.4). Invariably $N_{eg} < N$. The context in which one or other is used is critical because the best age distribution and sex ratio for minimizing genetic drift in a zoo may be those least appropriate for encouraging a natural population to break away from the trap of low numbers.

6.2   ENVIRONMENTAL STOCHASTICITY

Year-to-year variation in environmental conditions, particularly in weather and its effect on the resources that a population needs, affects a population's rate of increase in a way that differs notably from the influence of demographic stochasticity. The variance in r resulting from demographic stochasticity declines with increasing population size so that by $N = 300$ its effect is negligible. In contrast, the variance in r generated by environmental fluctuations is largely independent of the size of the population. It will lead to much the same proportionate increase or decrease in numbers whether the population is large or small. May (1973) introduced the term "environmental stochasticity" to describe this second source of unpredictability in rate of increase.

Previously, conservation biologist had considered year-to-year fluctuations in environmental conditions distinct in their effect from that of "catastrophes," which are large infrequent environmental perturbations. The rationale for the division was the belief that the trend of persistence time on population size differed in kind according to which effect was operating. Anticatastrophes, the large but favorable perturbations, did not rate a mention. Lande (1993) has shown that, contrary to the view of the previous consensus, the scaling rules governing the trend of persistence on population size do not differ between year-to-year fluctuations and catastrophes, and so the two are not differentiated here: the chance of a 100-year flood comes from the same probability distribution as does the chance of a 1-year flood.

Examples of the effects of environmental stochasticity on small populations include drought and the disappearance of the last large blue butterflies in England (Section 4.4). Another example relates the fate of the last three Laysan apapanes, *Himatione sanguinea freethii*— they disappeared after a sandstorm (Box 8.9). The Puerto Rican parrot was lucky—even when reduced to only four breeding pairs, the population survived Hurricane Hugo, although half the parrots disappeared in its aftermath (Section 4.9).

6.2.1 Persistence time under environmental stochasticity

It was previously thought (e.g., Ewens et al 1987; Goodman 1987; Pimm and Gilpin 1989; Soulé and Kohm 1989; Pimm 1991; Hedrick and Miller 1992; Stacey and Taper 1992) that environmental stochasticity necessarily led to an upwardly convex trend of persistence time on population size—that the effect on mean time to extinction of increasing population numbers by a given proportion is stronger at low numbers than at high numbers. Lande (1993) has shown that this result is not general, at least not for the simple models that he and those authors cited have used (truncated exponential, logistic). The shape of the trend is determined by the relative magnitudes of mean rate of increase $\bar{r}$ and the environmentally induced variance in r, V_e. With $\bar{r} < V_e$ the trend of persistence time on population size is convex; with $\bar{r} > V_e$ the trend is concave (as with demographic stochasticity).

The simple models used to provide generalizations about persistence time are often difficult to relate to real populations. The two most favored are the truncated exponential and the logistic. Each has but two parameters, r_m and K. We will examine briefly their applicability to real populations in the context of estimating a population's persistence time.

The truncated exponential model has a single rate of increase to which the population holds until it reaches a size of K, at which rate of increase becomes zero (Section 7.4.1). The population cannot exceed K. Parrots limited by nesting holes might produce similar dynamics. So might territorial species whose territories are incompressible. However, the model is entirely inappropriate ecologically for most other populations.

In the logistic model (Section 7.4.2) the realized rate of increase r is entirely a function of the population's size N, and it declines linearly as N increases, reaching zero when N reaches K. The only ecological system that would conform precisely to these rules is one in which the population depends on a renewable resource but can influence neither the standing crop of that resource nor the rate at which it is renewed (Caughley 1976a). Further, it cannot get at the standing crop but utilizes only the "interest" generated by the "principal" represented by that standing crop. Surprisingly enough there are such populations. Plants do not influence the rate of renewal of their major resources, water and sunlight, but compete for what comes their way, the competition among individuals intensifying with increasing density. Their population growth might therefore be expected to approximate a logistic curve, and it often does. Similarly, a population of scavengers conforms to logistic rules, and its growth might therefore be approximately logistic.

Most other populations are not like this. Their rate of increase usually depends on the standing crop of a renewable resource, and they are entirely capable of reducing its level. For a herbivore feeding on a logistically growing vegetation, the rate of increase of the animals is a function of the standing crop of its food plants, and the rate of increase of the food plants is at least in part a function of the standing crop of the animals feeding on them. The dynamics of these interactive multispecies systems are more complex than those of the single-species models outlined earlier. We must be wary of applying generalizations derived from the simple dynamics of single-species models to the complex dynamics of two-species or multispecies systems. It may often be possible, because large conclusions tend to be robust, but that must be demonstrated rather than assumed.

Although we would seldom admit it in public, we are continually frustrated by the penchant of many theoreticians to believe sincerely that simple population models actually describe the dynamics of real populations. We are particularly bothered by their cavalier use of the constant K as if it had a precise ecological meaning rather than being simply a mathematical convenience. In this we are not alone (Pimm 1991, p. 136): "Quite what K has to do with real populations is far from obvious. I have no idea whether it exists or how to calculate it. However, in simulating a large number of different models of species extinction, I find that K is closely related to average population size. Typically, average population size is about half K."

6.3 LOSS OF GENETIC VARIATION

One of the most voiced concerns for small populations is the threat imparted by loss of genetic variation. We briefly introduce genetics to explain the terminology and then describe measuring genetic variation. We explain the mechanisms by which genetic variation is lost and then possible consequences of that loss.

A chromosome comprises two strands of DNA which separate when gametes (i.e., sperm and eggs) are formed. A gamete carries only one of the strands from each chromosome. At fertilization, which forms a zygote (i.e., a potential embryo), the two strands come together again, one contributed by the sperm and the other by the ovum. A region of the chromosome that contains the genetic code or DNA sequence for a heritable trait is called a *locus*, and at each locus there will be two *alleles* of genetic material, one on each chromosomal strand. The *gene* is a general term for the heritable unit, but the allele is the specific form of that gene. The two alleles at a locus may be the same (*homozygous*) or different (*heterozygous*) in any individual. Within a population there may be many alleles occurring at a given locus (a *polymorphic* locus), although there can be no more than two allelic types occurring at that locus for a given member of the population. Many individuals will have two alleles of the same type at that locus even when the locus is polymorphic across the population. Alternatively, and more commonly, the locus will be *monomorphic* and all members of the population will, at that locus, have two copies of the one distinct allele type. That single allele type is then said to be *fixed* at that locus.

6.3.1   The Hardy-Weinberg equilibrium

If the progeny generation is relatively large, in the hundreds, the frequency of any particular allele and the genotypes that it produces can be predicted with precision. As long as the mating is random, the combinations of the alleles will be random, and genotypic frequencies at any one locus will reach equilibrium that holds constant between generations. This is the Hardy-Weinberg law. Hardy was an American mathematician, Weinberg a German physician. In 1908 they independently derived the law, but Hardy published in the journal *Science*, and initially the law was named after him. Weinberg had published in an obscure journal. It was not until 1943 that Stern (1943) suggested that Weinberg be linked to Hardy for the law's name.

Suppose that at a given locus the parental generation has two distinct alleles, A_1 with a frequency $p = 0.6$ (i.e., 60% of the alleles at that locus, measured across all individuals, are of the A_1 type) and A_2 with a frequency $q = 0.4$ (i.e., $1 - p$). Since the progeny generation is relatively large, there will be very little deviation of those frequencies as the genetic material is passed from parents to progeny, so the progeny generation will also exhibit allelic frequencies of A_1 and A_2 at that locus. An offspring will have one of three possible combinations of those alleles at that locus—A_1A_1, A_1A_2, or A_2A_2—and the binomial theorem predicts that those genotypes will occur with proportions

Genotypes:	A_1A_1	A_1A_2	A_2A_2
Probabilities:	p^2	$2pq$	q^2
Expected frequencies:	0.36	0.48	0.16

This relationship between allelic frequencies and genotypic frequencies is the Hardy-Weinberg equilibrium.

6.3.2   Measuring genetic variation

Once the role of DNA in inheritance was unraveled, the way was opened to directly measure genetic variation. The first development was from the relationship between a single gene and the protein that it codes for. Previously, genetic variation of individuals had been laboriously measured from breeding experiments to reveal variation in the inheritance of physical characters.

Allozyme electrophoresis is a relatively simple and cheap technique, and we have followed Evans's (1987) description. The technique depends on the fact that an individual with heterozygous alleles at a locus will produce two versions of a protein, but only one if homozygous. The electrical properties of soluble proteins (usually enzymes) cause them to migrate in a gel when an electric field is applied. The different migration rates are revealed as bands by histochemical staining and are characteristic for each form of the enzyme (allozyme): an individual homozygous for an enzyme such as alcohol dehydrogenase formed by two polypeptide chains will have a single band, or two bands if heterozygous.

The most common use of electrophoretic data is to estimate *heterozygosity* (the proportion of heterozygous loci in an average individual). The number and frequency of different alleles at a locus comes from electrophoresis, and if p_{ij} is the frequency of allele i at locus j in the population as a whole, the proportion of individuals heterozygous at that locus is estimated as

$$h_j = 1 - p_{ij}^2$$

If the number of individuals n_j examined for locus j is less than 30, h_j is underestimated but can be corrected by multiplying by $2n_j/(2n_j - 1)$. Thus the more alleles at a locus the higher the value of h_j, and the less diverse the frequencies of the alleles the higher is h_j.

Mean heterozygosity (H) is estimated as

$$H = \frac{\Sigma h_j}{L}$$

where L is the number of loci examined.

Electrophoresis has limitations. Relatively few enzymes are polymorphic, and even those are not all suitable because they have to be soluble and capable of taking a stain. Interpretation of the bands is a further suite of problems. Not all bands represent alleles: proteins denatured during sample preparation can form misleading (false) bands. An elaboration of techniques reveals that some bands actually represent several alleles. Alterations in the amino acid sequences have to change the protein's charge if they are to be detected, but this only happens for about a quarter of them. Sex and cohort can contribute variation to measures of heterozygosity, and with endangered species those sources of variation can be compounded by small sample sizes. Required sample sizes can be estimated in advance (see Sherwin et al 1991 for an example), but even then few loci or low levels of heterozygosity can limit statistical testing. Computer simulations and iterative sampling can overcome those problems, and Wayne et al (1991) illustrate this in their analysis of wolf genetics.

Labeling heterozygosity (or any measure of genetic variation) as low or high implies relative to some standard. Sherwin et al (1991) found zero heterozygosity in the small and isolated population of the eastern barred bandicoot, which has sharply declined. Before succumbing to the temptation to associate the population's problems with the zero heterozygosity, the researchers had used as a control a widespread and relatively large population of the same species in Tasmania. Its heterozygosity, as measured the same way, was also zero.

Electrophoresis deals only with qualitative characters. A single gene codes for a protein, or not, and those characters segregate in a Mendelian manner. Quantitative characters (body weight, tail length) vary in a continuous fashion and are the result of the effect of alleles within and between loci (additive genetic variance), dominance effects, and interactions between loci. The additive genetic variance is by far the most important and for present purposes can be discussed as if it constituted the total genetic variance. That is because dominance effects and interlocus interactions are not inheritable (only one allele at a locus is donated by each parent). Factors that influence additive genetic variance also influence heterozygosity, and they do so by similar amounts and in similar ways. We can use heterozygosity H as an index of additive genetic variance.

The value of H ranges between 0.00 and 0.26 for mammals and averages 0.04. It is zero (homozygous at all loci) in about 11% of mammals that were examined by electrophoresis (summarized in Caughley 1994). Comparable figures for birds are 0.00 and 0.13 with an average of 0.06, and only 1 of 86 species was homozygous for the 27 loci examined (summarized in Evans 1987). Invertebrates tend to be three times more heterozygous than vertebrates.

Species that scored with zero heterozygosity based on electrophoresis may not lack genetic variation. Electrophoresis only samples the 5% to 10% of DNA that codes for protein. The ability to directly sample more of the DNA (but not all of it) to measure genetic variation has markedly changed our ability to differentiate populations, their origin, and the relationships between individuals.

Mitochondrial DNA is highly polymorphic and has a higher rate of nucleotide substitution than nuclear DNA. It is maternally inherited and is particularly revealing about the relationships between subpopulations. Bacterial enzymes (known as restriction enzymes) are used to fragment the DNA at specific sites, and the fragments separate by size in an electric field. They are developed for view by hybridization with a radioactive DNA probe which is a specifically cloned sequence. The hybrid fragments appear as a series of bands after exposure to x-ray film. The fragments (restriction fragment length polymorphisms, RFLP) behave as alleles, and their frequencies are determined even though they are an arbitrary DNA sequence.

Much of chromosomal DNA is repeated sequences that vary in their length (i.e., the number of base pairs). The lengths of the repeated sequences, the number of repeats, and their location are unique to an individual and inherited equally from both parents. Development of probes which identified highly polymorphic DNA regions has increased the resolution of nuclear DNA. The probes that detect the specific DNA regions are known as microsatellites and minisatellites, and they can resolve a single hypervariable locus. The technique's label of DNA fingerprinting is evocative of its potential to identify individuals. To compare inbreeding in two populations of African lions, *Panthera leo*, the percentage similarity between pairs of individuals was estimated from numbers of DNA fragments in common and the number resolved in each pair. Those estimates of similarity were then calibrated by kinships based on long-term observations of the individual lions (Packer and Pusey 1993).

The northern elephant seal, *Mirounga angustirostis*, increased from about 100 individuals in the late 1800s to 125,000 by 1989. Through allozyme electrophoresis, no polymorphisms were found. DNA fingerprinting with minisatellite loci revealed high levels of mean pairwise similarity values, which does confirm the low genetic variation as measured by electrophoresis (Lehman et al 1993). It is too soon to determine what the low variation means for the species, but it did not impede its initial rate of increase.

A cautionary note is that DNA fingerprinting is not free of problems. The assumption that each band represents an independent locus was not found to be true in a study with Hispaniolan parrots (surrogates for management of the endangered Puerto Rican parrot, Section 4.9). The check on the independence of the bands came from the pedigrees of the parrots that were recorded for three generations. Despite the presence of the multibands, most bands were single alleles and can be used to estimate genetic variation even if not accurate estimates of relatedness (Brock and White 1992).

Determining the sequence of base pairs rather than detecting areas of similarity as in RFLP or fingerprinting was a slow and arduous technique until the discovery of polymerase chain reaction (PCR). The polymerase enzyme can be used to make millions of copies of a selected DNA area once the DNA has been unraveled by heat.

Practical applications of measuring genetic variation are identifying individuals and their parentage to verify *bona fide* captive breeding (as against illegal take from the wild). It is helpful in unraveling breeding systems and relationships in the wild and in captive populations, particularly when pedigrees are incomplete (Wayne et al 1994). An example of using the different techniques comes from Wayne et al (1991). Protein electrophoresis revealed a significant difference in the heterozygosity between the wolves in Isle Royale National Park, Lake Superior, and those on the mainland both in Canada and the United States. Mitochondrial DNA analysis revealed which mainland populations were most closely related to the Isle Royale wolves. Comparison of fingerprinting from known (captive) siblings, unrelated wolves, and the Isle Royale wolves identified the island wolves as siblings, a view which accords with the mtDNA results that they are descended from a single female.

Now that there have been anywhere from 3 to 40 generations since repatriations of white-tailed deer, *Odocoileus virginianus*, to the southeast United States, analyses of genetic variation are revealing their dispersal and mixing with local populations. The approaches were to use protein electrophoresis to compare genetic variation between populations (Ellsworth et al 1994; Leberg et al 1994) and mtDNA to estimate gene flow among female deer (Ellsworth et al 1994).

6.3.3   Genetic drift—the loss of alleles by chance

The Hardy-Weinberg equilibrium formally holds only for very large populations subject to various restrictions such as random mating and no inbreeding, but it is a close and robust approximation for populations of modest size (high tens rather than hundreds) even in the presence of random genetic drift. That does not hold, however, for very small populations (low tens). For purposes of charting the change in allelic frequency between such generations, it helps to think of all the gametes of one generation being mixed in a pool (the gene pool) from which a finite number are selected at random to provide the zygotes of the progeny generation. As with drawing a random sample from a bag of black and white marbles, the sample of alleles drawn from the gene pool will seldom reflect exactly the frequency of allelic types present in that pool. The disparity in gene frequency between the two generations is a matter of chance—it can be thought of as the "sampling error" of a random draw—and each locus is subject to that lottery. Ignoring details, there are only three important rules:

1.   The smaller the sample the greater the likely disparity in gene frequencies between the two generations
2.   The lower the frequency of an allele in the parent generation the more likely it is that it will be lost in the transfer of genetic material to the progeny generation
3.   The higher the frequency of a given allele the more likely it is to become fixed in the progeny generation

The loss of alleles by chance during the intergenerational genetic lottery is called *random genetic drift*. Ignoring mutation, the number of alleles at a locus can only reflect that of the parent generation, that is, decrease. It cannot increase. Thus there is a finite probability that a locus polymorphic in the parent generation will be occupied by a single fixed allele in the progeny generation. That effect is most easily measured as the change in mean heterozygosity between generations.

Despite our emphasis on the "three important rules" of genetic drift, the actual change in heterozygosity between generations is affected by a host of annoying minor influences which yet, when added, have a considerable cumulative effect on the rate of drift. The rate increases with disparity of sex ratio in the parental generation, with variation between parents in the production of offspring, with deviation from random mating, to name but a few factors. Chepko-Sade et al (1987) gave an excellent account of the effect of these secondary influences. A result is that it is almost impossible to work out the rate at which a population is losing heterozygosity because the necessary information spanning so many demographic parameters is virtually unobtainable from a single population at one time.

That Gordian knot was cut by Wright (1931, 1938, 1940), who took a Newtonian approach to the problem. What would be the rate of loss of heterozygosity for a population of a given size subjected only to the three main rules of sampling error? He invented an "ideal population" influenced only by the major determinant, the secondary determinants being set at their neutral values. Thus sex ratio is 50:50, mating is at random, the population is stable in numbers, and so on, the only influences on the rate of genetic drift being population size interacting with the laws of probability. In these circumstances the answer is breathtaking in its beauty and simplicity: between generations a population will lose a proportion of its mean heterozygosity H equal to $1/2N$, where N is the size of the progeny generation. The proportion of its mean heterozygosity bequeathed by the parent generation to its progeny is thus $1 - 1/2N$. Over that intergenerational exchange H therefore changes as

$$H_1 = H_0\left(1 - \frac{1}{2N}\right)$$

and over t generations, N being constant,

$$H_t = H_0\left(1 - \frac{1}{2N}\right)^t$$

Note that the size of the parental generation and the difference in numbers between parents and progeny have no bearing on the outcome. It is determined only by the size of the progeny generation, that is, the size of the *sample* of gametes drawn from the genetic material on offer from the parents, not the size of the pool of gametes from which that sample was drawn.

6.3.4   *Effective population size (genetic)*

The previous section on genetic drift treated all individuals as equal. It was the only way to extract a generalized relationship between population size and the rate of loss of alleles. That is all we need in most cases, but occasionally we seek a more accurate estimate of the rate at which alleles will be lost. Genetic management of the usually very small populations held in zoos is an obvious example. We have two choices: either derive the rate from first principles according to the special characteristics of the population or make the much easier calculation of the expected difference in rate of genetic drift between the real population and Wright's "ideal population." The second is the universal choice. We estimate N_{eg}, the *effective population size (genetic)*, the size of an ideal population that loses genetic variance at the same rate as does the real population. The population's effective size (genetic) will be less than its census size (Wright 1938), except in special and unusual circumstances, typically by a factor of 2 to 4 in birds and mammals and by rather more in fish and invertebrates. As a first approximation, effective population size is equal to or less than the number of breeding individuals.

The usual tactic is to estimate N_{eg} in steps. It is first calculated according to the most important discrepancy between the structure of the real population and the ideal population. Next, that value of N_{eg} is fed into the equation, correcting for the next most important discrepancy, and so on. We will outline the effect of some of these discrepancies, and more detailed treatments are to be found in Reed et al (1986), Chepko-Sade et al (1987), and Harris and Allendorf (1989).

Effect of sex ratio

Genetic drift is minimized when sex ratio is 50:50. Effective population size (genetic) in terms of sex ratio is given by

$$N_{eg} = \frac{4N_m N_f}{N_m + N_f}$$

(Wright 1940), where N_m and N_f are the numbers of males and females. The reason lies in the relatedness of the offspring. Using the example of Frankel and Soulé (1981), consider a population of zebra, *Equus* spp., comprising one male and nine females. The offspring would necessarily be related to each other as either half siblings or full siblings. But if the population were five males and five females, the offspring would on average be less closely related and hence would be less likely to lose an allele during transfer of genetic material from parents to progeny. The effective population size (genetic) is controlled by the number of the less common sex. For the example of one male and nine females, $N_{eg} = 1.6$. With one male and an indefinitely large number of females $N_{eg} = 4N_m = 4$ (Wright 1940).

Effect of family size

A second cause of the disparity between census size and effective size is the differences among individuals in the number of offspring they contribute to the next generation. In the ideal population their contribution has a Poisson distribution, the fundamental property of which is that variance equals mean. Should the variance of offspring production among individuals exceed the mean number of offspring produced per individual, the effective population size will be smaller than the census size. The effective population size N_{eg} corrected for differing numbers of progeny was first provided by Wright (1940), but it is given here in the form suggested by Lande and Barrowclough(1987):

$$N_{eg} = \frac{NF}{F + (s^2/F)}$$

where F is the mean lifetime production of offspring per individual and s^2 is the variance of production among individuals. It indicates that when mean and variance are equal $N_{eg} = N$ for this component of effective population size. The equation can be corrected for the bias imparted by small numbers by modifying it to

$$N_{eg} = \frac{NF - 1}{F + (s^2/F) - 1}$$

Since males and females sometimes differ in mean and variance of offspring production, those two equations are often solved for each sex separately and the sex-specific N_{eg} values summed.

Effect of fluctuating numbers

The effective population size (genetic) of a fluctuating population is not the arithmetic mean but is close to the harmonic mean (Wright 1940):

$$N_{eg} \approx \frac{n}{\sum_{i=1}^{n} \frac{1}{N_i}}$$

where n individual censuses each yield an independent estimate N_i, because the overall leakage of genetic variance is controlled disproportionately more by smaller than by larger numbers. Lande and Barrowclough (1987) give the exact formulation as

$$N_{eg} = \frac{1}{2}\left[1 - \left\{\prod_{i=1}^{n}\left[1 - \frac{1}{2N_i}\right]\right\}^{1/n}\right]$$

Other influences

N_{eg} is affected also by overlapping generations, dispersal, and the way in which the members of the population are distributed. The effect of these, and those discussed earlier, can all be linked in a stepwise manner to produce an estimated value of N_{eg} incorporating all the ways in which real populations differ in genetic behavior from the ideal population (see particularly the examples given by Chepko-Sade et al 1987; Rowley et al 1993). We are not convinced that this procedure has much utility given the sampling variances.

6.3.5 The estimate of inbreeding (F)

Inbreeding is mating between close relatives, and the probability of that happening increases as numbers decline. Even such an apparently simple definition turns out to have shortcomings in practice (Shields 1993). At the individual level, estimation of inbreeding for captive species can be deduced from pedigree tables (Ballou 1983). For species in the wild, direct observations of the pairings of marked individuals can be used to estimate inbreeding, but extra-pair matings will compromise those estimates. The solution is direct measures such as DNA fingerprinting (Section 6.3.2). At the population level, inbreeding is a relative measure that has to be compared to what could be expected by chance (i.e., a reference population with random mating).

Inbreeding is measured as F, formally the probability that the two alleles at any given locus on an individual's chromosomes will be identical by descent. For mating between full siblings of unrelated parents $F = 0.25$, and $F = 0.25$ also for the mating of a sibling with one of its parents. For mating between half siblings $F = 0.125$, and so on. The probability F also has an informal and more useful meaning: the proportion of heterozygosity likely to be lost by an average offspring of the mating with an inbreeding level of F.

6.3.6 Genotypic frequencies under inbreeding

The Hardy-Weinberg equilibrium (Section 6.3.1), which predicts the expected frequencies of genotypes in the progeny generation, is disrupted if the parents are inbred. For $p = 0.6$ the expected frequency of genotypes within their progeny will be (Frankel and Soulé 1981)

Genotypes:	A_1A_1	A_1A_2	A_2A_2
Probabilities:	$p^2 + pqF$	$2pq - 2pqF$	$q^2 + pqF$
Expected frequencies:	0.384	0.432	0.184

where for illustration F is set high for a wild population at $F = 0.1$. Thus both homozygotes increase in frequency beyond that of unrelated matings by pqF (by 0.024 in our example) at the expense of the heterozygote. The frequency of the recessive phenotype (that produced by genotype A_1A_1) increases by pqF, at the expense of the dominant phenotype (that produced by A_2A_2 and A_1A_2). Note that the result is the same when A_1 is declared the dominant and A_2 the recessive. Frankel and Soulé (1981, p. 63) give a more detailed explanation.

6.3.7 *Inbreeding depression (B)*

"Inbreeding depression" is the production of inherited deleterious traits in progeny as a consequence of a close relationship between their parents. Inbreeding and inbreeding depression are not the same thing. Some species can cope with a high degree of inbreeding without deleterious effects, the extreme being those species that are self-fertilizing.

Inbreeding depression shows up preferentially in the reduction of the so-called fitness characters (Frankel and Soulé 1981): those related closely to an individual's probability of surviving and reproducing (Section 6.3.9). Such attributes are typically controlled by dominant alleles that mask deleterious recessives. Often the dominant alleles interact with the recessives to give the heterozygote a selective advantage over either homozygote (termed *heterosis* or *overdominance*). Inbreeding, which boosts the frequency of the recessive phenotype over that of the heterozygote, thus has the potential to decrease fitness both by exposing the recessive and by reducing heterosis (Section 6.3.8). The relationship between inbreeding and inbreeding depression and their consequences are, however, by no means clear (for example, Shields 1993). Hamilton's (1993) elegant exposition of the questions raised by inbreeding and inbreeding depression admirably conveys those uncertainties.

In conservation biology concerns about inbreeding depression are based on captive populations in zoos. The debilitating effect of inbreeding in zoos on juvenile survival was not recognized until recently (Ralls et al 1979) but that belated recognition triggered a burst of pedigree analysis aimed at detecting potential inbreeding and the lowering of its rate by careful selection of mates. Those analyses have spilled over into a wave of concern for the pedigree of individuals released to start or augment populations of endangered species.

There are dissenting voices, notably Shields (1993), asserting that the debate on inbreeding depression has slid into dogma. Lack of data and careless use of terminology cloud the debate. Be that as it may, our concern here is more that there has been little equivalent research to determine the prevalence of inbreeding in the wild, the extent to which it leads to inbreeding depression, and its significance as a threat to wild species.

What has been studied is reviewed for most phyla in a compendium on the natural history of inbreeding (Thornhill 1993). The results are predictable in their range, given the diversity of breeding and dispersal systems that abound. They serve to illustrate some of the problems of pinning down this seemingly elusive topic. In Australia's superb wrens, *Malurus splendens,* the incidence of apparent inbreeding apparently increased to exceptionally high levels (21% of 136 pairs were closely inbred: mother-son, father-daughter, or full siblings) during the 9-year study (Rowley et al 1986), which should serve to sound as a warning against short-term studies. Even more significant, however, were the results of an amplification of techniques as the study continued. The initial description of inbreeding was based on observed pairings of individually marked birds. Once allozyme electrophoresis was used to identify paternity, it became apparent that 65% of the nestlings were fathered by wrens outside the group attending their nest (Rowley et al 1993).

Another pause for thought comes from a small island population (about 50 pairs) of great tits, *Parus major.* These small insectivorous birds had a reduction of 7.5% of egg hatching for each 10% increase in estimated inbreeding (inbred pairs were those with at least one great-grandparent in common—the study began in 1912, hence the lengthy genealogies). But those fledglings that survived had higher reproductive output (van Noordwijk and Scharloo 1981). Shields (1993) uses this example to make the point that there is no reliable evidence (yet) that inbreeding depresses *total* reproductive effort.

Zoo populations in contrast to wild populations are usually very small and remain small for a long time, they are shielded from year-to-year and season-to-season environmental variation, and they thus are maintained in circumstances that will extinguish a wild population. They accumulate abnormalities (e.g., drift in breeding season and retention of deleterious alleles) that seldom occur in natural populations. Their dynamic behavior may be used to generate hypotheses as to what is happening in the wild, but it must not be used as a model of natural population dynamics.

Most information on inbreeding depression comes from small populations held in captivity. Ralls et al (1988) give a particularly clear account of estimating inbreeding depression (B) in zoos. They relied on using juvenile survival as the measure of inbreeding depression. Most of the 38 species that they examined were susceptible to inbreeding depression, and therefore they concluded that parents must be chosen carefully to stave off that effect within the typically small breeding groups held in zoos. That conclusion is further reiterated in an experiment on mice. Survival of white-footed mice, *Peromyscus leucopus,* which had been bred from full-sib pairings for only three to four generations in captivity was significantly less than noninbred mice when they were released at the same site (Jiménez et al 1994). The reasons for the lower survival after release, based on trapping recaptures for 10 weeks, is unknown except that inbred mice lost more weight and the effects were more obvious in the wild than in captivity. The study says less about the fate of small populations in the wild than the implications for captive breeding, even for a short time, and subsequent releases.

6.3.8 *The effects of genetic drift and inbreeding*

No clear distinction can be made between the processes of random genetic drift and inbreeding. The two overlap but nonetheless can be separated conceptually, if a little artificially. For the purposes of this section we consider random genetic drift as the consequences only of random sampling. Its first major driver is the random segregation of alleles into gametes and the sampling effects that occur as they come together again as the zygotes of a finite population. Its second major driver is the inequality of reproductive effort among individuals, which puts some alleles at a numerical advantage over others in the next generation. These effects are progressively exacerbated with declining numbers. Inbreeding, on the other hand, is mating between close relatives. Individuals may breed with close relatives at both high and low population size according to the breeding system of the species and its pattern of juvenile dispersal.

Genetic drift purges rare alleles by random sampling and can therefore lead to fixation of alleles (either moderately favorable or unfavorable). In the absence of inbreeding, the recessive homozygote with frequency p^2 is outnumbered by the heterozygote with frequency $2pq$ when $p < 0.666$. That follows directly from the Hardy-Weinberg equation. Inbreeding increases homozygosity directly by distorting the Hardy-Weinberg frequencies. In the presence of inbreeding, the homozygote with frequency $p^2 + pqF$ is outnumbered by the heterozygote with frequency $2pq - 2pqF$ at lower levels of p. The breakpoint of $p < 0.666$ in the absence of inbreeding is reduced to $p < 0.555$ when $F = 0.25$ and to $p < 0.333$ when $F = 0.5$. As they are defined here, inbreeding and genetic drift act in different ways, although both decease the number of heterozygous loci in an average individual and both are accelerated by low numbers.

6.3.9 *Genetic variation and fitness*

Fitness is usually measured by the intrinsic rate of increase or net reproductive rate which sums up such characteristics as age at first breeding, fecundity, and juvenile survival. Those characteristics typically display dominance or overdominance and as such are reduced by inbreeding (Frankel and Soulé 1981). Loss of genetic variation then reduces fitness, but the relationship is not simple. Alleles are neutral, disadvantageous, or held together as balanced polymorphisms by overdominance, and those complexities explain why the effect of the loss of alleles is not entirely predictable. Single-locus characters and those quantitative characters based on a combination of loci do not have similar dynamics within small populations.

Figure 6.1 shows the relationship between fitness, heterozygosity, and the inbreeding coefficient F for five possible populations. Fitness decreases with inbreeding but only increases up to a point with cross-breeding. Then, although heterozygosity remains high, fitness decreases when crosses are between very different populations—this decline in fitness is called outbreeding. Well-documented examples

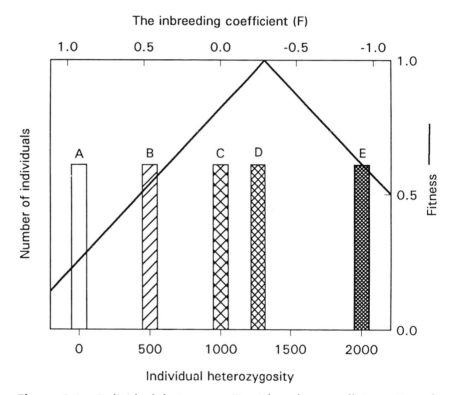

Figure 6.1. Individual heterozygosity, inbreeding coefficient F, and trend in hypothetical fitness for five populations. Population C is outcrossed and in Hardy-Weinberg equilibrium; B is a sample of individuals produced by one generation of selfing in population C; A is one of many possible inbred strains; D and E are crosses between inbred strains; D exhibits heterosis, while E exhibits outbreeding depression. (Modified from Mitton JB. Theory and data pertinent to the relationship between heterozygosity and fitness. In: Thornhill NW, ed. The natural history of inbreeding and outbreeding; theoretical and empirical perspectives. Chicago: University of Chicago Press, 1993:17–41.)

from the wild are conspicuously absent from the literature despite its importance for the translocations of animals between populations.

The one example that is often cited for outbreeding depression in vertebrates is Turcek's (1951) report on the introduction in the early 1900s of two Asian goat species to Czechoslovakia, which then inter-bred with ibex, *Capra ibex*. Turcek stated that the resulting hybrids maintained the seasonality of their Asian origin and lambed in mid-winter, whereupon "the offspring perished from generation to genera-tion." Despite this the population survived (with protection and feeding) until World War II. Other records of translocated ungulates have shifted their breeding cycles (Marshall 1937), which suggests that, in the absence of data, the inappropriate lambing season of the ibex is not a credible example of outbreeding depression.

Given that the relationship between genetic variation and fitness is not simple—and it is fitness of the individuals and populations that we are concerned about—what about measuring fitness? Mitton (1993) reviews the measurement of various components of fitness (viability, growth rate, physiological efficiency, and fecundity) in relation to heterozygosity. The expression of those components of fitness also depends on the environment, which complicates partitioning of mea-sured fitness by genetic or environmental influences. The long-term monitoring of individual red deer, *Cervus elaphus*, in a population limited by food lent itself to an examination of the relationship between fitness and genetic variation as the contribution of environmental ef-fects could be controlled in the data analysis. The fate of red deer calves was individually known, and their genetic variation was measured by protein electrophoresis (Pemberton et al 1988). Calves which were homozygous for two alleles fared poorly compared to calves hetero-zygous for both. Calves heterozygous for another allele survived less well, and that allele was also being lost from the population. It was either being replaced by another one, presumably superior, or was neutral until the current higher densities of deer and consequent food shortage caused it to now become disadvantageous. Complexity was further added by differences in the expression of two alleles in males compared to females and possible linkage between alleles (Pemberton et al 1988).

More often than not, we do not know enough about environmental effects on fitness compared to genetic effects. Difficulties of measuring aspects of fitness such as juvenile survival and relating them to genetic variation has led to indirect approaches. Fitness includes an animal's ability to buffer itself against environmental changes during its devel-opment. The effects of an adverse environment on development include reduction in symmetrical growth, and thus asymmetry of bilateral structures is assumed to reflect a reduction in fitness. As it turns out, asymmetry increases with increased homozygosity, al-though not invariably. Sarre et al (1993) review this. Fluctuating asym-metry has the merit that necessary measurements can be quickly and easily scored, although several characters should be tested. Quattro and Vrijenhoek (1989) found significant differences in three of seven bilateral meristic characters of Sonoran topminnows, *Poeciliopsis occidentalis occidentalis*. Fluctuating asymmetry is an index to fitness. There is nothing wrong with that so long as we remember that such indices are not complete measures of fitness.

There are grounds to argue that reduced genetic variance could increase susceptibility to disease and reduce fitness, and Black (1992) summarizes this for people exposed to a novel disease for them such as measles. Viruses and bacteria are constantly upgrading their immunogenic sequences as they replicate their RNA even within a single host. They are adapting to their host's immune system. Parasites from protozoans upward also show an astonishing diversity of mechanisms to outdo their host's immune defenses. Within an outbred population of hosts, the polymorphism of the immune proteins will be great, and each time a virus or bacteria infects a new host it is less likely to run into an antigen that it is already adapted to. Conversely, fewer alleles and their higher frequency will ensure that a virus will be dealing with a similar immune response in more hosts, and that could be the prelude to an epidemic.

The link between outbreaks of disease and low rates of polymorphisms of the proteins involved in the immune system will be difficult to establish in the absence of tightly designed experiments. Given the availability of inbred strains of laboratory animals and the depth of knowledge of their immune systems, more progress might be made initially with controlled laboratory experiments than with retroactive reviews of disease in captive or wild populations. Ascribling an outbreak of disease in a species that has low heterozygosity (based on protein electrophoresis) is coincidence and not causality.

We have to hang onto a sense of perspective about genetic variance. Small populations, if they escape demographic and environmental stochasticity, are the basis of intrapopulation variability, leading to adapation to local environments and eventually speciation. Berry (1974) wrote on genetic conservation before fears of the loss of variability in small populations had become the dominant theme of conservation genetics. His concluding generalities are

- Any environmental change (including management) may produce a genetical change.
- Adaptation is rapid and precise.
- Intrapopulation variation is large and resistant to loss.
- A local genotype cannot be preserved unchanged.

On the other hand, inbreeding and genetic drift can reduce genetic variance in small popoulations and that, in theory, could have deleterious effects. The quandary then becomes whether to manage small populations to supplement genetic variance by encouraging immigration. The use of corridors linking reserves springs to mind (Section 10.2.6). The alternative is to increase interpopulation variability by maintaining isolation of the populations. The answers are not immediately at hand in the sense that generalizations from our current understanding are premature and the range of possible problems is wide. Genetical management for captive populations and founders for newly established populations is unlikely to be transferable to established populations that happen to be small. For those, their existing genetic variability, history, breeding systems, and relationships to other populations all have to be taken into account before deciding whether to supplement genetic variation from another population through translocations or immigration via a corridor.

6.4 LOSS OF SOCIAL STRUCTURE

Many species have elaborate social breeding systems which range from the mating swarms of some insects to the clans of some mammals and birds. There are colonial breeders or those that feed and move in schools and swarms for reasons of defense or feeding efficiency. We know relatively little about when too few becomes not enough for those social behaviors to function. The collapse of the huge flocks of passenger pigeons, *Ectopistes migratorius,* is an oft-cited example of where numbers decreased to where there was no longer apparently safety in numbers and the passage to extinction was accelerated (Schorger 1955). Flocks became too small to locate the unpredictable fruiting of mast crops and too small to nest colonially. We could list other examples, such as the Puerto Rican parrot, where possibly the population is too small for "normal" social behavior (Walters 1991).

6.5 MINIMUM VIABLE POPULATION SIZE

At some stage we have to deal with what we mean by a small population. Others sought to answer that question (Franklin 1980; Lande and Barrowclough 1987) and their models of effective population size to avoid genetic problems led them to attach numbers to minimum population sizes as an illustration of the order of magnitude. The consequent 50/500 guideline became firmer at each reiteration even though that was not the original intention. A reaction has now set in, questioning unthinking adherence to the guidelines (for example, Walters 1991; Craig 1994).

Common sense tells us that there is no single number that tips a species into the category of small or minimum viable population. Simberloff (1986) suggested that the minimum viable population was an adequate metaphor for the size below which rapid extinction is likely and that we cannot assign a specific value. It is when proximate factors are replaced by ultimate factors that the final stages of a decline are precipitated. These ultimate factors are the trio that we have described: demographic and environmental stochasticity and reduced genetic variance. The temptation of focusing exclusively on small population size as such carries the risk of missing the processes that reduced the size in the first place.

6.6 SUMMARY

Small populations face the luck of the draw: sampling probabilities determine their persistence. Demographic stochasticity is the effect of chance on whether a population of a few animals increases or decreases over a year rather than depending on the age-specific probabilities of survival and reproduction. The effect of the instability of age distribution and sex ratio is referenced as the size of an "ideal population," producing the same net increment per year as the population of concern. This effective population size (demographic) may be greater than or less than the census population size N, both in terms of sex ratio and age distribution.

Year-to-year variation in environmental conditions, particularly in weather, is the second category of risks facing a small population. Environmental stochasticity affects a population's rate of increase in a way that differs from the influence of demographic stochasticity. The variance in r generated by environmental fluctuations is largely independent of the size of the population.

The third category of risk facing small populations is a reduction in genetic variation. The amount of variation can be sampled from proteins or, with a much higher resolution, from the DNA itself. Variability is lost by the random sampling of genes (random drift) and the increased likelihood of breeding with close relatives in a small population. How reduced genetic variation actually affects populations in the wild is unresolved, which leaves conservation biologists in a quandary. The effective population size (genetic) is the size of a genetically ideal population having the same loss of heterozygosity as the population of concern, which is invariably less than census size.

CHAPTER 7

Risk assessment

Risk assessment is estimating the probability of an adverse event. In the context of population biology it deals with the effect of various events on the chance that a population will become extinct. It requires two probabilities for each potential event: the probability that the event will occur over a specified time interval and the probability that it will lead to the extinction of the population. Marshaling together the events and their probabilities is what Gilpin and Soulé (1986) termed "population viability analysis," and it is the most common form of risk assessment used in conservation biology.

7.1   POPULATION VIABILITY ANALYSIS

Burgman et al (1993) provide an excellent outline of the techniques of population viability analysis (PVA). Most of the methodology available so far makes use of very simple single-species population growth models such as the logistic and the truncated exponential. These are approximations and must be treated with caution. They employ the assumption that rate of increase and present density are causally related, but that holds only when the species exhibits spacing behavior of some sort or when the limiting resource is inert and used preemptively (e.g., nesting holes). Density is used more commonly in these models as a surrogate for the level of availability of a resource, the rationale being that the level of that limiting resource will be high when few animals are using it and low when it is being used by many. It is dangerous to use present density in that way because the resource level is determined not by present density but by the history of densities. Nonetheless it works tolerably well if the species' intrinsic rate of increase is below about $r_m = 0.08$ on a yearly basis or when year-to-year environmental variation is small. Such models are often useful for extracting first approximations to expected persistence when the quantitative ecology of the species of concern is poorly known.

The efficacy of these shortcuts has seldom been tested. Worse, they are often used by biologists who assume that they are an adequate depiction of reality. But before accepting a shortcut, one must know how it measures up against the results of the long cut. An appreciation of how real systems work ecologically is a prerequisite to understanding where simple models may be adequate and where they will be disastrously inept. We therefore explore an elementary but real system and consider the causal relationships within it that those shortcut models summarize or circumvent.

There is a plethora of single-species ecological models in theory but no single species ecological system in actuality. The simplest of all ecological systems comprises a herbivore and the plants it eats. We choose the red kangaroo, *Macropus rufus*, and its food supply as the most elementary example we know of. It lives in an essentially nonseasonal environment and breeds throughout the year, the population having no seasonal structure. That greatly simplifies the mathematics. It is a common species whose ecology is tolerably well known.

7.2   SPECIFIC VIABILITY ANALYSIS: THE RED KANGAROO

The red kangaroo is ubiquitous across the arid and semiarid zones of Australia, its range covering approximately 5,000,000 km^2. A survey covering its range in 1981 returned an estimate of 8.4 million (Caughley et al 1983). The system of which it forms a part is driven by rainfall, the plants responding within days to any fall above 15 mm. The study borrowed here was on Kinchega National Park (440 km^2) and the sheep stations surrounding it in western New South Wales (Figure 1.1). We summarize how often the rain comes, how much of it comes, and how the system responds to it.

7.2.1   The rainfall

The area has a mean annual rainfall of 236 mm (9.3 in.) with a standard deviation of 107 mm. Table 7.1 summarizes its average distribution by season over the 100 years between 1883 and 1984. The rainfall is unpredictable: Serial correlation for the same season between successive years yields a coefficient of 0.05. In 3 of 10 years the annual rainfall is above or below the annual average by more than 50%, so floods and droughts are common.

Table 7.1.   Annual and seasonal rainfall (mm)
for Kinchega National Park

	Summer (Dec–Feb)	Fall (Mar–May)	Winter (Jun–Aug)	Spring (Sep–Nov)	Year
Rainfall	62	57	59	61	236
SD	59	47	34	44	107
CV%	94	84	60	73	47

Modified from Robertson G, Short J, Wellard G. The environment of the Australian sheep rangelands. In: Caughley G, Shepherd N, Short J, eds. Kangaroos: their ecology and management in the sheep rangelands of Australia. Cambridge: Cambridge University Press, 1987:14–34.

7.2.2 The plants

The plants take advantage of any but light falls of rain, responding with a burst of germination from the seed bank and a flush of growth from established plants. Despite a marked seasonal trend in temperature, the consequent production of plant biomass does not differ much between summer and winter. The increment to ungrazed plant biomass over any of the four seasons was estimated by Robertson (1987), and modified by Caughley (1987), to be

$$\Delta V = -55.12 - 0.0154V - 0.00056V^2 + 2.5R$$

where ΔV is the increment to plant biomass (kg/ha dry weight) over 3 months, V is the standing plant biomass at the start of those 3 months, and R is rainfall in millimeters over those 3 months (Figure 7.1).

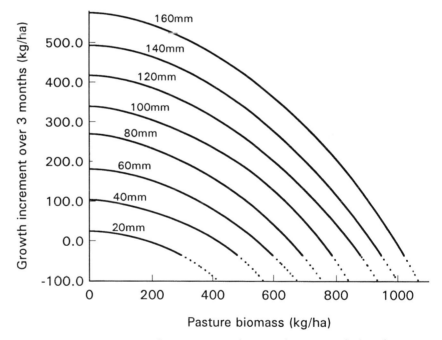

Figure 7.1. Plant growth over 3 months as a function of plant biomass at the beginning of the interval and rainfall during the interval, for pastures on Kinchega National Park. (Modified from Robertson G. Plant dynamics. In: Caughley G, Shepherd N, Short J, eds. Kangaroos: their ecology and management in the sheep rangelands of Australia. Cambridge: Cambridge University Press, 1989:50–68.)

7.2.3 The herbivore

The kangaroos react to the standing crop of vegetation in two ways: they reduce its standing biomass by grazing, and they use that harvested energy to produce more kangaroos. How much an animal eats over a day is determined by the density of plants in front of its nose. Short (1987) determined from grazedown trials that an average kangaroo (35 kg) will eat $0.95\,(1-e^{-V/34})$ kg dry weight of pasture per day when the vegetation is composed only of grasses and forbs, and $0.89\,(1-e^{-V/84})$ kg per day when the grasses and forbs share the area with scattered shrubs. The terms outside the parentheses (0.95 and 0.89 respectively) are much the same and represent the satiating diet: the amount of plant material (kg/day) a kangaroo eats when faced by a superabundance. Hence that value will be about the same whether the preferred food occurs alone or whether it is interspersed with inedible shrubs. Figure 7.2 shows what these curves look like. They depict the relationship between intake per individual and the standing crop of food on offer. It is called the *feeding response* (or functional response) of the herbivore to plant biomass.

The rate of increase of a population as a function of available food (or of any other limiting resource) is called its *numerical response*. Some response functions presented in the literature under that name give the relationship between rate of increase and density. We do not use that relationship because there is no instantaneous causal relationship between the two—in a fluctuating environment the two are seldom, if ever, at equilibrium—and because the relationship between a rate of increase determined by the level of a resource, on the one hand, and the density of the species under study, on the other, is distorted by other competing species utilizing that resource. That distortion is eliminated when the numerical response of the species under study is estimated in terms of the level of its own limiting resource; it should be the same on average whether the species occurs alone or in the presence of other species using that same resource.

Bayliss (1987) estimated the numerical response of red kangaroos from 3-monthly aerial surveys and concomitant measures of pasture biomass on Kinchega National Park and on the abutting sheep station of Tandou (Figure 1.1). On Kinchega the red kangaroo competes for food with the southern gray kangaroo, *Macropus fuliginosus*. It competes on Tandou with that species and with merino sheep. Consequently the red kangaroo is considerably more abundant in the national park than on the sheep station.

Bayliss used Ivlev's (1961) model:

$$r = -a + c\,(1 - e^{-dV})$$

in which r is the realized exponential rate of increase on a yearly basis, a is the maximum rate of decrease (the rate at which the population

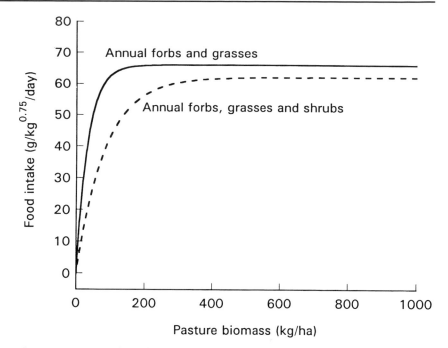

Figure 7.2. Food intake per individual per day for kangaroos at varying levels of food availability on two plant communities. (Modified from Short J. Factors affecting food intake of rangeland herbivores. In: Caughley G, Shepherd N, Short J, eds. Kangaroos: their ecology and management in the sheep rangelands of Australia. Cambridge: Cambridge University Press, 1989:84–89.)

crashes when pasture biomass is zero), c is the extent to which that rate of decrease is ameliorated when pasture biomass is more than enough to allow a satiating diet (hence $r_m = c - a$), d is an index of grazing efficiency (the ability of an animal to find food when food is scarce), and V is the average biomass of pasture (kg/ha dry weight). Figure 7.3 gives numerical response curves for red kangaroos on Kinchega National Park (sheep absent) and on Tandou (sheep present). As predicted, they are much the same despite coming from populations differing in density. Caughley (1987) slightly modified Bayliss's estimates to include an observed, rather than extrapolated, maximum rate of decrease to yield

$$r = -1.6 + 2.0(1 - e^{-0.007V})$$

By this formulation the population's realized rate of increase on a yearly basis varies between a maximum of $r = r_m = 2.0 - 1.6 = 0.4$ and a minimum of $r = -1.6$, according to the level of pasture biomass V.

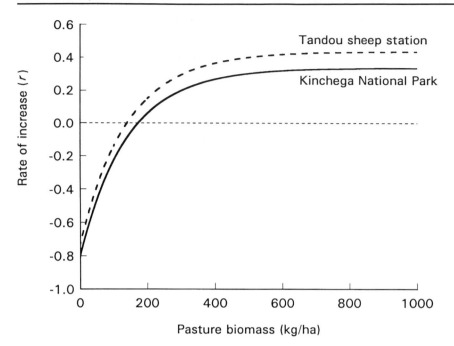

Figure 7.3. Rate of increase on a yearly basis of a population of red kangaroos at varying levels of food availability and in the presence of sheep (Tandou sheep station) or absence of sheep (Kinchega National Park). (Modified from Bayliss P. Kangaroo dynamics. In: Caughley G, Shepherd N, Short J, eds. Kangaroos: their ecology and management in the sheep rangelands of Australia. Cambridge: Cambridge University Press, 1989:119–134.)

7.2.4 *System dynamics*

We now have estimates of each of the parameters, allowing construction of a model of the red kangaroo's population dynamics. Each season was assigned the average expected 3-month rainfall and sampled from a normal distribution with a mean of 59 mm and a standard deviation of 46 mm (Table 7.1). The QuickBASIC subroutine given in Box 7.1 will do that.

That allowed an estimate of the increment to pasture biomass, according to the estimated relationship between plant growth and rainfall, which was added to the biomass present at the beginning of the season. That then allowed an estimate of the amount by which the red kangaroos would increase or decrease (from the numerical response equation) and that increment was added to their standing population. Finally, the plant biomass removed by the grazing of kangaroos was estimated per individual from the feeding response equation and multiplied by the size of the population. That loop of calculations was applied once a week to generate trajectories of pasture biomass and

---

Box 7.1.   QuickBASIC subroutine to make a random draw from a normal distribution with specified mean and standard deviation

---

NORMAL:
Subroutine to calculate a random draw X from a normal distribution with mean EX and standard deviation STD. Values of EX and STD are imported from the main program and the value of X is returned to it.

```
SUM = 0
FOR I = 1 TO 12   The 12 is critical. Do not change it.
RR = RND
SUM = SUM + RR
NEXT I
X = STD*(SUM − 6) + EX
RETURN              Return to the main program bearing a
                    normal deviate X
```

kangaroo density over time. The behavior of the modeled system is unaffected by initial conditions after 10 years (Caughley 1987). Figure 7.4, showing a typical outcome, indicates that the rainfall of the region takes the form of high-amplitude, high-frequency fluctuations, and the biomass of the plants generates a similar pattern as it reacts speedily to the rainfall or lack of it. The kangaroo's rate of increase follows the trend in plant biomass but at a lower frequency.

7.2.5   Population viability analysis for red kangaroos

We now have a model that purports to duplicate the internal workings of a weather-plant-kangaroo system. Each of the constants therein has a precise ecological meaning and the value of each is estimated. It is described as a "consonant model" to distinguish it from the simple "predictive model" (e.g., the logistic or the polynomial regression). The function of a predictive model is simply to duplicate an observed outcome. Its constants need have no necessary correspondence with ecological processes.

We will use this model to estimate how large a reserve must be (and hence how large the average population) to conserve a population of red kangaroos for a specified number of years. To conform with the circumstances under which the constants of the model were estimated, we must stipulate that the reserve has the same climate as Kinchega National Park, that kangaroos neither enter nor leave the reserve, that

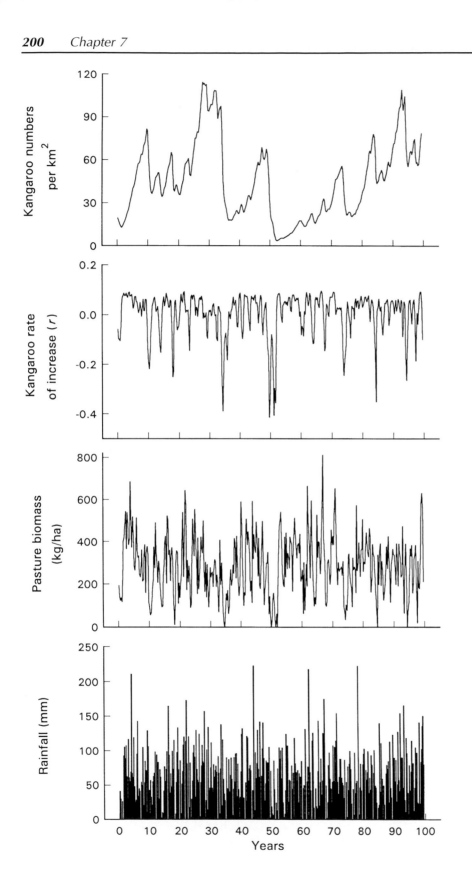

it is surrounded by a firebreak such that wildfire does not affect the level of pasture biomass, that it contains neither predators nor competitors of red kangaroos, and that surface water is never in short supply. It could be envisaged as an island in Lake Menindee which borders Kinchega National Park (Figure 1.1).

Average density under this model is 0.45 red kangaroo per hectare and so a reserve of 1 km² will contain 45 kangaroos on average, although the actual numbers will fluctuate markedly around that mean. Our first analysis will answer the question: "How long will a population persist if restricted to a reserve of a given size?" The size of the reserve is indexed by the extinction threshold, the density at which the population dies out. The extinction threshold in terms of population size is clearly one kangaroo, and so the threshold in terms of density is 0.1/ha (i.e., 1/10) for a reserve of 10 ha and 0.01/ha (i.e., 1/100) for a reserve of 1km² (= 100 ha). Since the model is written about pasture biomass per hectare and kangaroos per hectare, the persistence of the population on a reserve of any given area can be gauged by noting how long the system runs before the density extinction threshold, specific to that area, is crossed. Figure 7.5 shows average persistence as a function of reserve size. Each point is an average of 100 runs. The trend conforms closely to the fitted regression of

$$T = 0.944A^2$$

where T is mean persistence time in years and A is the area of the reserve in hectares. Figure 7.5 shows the trend.

How precise are those estimates? It is all very well to suggest that a reserve of 20 ha can be expected to hold a population of red kangaroos for about 400 years, but that is an average. Precision is measured as the standard deviation of the outcome of replicated runs. A regression of the standard deviation of times to extinction on their mean provided a linear trend through the origin with a slope of $b = 0.96 \pm 0.007$. Hence the standard deviation of times to extinction is about 96% of the estimated mean time to extinction. As a close approximation, the standard deviation equals the mean. That has two implications. First, although

Figure 7.4. The bottom histogram shows 3-month random draws for rainfall from an Australian arid-zone climate. The next graph up is the biomass of grasses and forbs that would be generated by that rainfall and reduced by the grazing of kangaroos. Above the plant biomass graph is the rate of increase of red kangaroos, on a 3-month basis, as it reacts to plant biomass; and the top graph is the trend of kangaroo numbers generated by the run of rates of increase directly below it. (Modified from Caughley G, Gunn A. Dynamics of large herbivores in deserts: kangaroos and caribou. Oikos 1993;67:47–55.)

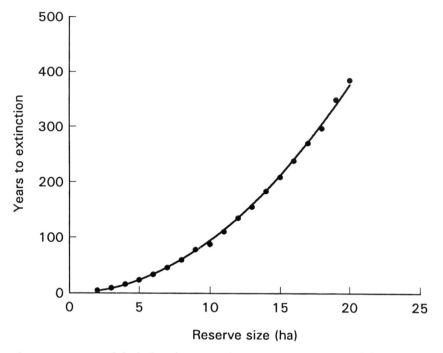

Figure 7.5. Modeled distribution of time to extinction of kangaroos inhabiting a reserve of varying size.

the mean time to extinction can be predicted accurately, the time to extinction for a given run is unpredictable. Second, the parity of mean and standard deviation suggests that the individual runs, each yielding a time to extinction, are drawn from something like an exponential distribution. That suggests further that we should have a look at the shape of the distribution because an exponential distribution is concave with a modal value at zero (i.e., it conforms to what mathematicians call a J-shaped curve, although what they actually mean is a reversed J curve). Such a shape has worrisome implications for conservation.

Figure 7.6 shows the frequency distribution of replicate times to extinction from the red kangaroo model for areas of 4 ha, 10 ha, and 16 ha. They are indeed J-shaped and closely approximate exponential distributions. We will therefore assume as an approximation that we are looking at exponential distributions in fact.

The exponential distribution has a single parameter α which is the reciprocal of the mean of the distribution and of its standard deviation. The frequencies of years to extinction (where years are $x = 0, 1, 2, \dots$) are given by

$$f(x) = \alpha e^{-\alpha x}$$

and the cumulative frequency distribution is

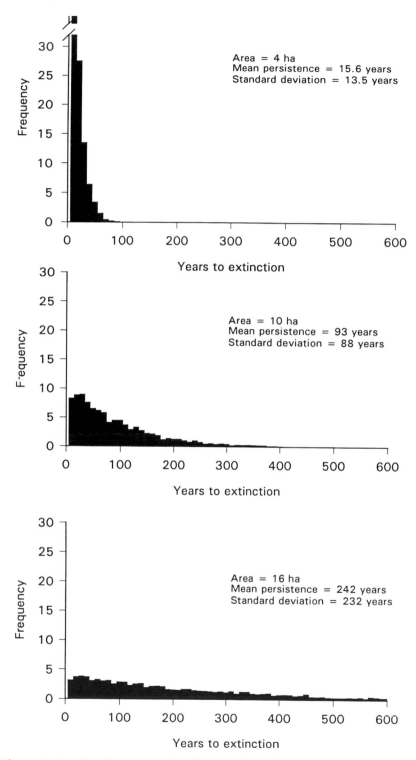

Figure 7.6. The frequency distribution of replicate times to extinction from the red kangaroo model for areas of 4 ha, 10 ha, and 16 ha.

$$F(x) = 1 - e^{-\alpha x}$$

Median time to extinction can be solved from that equation by substituting $\bar{x}$ for $1/\alpha$, setting $F(x)$ to 0.5, and solving for x:

$$F(x) = 1 - e^{-\alpha x} = 1 - e^{-x/\bar{x}}$$

$$0.5 = 1 - e^{-x/\bar{x}}$$

and so

$$-(\ln 0.5) = \frac{x}{\bar{x}}$$

to give median x as

$$x = (-\ln 0.5)\bar{x} = 0.69\bar{x}$$

Thus the median x marking the time up to which 50% of populations would have died out is only 69% of the mean time to extinction. We can also calculate the percentage of populations that would have died out before the mean time to extinction by solving for $F(x)$ corresponding to $x = \bar{x}$:

$$F(x) = 1 - e^{-\alpha x}$$

and so

$$F(\bar{x}) = 1 - e^{-\alpha\bar{x}}$$

$$= 1 - e^{-1}$$

$$= 0.63 \quad \text{(i.e., 63\%)}$$

Another way of expressing a population's expected persistence, and rather faster to compute, is the probability of its surviving for a given number of years. We choose 100 years for illustration. (A century is a good interval because by then your great-grandchildren will be old enough to judge you.) The same strategy as for estimating T is used, except that each run is terminated at extinction as before or at 100 years if the replicate population is still extant. Figure 7.7 shows the results derived from 5,000 runs for each specified area A measured in hectares. A very close fit is supplied by the regression equation

$$P_{100} = 1.0234e^{-110.9/A^2}$$

Those are two ways of depicting persistence: mean time to extinction T and probability of extinction over a specified period P_x. Their relationship is given in Box 7.2.

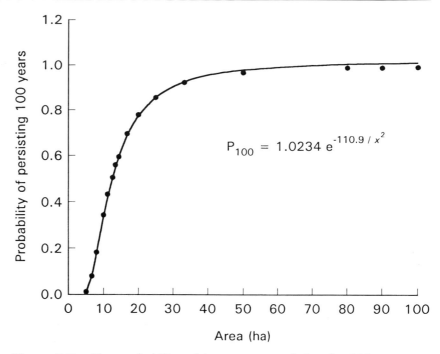

Figure 7.7. The probability of kangaroos persisting for 100 years on varying sizes of reserves.

Box 7.2. Converting a probability of persistence over a specified time P_x to an expected mean time to extinction T

Probability of persisting for 100 years is (say)	$P_{100} = 0.7800$
Probability of persisting for 1 year is	$P_1 = 0.7800^{1/100} = 0.9975$
Probability of dying out over 1 year is	$(1 - P_1) = 0.0025$
Mean time to extinction is	$T = 1/(1 - P_1) = 1/0.0025 = 400$ years

---

A reason for running these analyses might be to estimate the minimum size of a reserve needed to hold a self-sustaining grazing system for which the red kangaroo is the sole herbivore. The equation estimating P_{100} is reversed to allow a calculation of the area required for a given probability of persistence. We choose $P_{100} = 0.999$. Since $P_{100} = 1.0234 e^{-110.9/A^2}$,

$$A = \sqrt{\dfrac{-110.9}{\ln P_{100} - \ln 1.0234}}$$

$$= \sqrt{\dfrac{-110.9}{-0.0001 - 0.0231}}$$

$$= 69 \text{ ha}$$

We estimate that a reserve of 69 ha will hold a self- sustaining population of red kangaroos for a century in 99.9% of cases (at a mean density of 0.45 kangaroo per hectare).

What margin of error should we attach to that estimate? The standard deviation around the regression does not help with that decision because it tells only how well the regression describes the output of the model. It does not indicate how closely the model fits reality. A civil engineer designing a bridge estimates how robust it must be to withstand any stress he can imagine and then doubles that estimate. But we are dealing with information less reliable than is available to the engineer, yet the need to get it right is quite as critical. Consequently, a comfortable margin of error is called for. We recommend the following rules of thumb. Double the estimate to allow for possible influences on the system not included in the model: nonlinear relationships assumed incorrectly to be linear, demographic stochasticity, climatic trend, and various unexpecteds and unknowns. Double it again to allow for absurd mistakes in field measurement and computer programming. Double it yet again to allow for future inappropriate management by inept or inexperienced biologists and rangers. Thus the original estimate of 69 ha needs to be increased, by a margin of error of about 8, to about 550 ha (5.5 km²) before one would be satisfied that an adequate buffer against misfortune had been built into the recommendation.

7.3 GENERALIZED POPULATION VIABILITY ANALYSES

A generalized population viability analysis is one that can be applied to any species. The structure of the model remains the same, but input—mortality rates, fecundity rates, carrying capacity, and environmentally induced variance in r—differs between cases.

If we assume that the dynamics of the species of interest can be described by a single-species model (i.e., its birth rate, death rate, and hence rate of increase are predicted by its density), then its trend in numbers can be described by a birth–death stochastic model. That in turn can be manipulated to estimate mean time to extinction for any starting value of N. Richter-Dyn and Goel (1972, Eq. 2.3a) gave the general case, here presented in Goodman's (1987) notation, as

$$T_{(N)} = \sum_{x=1}^{N} \sum_{y=x}^{N_m} \frac{1}{yd_{(y)}} \prod_{z=x}^{y-1} \frac{b_{(z)}}{d_{(z)}}$$

where $T_{(N)}$ is mean time to extinction for a population presently of size N, b is its birth rate, d its death rate, those rates being a function of population size, and N_m is maximum population size and would be written as K if the relationship between rates and population size were as in the logistic model. In fact, this equation was used previously in logistic form (i.e., b and d linear on N) by MacArthur and Wilson (1967) to estimate the chances of a species successfully colonizing an island.

Goodman (1987) took advantage of the relationship of stochastic b and d to the variance in rate of increase r, rewriting the equation with parameters more easily estimated in the field:

$$T_{(N)} = \sum_{x=1}^{N} \sum_{y=x}^{N_m} \frac{2}{y[yV_{(y)} - r_{(z)}]} \prod_{z=x}^{y-1} \frac{V_{(z)}z + r_{(z)}}{V_{(z)}z - r_{(z)}}$$

in which r and V are respectively exponential rate of increase and its variance, each as a function of population size.

It can be simplified further if one is willing to guess at the relationship between r and its variance, on the one hand, and population size, on the other. Belovsky (1987), for example, assumed that r and V were independent of population size except at N_m where r is zero:

$$T_{(N)} = \sum_{x=1}^{N} \sum_{y=x}^{N_m} \frac{2}{y[yV - r]} \prod_{z=x}^{y-1} \frac{Vz + r}{Vz - r}$$

converting the model to a stochastic truncated exponential where the population increases on average at a rate of $r = r_m$ until reaching the ceiling of N_m. Lande (1993) gives a further simplification of the truncated exponential model by way of the diffusion equation. For such a model, the expected time to extinction from its maximum size is approximately

$$T_{(N_m)} = \frac{2}{Vc} \left(\frac{N_m^c - 1}{c} - \ln N_m \right)$$

in which $c = 2r/V - 1$.

7.3.1   *Generalized population viability analysis by computer simulation*

Simulation modeling is an alternative way of gauging a population's viability. The model "grows" the population according to rules set by the structure of the model and the input values of fecundity, mortality, carrying capacity, and so on, supplied by the user. The effects of demographic and environmental stochasticity are simulated by random draws from binomial (Box 5.1) and normal (Box 7.1) distributions. The population is grown many times with the same input values, but the resultant replicate trajectories differ because of the pseudostochasticity imparted by the computer's random number generator. The probability of persistence over any nominated period is then estimated as the proportion of replicates still extant at the end of that period. That value can be converted, if necessary, into a mean persistence time (Box 7.2).

The most widely used PVA computer program is VORTEX (Lacy and Kreeger 1992; Lacy 1993), which will serve here as a representative example. The program interrogates the user as to the characteristics of the population of interest and replies with the probability of extinction over intervals of time. It assumes that rates of fecundity and mortality are independent of age after first breeding, it accepts mortality rates for each age class before that, and it is interested in the mating system of the species. The user specifies the proportion of females breeding and the proportion of breeders that produce litters of $1, 2, \ldots, 5$. The program insists upon receiving an estimate of carrying capacity K. Genetics (optional) are considered as an infinite-allele model of genetic variation, each founder having two unique alleles. The user may tell the program how much inbreeding depression to expect from loss of heterozygosity. A linear decline in available habitat (treated as a falling K) may be specified where appropriate. Reduction in numbers by harvesting, or supplementation of numbers as from a captive breeding program, are allowable options. Metapopulation structure is catered for. Environmental variation may be imparted optionally to the carrying capacity and to rates of mortality and fecundity. "Catastrophes" of specified frequency and intensity may be invoked to act on the specified rates of fecundity and mortality. Fecundity (but not mortality) may be rendered density dependent by ascribing values to the constants of a fourth-order polynomial relating fecundity to population size. At default values these characteristics add up to a stochastic model, partially age structured, of exponential increase truncated at a reflecting boundary ("carrying capacity") specified by the user.

VORTEX and like programs do exactly what they are told to do, as constrained by the static single-species models that provide their structure. They can be useful for various purposes so long as the user understands what the programs are doing and how the ecology implied by the model differs from the ecology of real populations. Even if the parameter values fed into them are correct, their structure may be inappropriate and, hence, their output may be of dubious usefulness. A further problem is that the detailed information they seek as input is almost never available from an endangered species. Hence they often get fed with guesses, or "consensus views of experts familiar with the species," which amounts to the same thing. This is a fault not of the software but of the user.

7.4 ASSESSMENT OF GENERALIZED POPULATION VIABILITY MODELS

We will examine the characteristics of the single-species population models on which most PVAs are based and determine the extent to which they duplicate the more elaborate model provided by the red kangaroos. In particular, we ask how they match up with real ecological systems and how their parameters are estimated from real populations.

We restrict ourselves to the truncated exponential and the logistic models of population growth. Both are density-dependent models based on the premise that rate of increase is predicted by density. An excellent evaluation of alternative density-dependent models, along with the relationship of the model structure to the nature of the interaction between the members of the population and how these impinge the use of resources, is provided by Burgman et al (1993). The density dependence is restricted in the truncated exponential model to the population level N_m, where rate of increase becomes zero. Below that level the population increases at its constant intrinsic rate r_m. The logistic is density dependent at all levels of N, the realized rate of increase declining linearly from a maximum of r_m at insignificant population size to a minimum of zero when $N = K$, the "carrying capacity." Neither model in its deterministic infinitesimal form allows for a negative rate of increase.

We have emphasized that the rates at which real populations increase bear a tenuous relationship to their current densities, for good ecological reasons, except in the special cases of populations with low r_m in a near-constant environment, or populations whose upper limit is set by a fixed number of breeding sites, or populations whose members exhibit strong spacing behavior. These are a distinct minority. Figure 7.8 shows for red kangaroos the modeled relationship between rate of increase and density, on the one hand, and rate of increase and level of the limiting resource (grasses and forbs), on the other. In this modeled system the rate of increase graphed against plant biomass is a tighter

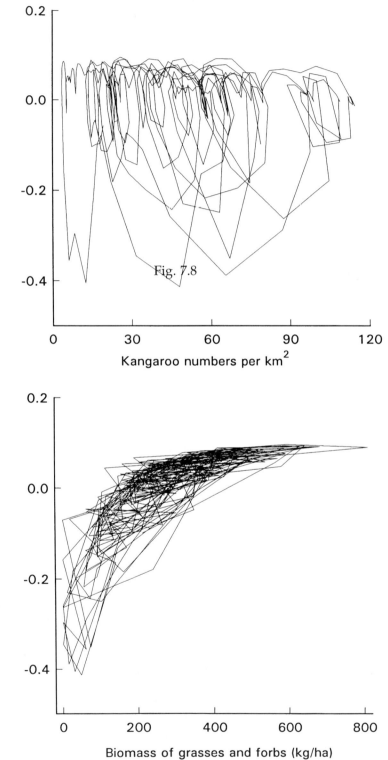

Fig. 7.8

predictor of herbivore rate of increase. It is apparent that density tells us little about their rate of increase and therefore is not a particularly useful parameter to include in a model of their population dynamics.

7.4.1 The truncated exponential model

In its deterministic form the truncated exponential has only two parameters: an intrinsic rate of increase r_m at which the population grows in all circumstances until numbers reach an upper limit, which is labeled variously as K or N_m (the subscript standing for "maximum"), after which the population stops growing and numbers hold to that constant value.

This model provides the skeleton of the equations (Section 6.2) for population persistence (Richter-Dyn and Goel 1972; Goodman 1987; Lande 1993) that have been used to investigate the influence of environmental stochasticity on the persistence of populations. In most cases they are used to provide qualitative information on the relationships among persistence, population size, and environmental fluctuation (Section 6.2.1), but Belovsky (1987) tested such a model against real data. He used information (Brown 1971; Patterson 1984) on persistence in the American southwest of species stranded on mountain tops by post-Pleistocene warming.

Boyce (1992) pointed out forcefully in the context of specific PVAs that the main problem with the truncated exponential is that it has no ecology. It has two parameters, r and N_m, neither of which is easily identified with parameters of real populations. Belovsky (1987) simply equated r with r_m, and N_m with reported estimates of density multiplied by area. This will not do when environmental stochasticity is imparted to r. By definition r_m is the population's maximum rate of increase and so, although r can be perturbed below r_m, it cannot be perturbed above (except for the special case of an unstable age distribution). Alternatively, environmental stochasticity can be imparted to N_m, but that makes little biological sense unless r varies in sympathy with it. We predict that the truncated exponential will ultimately be proven unsuitable as a skeleton for species-specific PVAs, although possibly adequate for answering qualitative questions about the relationship between environmental stochasticity and population persistence.

Figure 7.8. Rate of increase of red kangaroos given every 3 months for the 100-year simulation presented in Figure 7.4. The trends are graphed first against kangaroo density and then against the biomass density of grasses and forbs. (Modified from Caughley G, Gunn A. Dynamics of large herbivores in deserts: kangaroos and caribou. Oikos 1993;67: 47–55.)

7.4.2 The logistic model

We will approach the logistic from the exponential (geometric) model where the population simply increases at its intrinsic rate of increase. It takes the form

$$\frac{dN}{dt} = r_m N$$

where r_m is the intrinsic rate of increase. The realized rate of increase r is then

$$r = \frac{1}{N} \times \frac{dN}{dt} r_m$$

The realized rate of increase is reduced progressively as N increases, such that when N reaches K, r declines to 0. K is usually considered as a population size which saturates the area holding the population. It is often called the "carrying capacity." The logistic equation may be written

$$\frac{dN}{dt} = r_m N \left(1 - \frac{N}{K} \right)$$

and terms of the realized rate of increase r as

$$r = \frac{1}{N} \times \frac{dN}{dt} = r_m \left(1 - \frac{N}{K} \right)$$

We have seen (Section 6.2.1) that the logistic model is likely to approximate the population growth only of scavenging species. However, the model changes form when modified to allow for variation in K, and it then stands a chance of simulating the growth of a population dependent on a depletable and renewable resource. If environmental stochasticity is imparted to K alone, the model reveals two more characteristics that are ecologically sensible. First, as K rises and falls (most easily envisaged as a rise and fall in the standing crop of food), the realized rate of increase r varies in sympathy with it but at a lesser rate. That adequately describes the dynamics of most real populations. Second, r can never exceed r_m and so the r_m of the logistic equation is the same thing as the rate r_m that can be measured for a real population.

A major problem remains, however: what to measure from a real population that might correspond to the K of the unmodified logistic equation or the mean $\bar{K}$ of the stochasticized logistic. We suggest further on that this might not be the problem it appears if $\bar{K}$ is treated simply as a constant of fit.

Given any population size N_0 of a population growing logistically, its size t time units further on is solvable as

$$N_t = \frac{K}{1 + [(K - N_0)/N_0]e^{-r_m t}}$$

There are often advantages in using a discrete time version of the logistic equation, particularly when the population has a life history that moves in yearly jumps. Any seasonally breeding species meets that criterion. Even if not, the discrete-time version is useful when one wishes to run up on a computer the trajectory of a population growing logistically. There are a number of algorithms, but we recommend the Ricker function, which takes the form

$$N_{t+1} = N_t e^{r_m - \beta N_t}$$

and which approximates a discrete logistic when β is set to r_m / K. Another possibility is the general equation introduced by May and Oster (1976) and Hassell et al (1976) for density-dependent population growth in discrete time:

$$N_{t+1} = \frac{e^{r_m} N_t}{(1 + aN_t)^b}$$

where $a = (e^{r_m} - 1)/K$. Setting $b = 1$ renders the growth logistic.

Another algorithm in common use in ecology and conservation ecology, but not recommended, is

$$N_{t+1} = N_t \left[1 + r_m \left(1 - r_m \frac{N_t}{K} \right) \right]$$

It looks very much like its continuous-time analogue, and one might easily be fooled into thinking that it will therefore produce an appropriate discrete-time version of the logistic equation. However, it acts very differently. This discrete form of the logistic can produce complicated abundance curves, its behavior depending on the size of the time step and the magnitude of r_m. Very small changes in the initial population will lead to wildly different outcomes (Burgman et al 1993). This kind of unstable behavior is typical of many discrete-time models. The safest option is the Ricker function with $\beta = r_m/K$. Its parameters can be calculated from a string of population estimates by simple linear regression:

$$\ln\left(\frac{N_{t+1}}{N_t}\right) = r_m - \beta N_t$$

A highly recommended discussion of the various forms of density-dependent models in discrete time and their strengths and drawbacks is provided by Burgman et al (1993).

Figure 7.9 shows characteristics of the red kangaroo model as a function of density and compares them with those of the analogous logistic model. The kangaroos have a mean density of $46/\text{km}^2$, an r_m of 0.4 on a yearly basis, and a variance in r of 0.16 imparted by environmental stochasticity. The logistic model was set up to duplicate those parameters by assigning it an r_m of 0.4 and by setting $\bar{K} = 55/\text{km}^2$ with a normally distributed standard deviation of 14.5. Although the logistic population was "grown" every 3 months, the K was changed only once a year. Note that there is nothing in the kangaroo frequency-density histogram that could be construed as a carrying capacity K that could be borrowed for use in the logistic model. Maximum kangaroo density obeys the same rules as maximum flood level, and for the same meteorological reasons; you have not seen the maximum yet. Thus the $\bar{K}$ assigned to the logistic was chosen simply to align the other parameters of the two models.

There are still differences outstanding despite the alignment of three parameters between the two models. The distribution of densities is positively skewed (the tail points to the right) for the kangaroos and slightly skewed negatively for the logistic. Rate of increase r trends slightly upward against density for kangaroos but down for the logistic. The mean rate of increase per time interval $\bar{r}$ is slightly negative for kangaroos and slightly positive for the logistic. (A negative $\bar{r}$ does not mean that the population is necessarily decreasing. Only a negative long-run rate of increase $\tilde{r}$ implies that, and $\tilde{r}$ is, among other things, determined by the relationship of r to density.) Variance in r against density trends down for kangaroos and up for the logistic.

We will now run a PVA on kangaroos, using the logistic approximation, to see how closely it duplicates the PVA based on parameters estimated directly from the weather-plant-herbivore system of Kinchega National Park. We ask how the probability of red kangaroos persisting for a century changes according to the size of a reserve.

Figure 7.10 compares the probability of persistence of red kangaroos

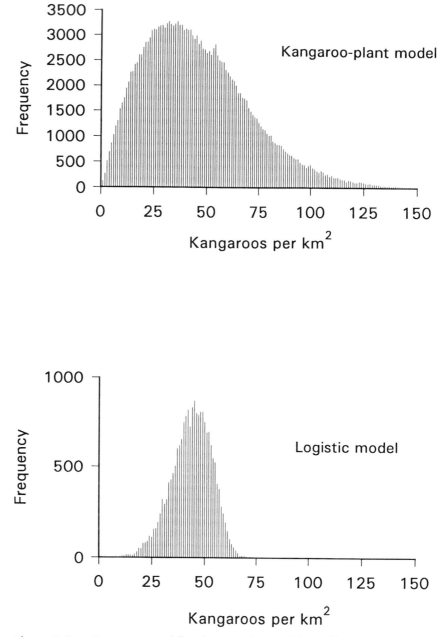

Figure 7.9. Comparison of the characteristics of the red kangaroo model as a function of density with those of the analogous logistic model.

as a function of reserve area, between an estimate based on measured parameters of the rainfall-plant-herbivore system and one based on a logistic approximation to it. The logistic gives the more optimistic estimates of persistence. The estimates of P_{100} from the logistic model are fitted reasonably well by the pragmatic regression

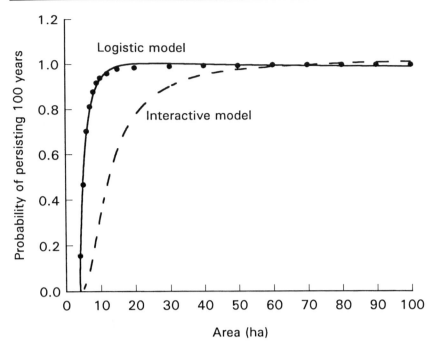

Figure 7.10. Probability of a kangaroo population persisting 100 years on reserves of various size according to the interactive model where rainfall, plants, and herbivores interact, and the logistic approximation to it.

$$P_{100} = \frac{0.9768 + 1.520}{\dfrac{A - 19.391}{A^2}}$$

where A is area in hectares, implying that an area of 17 ha is sufficient (with a probability of $P_{100} = 0.999$) to hold a self-sustaining population of red kangaroos for a century. That can be compared with the estimate of 69 ha from the more realistic model.

It is informative to compare estimated mean persistence according to more than one single-species model. Burgman et al (1992) ran a logistic PVA on shrews, *Crocidura russula*, in Préverenges, a suburban area of Switzerland, coming up with the surprising result that the populations were likely to die out in less than 20 years. Burgman et al (1993) repeated that exercise using the same data fed through the Ricker function of the logistic. By that model, the mean persistence time was estimated as greater than 50 years.

Comparisons between a model's predictions and reality may reveal missing factors, which confound any predictions. A viability analysis using the population's demographic data for a small population of acorn woodpeckers, *Melanerpes formicivorus*, predicted a median persistence of 16 years (Stacy and Taper 1992). The field evidence, however, was that the population persisted beyond that prediction by immigration even though it was isolated by some 45 km from the next nearest population.

7.5 THE USES AND ABUSES OF POPULATION VIABILITY ANALYSES

The major problem with generalized PVA methodology is that it treats the population at risk in the context of a density-dependent single-species model. Thus the effect of environment is left out of the calculation. Environment appears only indirectly as a number of animals K that saturates the area in which the population lives. The reciprocal interactions between the population and the environment are depicted in one form or another only by a reduction in realized rate of increase as population size approaches K. In brief, the models contain a minimum of ecology and cannot cope with the complex ecological interactions that constitute a standard conservation problem.

However, for a minority of species, viability analyses using simplified density response functions may be more appropriate than complex ecological models. Those species have populations with low reproductive potential, and nonlinear density effects may typically persist at levels below those that trigger density effects. Numbers of species with very high recruitment rates may fluctuate due to density-independent factors, and for them complex ecological models may not be necessary.

In a small population the fate and breeding success of individuals becomes increasingly important as population size decreases (Section 6.1). The breeding system of the species in trouble may accentuate this effect if, for example, the survival of breeding females depends on the survival of their partner. When these effects are known, they can be used to fine-tune population viability analyses. McCarthy et al (1994) refined a model of population persistence for helmeted honeyeaters, *Lichenostomus melanops cassidix*, an endangered bird in Victoria, Australia. The modifications were for estimates of survival and took into account the breeding system where the survival of individuals and their offspring were not independent events.

PVA operates on a set of static population parameters to determine how long the population will persist if nothing unusual happens. However, there are minor exceptions: VORTEX, for example, allows for a linear decline in carrying capacity and includes as an option a user-specified coefficient of inbreeding depression (Section 6.3.7). By and large PVAs deal with small static populations subject to the effects of demographic stochasticity reinforced by the effects of environmental stochasticity, with or without some attention to possible genetic mal-function. PVAs as presently constituted deal with the "extinction vortex," that interaction of positive feedback loops by which a small population becomes smaller. Thus, most PVA methodology addresses the perils of smallness, as such, for a population in steady state.

A major utility of PVAs is in designing reserves of adequate size rather than in dealing with populations in trouble. The stochastic events that a PVA portrays are exactly those of most concern to a park manager. PVA can be used to secure first approximations to minimum reserve size, so long as the structure of the model is realistic and the data are not wild guesses. Shaffer (1983) used a viability analysis in this way to suggest that the grizzly bear, *Ursus arctos*, of the Yellowstone National Park (8,992 km^2) needed an area of 2,500–4,000 km^2 to be rea-sonably secure. Armbruster and Lande (1993) did the same for African elephants, suggesting from the output of a model approximating a stochasticized logistic that an area of 2,500 km^2 was needed to maintain a self-perpetuating population in semiarid regions.

A second use of PVA is in assessing the vulnerability of a small population to stochastic events. An excellent example of this use is provided by Burgman et al (1993), who used data presented by Conway and Goodman (1989) on the vital statistics of white rhino, *Ceratotherium simum*, to assess the risks inherent in two harvesting regimens applied to a small population in a reserve. A third legitimate use of PVA is the playing of heuristic games. PVA software allows a user to see how a capped population behaves at low density under various specified regimes of fecundity and mortality.

That is the good news. The bad news is that PVA can hamper diag-nosis of a conservation problem because it begins from the assumption that the problem is low numbers. It is not concerned with why the population or species declined to low numbers, so it does not prescribe how to get numbers back up to a healthy level. Nonetheless, it has been represented as a general means of adjudicating between management options (e.g., Seal 1991; Lindenmayer et al 1993; Possingham et al 1993), even though the circumstances in which it can do so are severely constrained.

An example of PVA leading to acceptance of the status quo, rather than to diagnosis and treatment of the problem, is provided by a study of the endangered Mediterranean monk seal, *Monachus monachus* (Durant and Harwood 1992). These investigators presented a set of demographic indicators (e.g., "0 pups born in 1 year, population <20") that, according to their model, would identify a small population about to get smaller. "If a predictor of extinction is observed in a protected population then every effort should be made to increase juvenile sur-vival rather than fecundity. Methods of increasing juvenile survival include the establishment of a rescue scheme to increase the survival of abandoned pups and those entangled in fishing gear." However, the critical question not answered was, why did numbers decline?

7.6 SUMMARY

Risk assessment in conservation biology is mostly used to determine the probabilities of a population becoming extinct. The two steps of what has come to be termed population viability analysis are estimating the probabilities that various events will occur and then the probabilities that those events will drive a population to extinction. Most such analyses are based on single-species models such as the logistic and truncated exponential. Thus their output should be viewed with caution and should attract a substantial margin of error. Viability analysis has often been used as if it were a diagnostic tool, it has been used to justify management recommendations on how to cure a problem that it cannot in logic diagnose, and its use has sometimes led to essentially status quo management when a recovery operation was urgently needed. The viability analyses are at their most useful in estimating the vulnerability of small populations to stochastic events. Populations in reserves are often small and face the risks of stochastic events, and thus it follows that viability analyses are useful to estimate the size of reserves.

Diagnosis of declines

How to determine which agent or agents push a species toward extinction is pivotal to conserving it. We state this because, despite it being self-evident, too often the crisis of the moment obscures such fundamental logic. Our plea in this chapter is for a scientific approach to be adhered to and assumptions shunned insofar as this may be possible. Logically arriving at a diagnosis stated as a hypothesis to be tested by a designed experiment helps us avoid preconceived ideas. We need to heed Underwood's (1990) warning that "it is immensely easy to confuse observation with explanation." We can easily fall into the trap that, because an agent such as an introduced predator is present, it is the problem.

Although this chapter might be seen as restricted to the declining-population paradigm, that is not the way we view it. Take a species threatened with extinction through demographic stochasticity because its numbers are low. The antidote lies in treating the proximate factor—the reason why the population declined. We ask not what will happen if the species continues at low numbers but how it got to that state and whether the process might be reversed. Thus diagnosing the cause of decline is equally pertinent to species that have already declined to very low numbers.

First we have to decide that a species is in trouble and then we have to identify what is causing its problems. Most diagnoses start with observing an effect (poor survival or low fecundity) coinciding in time or space with an anticipated agent. But that is not enough. Coincidences alone are insufficient validation, even though they may seem overwhelming. The next step is to systematically weigh the evidence to accept or reject a hypothesis. We outline the diagnostic steps as they apply to the main classes of agents implemented in pushing species toward extinction to illustrate the types of effects that can be expected and, from examples, hints as to how we can detect them. Most methods are within the usual repertoire of ecologists as they deal with a standard ecological problem—what is limiting a species.

8.1 ENDANGERED SPECIES RECOVERY ANALYSIS

We have grouped the actions to determine what is driving a species toward extinction as an endangered species recovery analysis to emphasize that successful diagnosis depends on a logical series of steps (Caughley 1994).

1. Confirm that the species is presently in decline or that previously it was more widely distributed or more abundant.
2. Study the species' natural history for knowledge of and a feel for its ecology, context, and status.

3. When confident that this background knowledge is adequate to avoid silly mistakes, list all conceivable agents of decline.
4. For each agent, measure its level where the species now is and where the species used to be in time or space. For example, has a harvest increased? Does the species overlap with an introduced predator? Test one set against the other. Any contrast in the right direction identifies a putative agent of decline and thus a hypothesis to be tested. Do not assume that the answer is already provided by folk wisdom, whether lay or scientific.
5. Test the hypothesis by experiment to confirm that the putative agent is causally linked to the decline, not simply associated with it. Treatments can often be used for this.

This account of diagnosing conservation problems concentrates more on logic than on arithmetic. A frequent mistake in conservation biology is that because a species is common it has no conservation problem. Yet, a species in decline can seem relatively common until only a short time before it becomes rare. The mark of a species in trouble is not rarity as such but its rate of decline. We have given an example of the unpleasant shock that awaits when a species thought to be common declines exponentially (Section 4.8).

There are attributes that signal a species as being vulnerable to extinction—low numbers, colonial breeding at a few sites, restricted distribution, and so on—but these are not in themselves necessarily problems. They may prompt monitoring which will identify a decline, if there is one.

Once a population is of small size, diagnosis of agents forcing the decline may be difficult. Contrasting the conservation of two cavity-nesting birds makes this point. The scramble to protect the last handful of Puerto Rican parrots (Section 4.9) was less productive than the experimental approach to the reasons why the red-cockaded wood-pecker was declining even though it still numbered 7400 in 1992 (reviewed by Walters 1991). Waiting too long to initiate conservation action not only hampers diagnosis, but the species may disappear during the efforts to determine its problems, as happened with the large blue butterfly (Elmes and Thomas 1992).

We add one more example to emphasize the importance of getting the diagnosis right. The Aleutian Canada goose, *Branta canadensis leucopareia*, breeds on most of the outer Aleutian Islands in the Bering Sea (Figure 4.4) and winters in California and Japan where they are hunted. Throughout this century numbers of geese had declined. The decline was attributed to Arctic foxes, *Alopex lagopus*, taking eggs and goslings, and foxes were eliminated from four islands between 1949 and 1976. The elimination did not stem the decline, and by 1975 only 1,150 geese were left breeding on the one island which foxes had not reached. Captive-bred birds did not survive, and translocation of wild birds to fox-freed islands did not reverse the decline either. Only when hunting was closed in 1975 on the wintering grounds did the goose population triple (Springer et al 1978).

8.1.1 Detecting trends

The first step is to decide if a decline is taking place and then to determine the rate of decline. Knowing the rate of decline establishes the time available in which to take remedial steps. The greater the rate, the more urgent the task to determine what caused it. Declines detected while a species is relatively abundant may be easier to rectify than waiting until the species is no longer abundant. Two examples are the African elephant (Section 4.8) and the black rhino, *Diceros bicornis*. Had authorities realized the extent of declines before sharp rises in the value of ivory and horn, subsequent policies and actions might have been different. Herein lies the case for regular monitoring of a selected species. Sparrow et al (1994) describe their approach to techniques and species to monitor neotropical butterflies. Alternatively, determining the trend may avoid precipitating action before it is needed. The Jamaican hutia was taken into captive breeding before the extent of its decline was known (Section 4.17).

Detecting a trend is a matter of sampling design. A biometrician is useful here, as sampling is a troublesome prospect for detecting changes in very small populations. Green and Young (1993) demonstrate the Poisson distribution to allocate sampling effort for rare species. The method is general but does require predetermining what density is the threshold of rarity for the particular situation.

The threat of Type 2 errors looms large because failing to detect a trend is serious for a declining species. Suppose we found that two estimates of population size were not significantly different, we would conclude that the population had not decreased. That would be risky, as the probability of failing to detect the difference depends on the difference between the estimates as well as variability in the data and sampling design. As populations become smaller, the precision of estimates decreases and the likelihood of detecting that change in size then shrinks (Taylor and Gerrodette 1993). Precision can be increased by intensifying survey coverage (which is not the same as increasing survey frequency).

Often the data are restricted to two censuses only. If they come as total counts, as might be feasible with a penguin colony, there is no problem. The two counts are sufficient to indicate whether the population has increased or decreased over the intervening interval, and they can be analyzed to estimate the average rate of increase or decrease over that period. If the first count returns N_0 individuals and the second N_t, the two are related by

$$N_t = N_0 e^{rt}$$

where r is the exponential rate of increase (or decrease if it comes out negative), e is the base of natural (Naperian) logarithms, taking the value 2.7182817, and t is the time between the censuses. The exponential rate of increase is

$$r = \frac{\ln(N_t / N_0)}{t}$$

If, for illustration, $N_0 = 320$, $N_t = 95$, and $t = 7$ years,

$$r = \frac{-1.2144}{7} = -0.1735$$

The exponential rate of increase (in this case negative and hence a decrease) can be changed directly from one unit of time to another: -0.1735 on a yearly basis is $-0.1735/365 = -0.000475$ on a daily basis. Rate r on a yearly basis can be converted to a proportional decline D per year by $D = 1 - e^r = 1 - e^{-0.1735} = 0.159$ per year, which is a decrease of 15.9% per year.

Two points do not provide much information on a population's trend. The difference between the estimates may reflect only year-to-year variation in rate of increase, or, even more likely, it may reflect differences in census technique or (if they come from an aerial survey) viewing conditions. A decline depicted by just two points alerts us to a potential problem, and the more precipitate the decline the more worried we would be, but that is all. Further data are needed to confirm that the measured decline is part of a continuing trend rather than just a demographic incident or surveying glitch.

Analyses do not have to be complicated to separate out the type of agents causing the decline. We demonstrate that point with regression analyses of estimates by aerial survey of the Peary caribou, *Rangifer tarandus pearyi*, on the Queen Elizabeth Islands in the Canadian Arctic (Figure 4.4). Table 8.1 shows that estimated numbers dropped from over 20,000 in 1961 to a couple of thousand in 1986. On that basis, the Canadian government declared the subspecies endangered.

We fit a simple linear regression (Figure 8.1) of the form $y = a + bx$, the steps of which are given in Table 8.1. It provides a slope of $b = -738.5$, which indicates that the average trend is a net loss of 738.5 caribou from the population each year. The expected year of extinction T_E, if this linear model is correct and the trend continues, is the estimated value of x when population size y has declined to one animal: $T_E = (1 - a)/b$ $= -61,772/-738.5 = 84$ (i.e., [19]84).

Fitting a linear regression to these data assumes that the population loses a constant net number of animals per year. That might occur if habitat is being modified by a commercial company whose efforts are constrained by a fixed annual budget, for example, clear-felling forestry operations. However, for caribou on Arctic islands we would

Table 8.1.   Trend of numbers of Peary caribou on the Queen Elizabeth Islands of Canada

Year (*x*)	Estimated numbers (*y*)	Logged numbers (ln *y*)
[19]61	20,308	9.919
73	5,143	8.545
74	2,591	7.860
86	1,917	7.559

$$n = 4$$
$$\bar{x} = 73.5 \qquad\qquad \bar{y} = 7{,}490 \qquad\qquad \bar{y}_{\ln} = 8.471$$
$$\Sigma x = 294 \qquad\qquad \Sigma y = 29{,}959 \qquad\qquad \Sigma \ln y = 33.88$$
$$\Sigma x^2 = 21{,}922 \qquad\qquad \Sigma xy = 1{,}970{,}823 \qquad\qquad \Sigma x \ln y = 2{,}461$$
$$(\Sigma x^2)/n = 21{,}609 \qquad\qquad (\Sigma x)(\Sigma y)/n = 2{,}201{,}987 \qquad\qquad (\Sigma x)(\Sigma \ln y)/n = 2{,}490$$

$$\text{SS}_x = 313 \qquad\qquad \text{SS}_{xy} = -231{,}164 \qquad\qquad \text{SS}_{x\ln y} = -29$$
$$b_{y.x} = \text{SS}_{xy}/\text{SS}_x = -738.5 \qquad\qquad b_{(\ln y).x} = \text{SS}_{x\ln y}/\text{SS}_x = -0.095$$
$$a_{y.x} = \bar{y} - b_{y.x}\bar{x} = 61{,}773 \qquad\qquad a_{(\ln y).x} = \bar{y}_{\ln} - b_{(\ln y).x}\,\bar{x} = 15.48$$
$$r^2 = 0.76 \qquad\qquad r^2 = 0.86$$

Modified from Miller FL. Peary caribou status report. Ottawa: Canadian Wildlife Service, 1990.

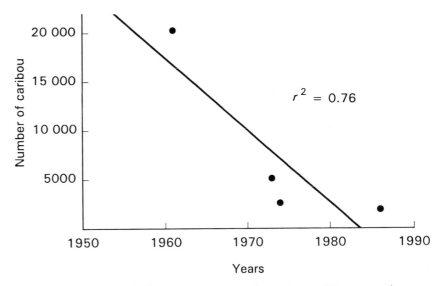

Figure 8.1.   A simple linear regression of numbers of Peary caribou on Canada's Arctic Islands. (Modified from Miller FL. Peary caribou status report. Ottawa: Canadian Wildlife Service, 1990.)

expect an agent of decline to reduce a population by a fixed proportion each year. The model reflecting that expectation is a linear regression of the logarithm of numbers on time. The last column of Table 8.1 gives numbers as logarithms, and below them are the steps of the regression analysis on those log-transformed estimates. Figure 8.2 shows the regression fitted to the data. It has a y-intercept of $a = 15.48$ and slope of $b = -0.095$, which is an estimate of the exponential rate of increase r, not to be confused with the r of r^2 we have been using to symbolize the proportion of total variance in numbers, or in logged numbers, accounted for by the regression. Raising $r = b$ to the power of e converts it to the net survival rate per year, $e^r = e^{-0.095} = 0.91$, the factor by which population numbers are multiplied each year to give an estimate of numbers in the next year. Thus we can estimate expected numbers in 1986 by multiplying the 1961 estimate 25 times by 0.91: $20,308 \times 0.91^{25} = 1,922$, which compares nicely with the observed 1917. The mean net rate of loss per year is estimated as $1 - e^r = 1 - e^{-0.095} = 0.09$ or 9% of the population per year. The expected year of extinction T_E, if this exponential model is correct and the trend continues, is the estimated value of x when population size y has declined to one animal (i.e., when $\ln y = 0$):

$$T_E = \frac{-a}{b} = \frac{-15.48}{-0.095} = 163$$

Since years have been scaled from 1900, the estimated year of extinction is $1,900 + 163 = 2,063$. That is quite a difference from the 1984 calculated by the linear regression and emphasizes the necessity of selecting the right regression model.

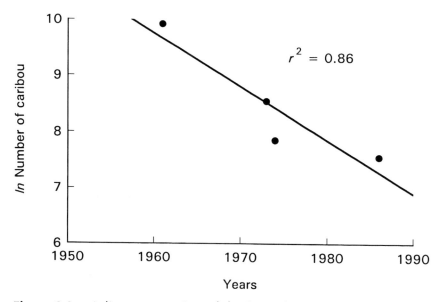

Figure 8.2. A linear regression of the logarithm of numbers of Peary caribou on Canada's Arctic Islands. (Modified from Miller FL. Peary caribou status report. Ottawa: Canadian Wildlife Service, 1990.)

The variance accounted for by this regression is $r^2 = 0.86$, in contrast to the $r^2 = 0.76$ of the regression of untransformed numbers on time. So we conclude that the model of a constant net proportion being lost each year fits better than that implicating a constant net number being lost each year. That implies in turn that the agent of decline is acting on the survival and/or fecundity of the population with a severity independent of population size.

These two regression analyses are the irreducible minimum for investigating a trend. In practice we would investigate more closely by analyzing for trend separately on each of the 10 islands, checking whether the slopes are the same or different, and then relating any detected disparity to differences in weather patterns, harvesting pressure, and predator density over time and among islands. The appropriate techniques are analysis of covariance and multiple regression. More on procedures to determine trends are found in Caughley and Sinclair (1994) and Krebs (1989).

Estimating Laysan duck, *Anas laysanensis*, numbers (Box 8.9) makes the point to match survey technique to the species' ecology and then to standardize techniques. Storms, inbreeding, and food availability were offered as explanations for the wide fluctuations in annual numbers. An evaluation of survey techniques painted a different picture. Comparing estimates within a year revealed more variation than between years and that the estimates depended on timing. Counts during nesting or molting when the birds were secretive returned low estimates (Marshall 1992).

Detecting long-term trends is tricky when a population responds quickly to environmental variations by fluctuating widely in size. The problem then becomes avoiding unnecessary treatments or, worse, failing to take the necessary steps. One solution is to raise the frequency of estimates by using the scale and frequency of environmental variations to decide on a suitable sampling strategy.

An alternative approach to determining trend in population size from repeated surveys is to use estimates of survival and birth to estimate annual geometric growth rate λ (Lande 1988a, 1988b). Taylor and Gerrodette (1993) used an analysis of statistical power to compare the two approaches for determining trend in population size. At low densities the demographic method is more powerful, but after the population has increased to a certain size, repeated surveys become the more powerful method. The demographic method is particularly suitable for long-lived species as annual variations in survival and productivity balance out. The problem is that measuring survival and birth is often difficult in the field (Caughley 1977).

In many cases, reality forces us to forgo a statistically satisfying series. The bottom line is when there are no estimates of population size and the only information is local people's knowledge. Unsatisfactory as that may seem, it should *never* be ignored, but treated positively as the best information available.

8.1.2 *Contraction of range*

More often than not we have only guesses for the number of individuals. For example, we have no estimates for the initial population sizes of most endangered birds of the Americas, let alone estimates from which we can determine a rate of decline (Figure 8.3).

A sign of potential trouble is range contraction, and it may start with the extinction of a local population. Local extinctions may not be uncommon depending on the species' population structure (Section 5.6), and they are not diagnostic in themselves but they signal the need to determine the cause. A species may rehearse its extinction many times before the main event. An association between a disappearance or reduction of a local population and a possible cause is not in itself a diagnosis. It is a speculation leading to hypotheses to be confirmed by testing.

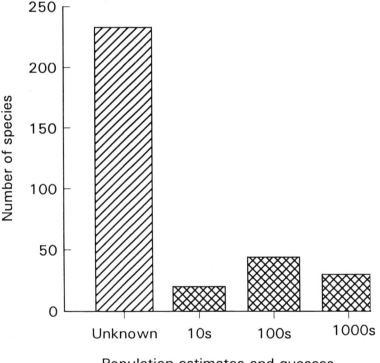

Figure 8.3. Population sizes of the endangered birds of the Americas. (Modified from Collar NJ, et al. Threatened birds of the Americas. London: ICBP, Washington DC: Smithsonian Institution Press, 1992.)

Range contraction may show up in the species' distribution as either a geographic trend—a partial erosion of the distribution's boundary— or populations extinguished within the distribution to produce range fragmentation. In range contraction, the agent of decline can often be identified by knowing something about those factors that position the edge of the range. Caughley et al (1988) give a protocol for identifying those factors.

8.1.3   Natural history investigations

It may seem trite to warn against tackling a conservation problem before gaining a basic knowledge of the species' natural history, but it happens. Diagnosing the decline of the large blue butterfly (Section 4.4) remained incorrect until the butterfly's ecology was understood. Another aberration is advising on a conservation problem without looking at it on the ground. Related to this is the folly of not tapping information from people who live in the same area as the species at risk.

Even though we term this first step natural history investigations, that does not justify any departure from a standardized and systematic approach. The methods do not have to be elaborate, but much follows from this first step, and clearly recording those preliminary ecological observations is important.

Natural history investigations are necessary, but they should not get out of hand. Time spent measuring growth rates of juveniles, banding everything in sight, or estimating age-specific rates of fecundity and mortality is unhelpful at this stage. The purpose is to gain the information necessary to construct a hypothesis as to the cause of the decline. More detailed investigation comes after identification of the relevant questions.

The importance of knowing the species' natural history and a "feel" for its ecology continues throughout all stages of analyzing and treating declines. Linked to this is the need for competent field biologists, and we note the intangibility of what it is that qualifies them as such.

8.1.4   Diagnosing the cause of the problem

By the time the basic natural history is understood, a list of possible agents driving the decline will have emerged. Agents often act in concert: for example, forest fires may, over the short term, destroy fallen logs in which numbats avoid predators (Section 4.14); without the fires the foxes and cats may not have caused the numbat decline; without predators the fires would not have caused the decline, and so on. Any interaction complicates but does not preclude accurate diagnosis. Box 8.1 describes the diagnosis of a single cause, the introduced brown tree snake, *Boiga irregularis*, affecting several endemic forest

Enormous effort required.

---

birds on Guam (see Figure 1.1). Then Box 8.2 describes the several problems facing a single species—the palila, *Loxioides bailleui*, an endangered endemic honeycreeper on Hawaii. The palila's story is not finished as research is still under way, but the way the factors were examined and discarded or accepted is fairly typical of reality.

We can take advantage of other techniques to explore possible explanations for a decline. Some species or situations are harder to tackle than others—large mammals or ocean dwellers come to mind. We can look for correlations between changes in numbers and environmental variables, or we can use population models as an investigatory tool.

Box 8.1. Diagnosis of the effect of the brown tree snake on Guam's endemic birds and lizards

---

Problem: the decline of forest birds
Ten species of endemic forest birds were common on the Pacific island of Guam (550 km²) until they started disappearing from their southern ranges in the 1960s. Their distribution contracted to a small (160 ha) area on the northern tip, and three species became extinct.

Diagnosis of factors driving the decline
Step 1.   Assemble evidence for and against possible causes.
Descriptive evidence for and against pesticide contamination, hunting, competition with introduced species, habitat change, disease, and predation was compiled and scrutinized. Evidence was lacking for all possible causes except predation. The brown tree snake, introduced in 1967, was the only predator unique to Guam which itself was the only island where the forest birds had virtually disappeared. Diagnosis then focused on predation by the snake.

Step 2.   Test predictions that the tree snake has driven the declines.
The hypothesis that predation had driven the declines led to three predictions:

1. Expansion of the snake's range will coincide with contraction of the bird's range.
2. Predation rates as indexed by live-bait trapping should be higher where the forest birds were declining than where their numbers were stable.
3. Small mammals would have also declined as the snake preys on them.

The bird's and snake's distributions were compared through interviews and records. The snake's diet was described by trapping using live surrogate birds and eggs.

Results and conclusion
The timing and geographic spread of the snake coincided with the contraction of the bird's distribution. Trapping demonstrated the effect of the snake (Table 8.5).

(Continued)

Box 8.1.   (Continued)

All three predictions were verified, and the conclusion that the introduced brown tree snake was the cause of the decline of the forest birds was supported.

Problem: the decline of endemic lizards
The introduced tree snakes, especially juveniles, preyed on endemic lizards, but the effect on the lizard populations was unknown. Complications were the snake having exterminated a kingfisher predator of the lizards, the introduction of a parasitic skink, an irruption of shrews, and the decline of a carnivorous gecko.

Diagnosis of factors driving the decline
Step 1.   Assemble evidence for and against possible causes.
Information on the lizard numbers and distribution was not collected systematically until 2 years after the snake's introduction. This prevented a comparison of the rate of spread of the snakes and the contraction of the lizard's range. The role of disease and pesticides could not be evaluated. Habitat change was discarded as a possible cause as none of the lizards were obligatory forest dwellers.

Step 2.   Test predictions that the tree snake has driven the declines.

1.   Numbers and species of lizards would be different on snake-free islands.
2.   Numbers and species of lizards would have changed between 2 and 24 years after the introduction of the snake.
3.   Snakes would be the principal predator of lizards.

Museum records were scrutinized and lizard numbers were estimated by counts. The rate of predation on lizards was estimated from using geckos as live bait in traps.

Results and conclusion
All three predictions were verified, leading to the conclusion that the introduced brown tree snake was the cause of the decline of at least three and possibly six species of geckos; but the cause of the decline of the skinks was more complicated.

(Modified from Savidge JA. Extinction of an island forest avifauna by an introduced snake. Ecology 1987;68:660–668; Rodda GH, Fritts TH. The impact of the introduction of the colubrid snake *Boiga irregularis* on Guam's lizards. J. Herpetol. 1992;26:166–174.)

Pascual and Adkison (1994) used these indirect approaches to close in on the most likely explanations for a decline in Steller sea lion, *Eumetopias jubatus*. Such explorations of possible causes are exactly that—explorations and not explanations, and further work is needed to support or reject them.

Box 8.2.   Diagnosis of the problems facing the palila on Hawaii

Problem
The palila was widely distributed on Hawaii in mamane, *Sophora chrysophylla*, forests where the finchlike honeycreeper feeds on pods and flowers. Clearing lowland forests left the palila isolated in remnants of forest on the slopes of Mauna Kea, an extinct volcano. Cattle, until 1931, and then sheep, until 1982, grazed the understory and browsed trees. In the early 1980s, no trend was discernible in counts of palila, which were numbered at 2,000–6,000. Feral cats, rats, and mongooses are abundant. The range of mosquitoes—vectors of avian cholera—reaches one area of forest. Palila were unevenly distributed among the patches of forest and occupy only a quarter of the available forest. Densities were lowest in the long narrow fragments of forest and the narrow patches on north and east slopes that are isolated from larger patches on the southwestern slopes.

Diagnosis of factors driving the decline
Step 1.   Assemble evidence for and against possible causes.
Initially, research focused on the palila's ecology and its annual variation in habitat use to screen factors limiting distribution.

1.   Habitat modification: the range contraction was determined from the palila's subfossil and present distribution. The palila's dependence on tree flowers and pods suggested why it is restricted to the forests. Densities of birds were lower in forests with exotic understory. Stock browsing on trees reduces the palila's foraging success as browsed trees are multistemmed with smaller branches and the birds have trouble getting pods among smaller twigs.
2.   Breeding failure: preliminary observations were that breeding success was low.

Step 2.   Test predictions that breeding success is low.
The next stage focused on larger sample sizes to confirm breeding success was low and which factors were reducing it. Palila nests were monitored for 1 year. The results were:

1.   Breeding success was low—only 25% of nests fledged at least one chick.
2.   Hatching failure was the greatest cause of breeding failure: 49% of eggs failed to hatch or were deserted. About 18% were infertile eggs or the embryo died.
3.   Nesting success significantly decreased during the breeding season.
4.   Predation of eggs was rare and of nestlings, uncommon.

   Conclusions.   The results pointed to food unavailability causing low breeding success, but the data did not rule out inbreeding depression or predation.

Step 3.   Test predictions that predation causes low breeding success.
The cause of the low breeding was unresolved. Food unavailability was a likely explanation, but ruling out predation was apparently more amenable to experi-

(Continued)

Box 8.2. (Continued)

mentation. The design was an unreplicated three-factor experiment. Factor 1 (density of dummy nests containing quail eggs) had two levels, factor 2 (density of palila) had two levels, and factor 3 (season) had two levels. The control was predation on actual palila nests. Analysis of variance was used to test the predation frequencies for effects of season, nest, and palila density. The results were:

1. Predation rates were low and the black rat was the only predator.
2. Predation was significantly higher during than before the breeding season.

Conclusions. Predation of eggs was not causing the low breeding success, which ruled out the need for predator control. The next step would be whether food availability was limiting productivity and whether it was lack of a resource such as food or dispersion between forest patches that explained why so little of the available habitat was unused.

(Modified from Scott JM, et al. Annual variation in the distribution, abundance, and habitat response of the palila (*Loxioides bailleui*). Auk 1984;101:647–664; Pletschet S, Kelly JF. Breeding biology and nesting success of palila. Condor 1990;92:1012–1021; Amarasekare P. Potential impact of mammalian nest predators on endemic forest birds of western Mauna Kea, Hawaii. Conserv. Biol. 1993;7:316–324.)

8.1.5 Overharvesting by people

Harvesting here means removing individuals from a population by hunting or fishing, and it is a common factor in driving declines. Once the point of economic extinction is reached (harvesting is no longer profitable) and harvesting stops, populations may rebound, which reinforces, but does not prove, overharvesting (Figure 8.4). This figure illustrates small populations that did rebound after overharvesting. Section 3.2.3 tells of the species that did not and became extinct. Harvesting may be unselective and unintentionally force other species into decline. The by-catch of marine turtles by shrimp boats (Box 9.10) and seabirds by long-lining for tuna, *Thunnus* spp. are examples.

The simplest and most secure diagnosis is to demonstrate that the harvest exceeds the maximum sustained yield. But only occasionally is it possible to collect data allowing an estimate of annual offtake Y and population size N. An alternative approach if r_m is known or can be approximated for the species is then the demonstration that $e^{r_m} < Y/N$ indicates that the population is being overharvested. Rate r_m can be approximated for herbivorous mammals as $r_m = 1.5W^{-0.36}$ (Caughley and Krebs 1983), where W is mean adult live weight in kilograms. Caughley et al (1990) used a variant of that method to show that the elephant population of East Africa was being hunted to extinction outside high-security areas (Section 4.8).

Harvesting suffers from the lure of tangible evidence. It is easy when seeing hundreds of caged birds or stacks of pelts to conclude that harvesting is excessive. The annual offtake is not sufficient evidence on its own, however, for a diagnosis of overharvesting, although it can be a strong hint that a species is heading for trouble. More often than not, determining harvest levels is difficult or becomes an end itself.

Harvest type affects determining harvest levels. Most international trade of species already in trouble is monitored through the Convention on International Trade in Endangered Species of Wild Fauna and Flora (CITES) (Section 12.1.1). It is usually only part of the annual offtake, but it is at least a starting point for estimating harvest levels, even if not always an easy one to interpret. Leader-Williams (1992) has scrutinized the trade in rhino horns and revealed much about difficulties in using trade figures.

Greater difficulties lie with determining national levels of subsistence and commercial harvesting, as most countries do not use a quota and tag controls typical of North America. A common approach to gauging harvesting levels is to interview hunters and gatherers. Osemeobo (1992) took this approach to estimate the consumption and sales of African giant snails, *Archachatina marginata*, in Nigeria (Figure 4.9), which may be declining.

Underreporting of harvests is common: losses (wounding, failure to retrieve, and misidentification) are rarely reported. Deaths in captivity for the live trade are often high—60% of Mexican parrots die before export (Iñigo-Elias and Ramos 1991). For discussion on why the losses in the international live trade of birds are high, see Section 11.5 and Box 11.1. Illegal harvests can be gauged from interviews either of the harvesters or at the market end. Silva and Strahl (1991) interviewed hunters in Venezuela about the harvest of guans, *Penelope*, and curassows, *Crax*, even within national parks. The openness of those hunters in admitting to illegal hunting is probably atypical. Investigators for Trade Record Analysis of Flora and Fauna in Commerce (TRAFFIC) (Section 12.1.2) sometimes resort to masquerading as potential purchasers to determine if illegal wildlife products are still being marketed. Such surveys are like the seizures of illegal wildlife: they are evidence that there is illegal trade, but they convey little about the volumes or their effect on the population.

Illegal harvesting can, for obvious reasons, be extraordinarily difficult to deal with. Spix's macaw, *Cyanopsitta spixii*, already conspicuous by its bright blue plumage and long tail, is further distinguished by, as Collar et al (1992) put it, "errors, contradictions, silence and disinformation . . ." which have dogged efforts to establish the macaw's status. Unfortunately by the time it was realized how few macaws were left, they were under intense pressure from illegal trapping. Captures in the 1980s reduced the last 40 parrots to four and then by 1991, only a solitary male remained—the last of his species in the wild (scattered in various captive collections are 25 male and female

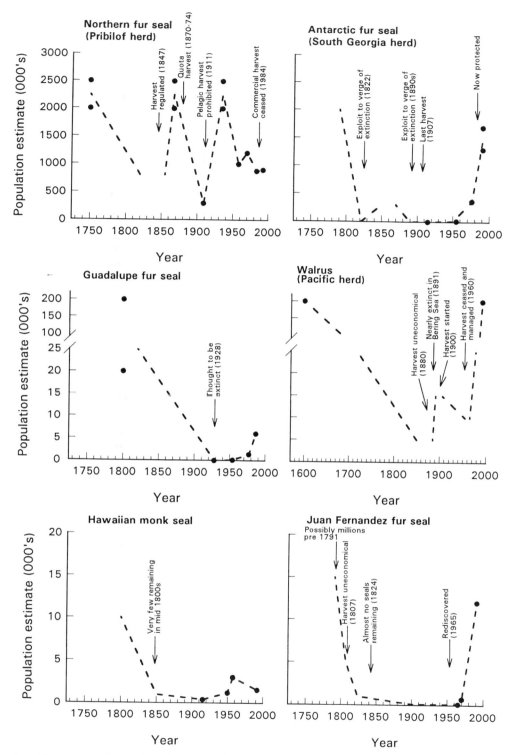

Figure 8.4. The status of eared seals, true seals and walrus species after their near extinction from commercial harvesting for their hides, meat, oil, and ivory. (Modified from Dampier 1729 [quoted in Hubbs and Norris 1972]; Morrell 1832; Hubbs 1956; Fay 1957; Baker et al 1970; Hubbs and Norris 1971; Johnson 1975; Payne 1979; Bonner 1982; Lander and Kajimura 1982; Busch 1985; Fleischer 1987; York 1987; Fay et al 1989; Anon 1993; Francis in Reijnders et al 1993; Gallo in Reijnders et al 1993; Fowler [1993] [quoted in Reijnders et al 1993]; and Reijnders et al 1993.)

parrots). Loss of its preferred habitat, caraiba woodland, to agriculture had probably been happening for 300 years, but by the 1970s live captures had virtually become the ultimate factor for the species.

Collecting as a particular form of overharvesting is more often blamed than diagnosed as causing population declines. Bourn and Thomas (1993) experimentally removed a proportion of an adult butterfly population to simulate collecting. They concluded that overcollecting was unlikely except for extremely rare species restricted to a single colony.

Without estimates of population size, diagnosing overharvesting is sometimes attempted from age and sex ratios when a population is classified into age groups such as young of the year, other subadult and adult, or more commonly just juveniles and adults, often divided further into males and females. These frequencies may then be combined to provide sex and age ratios. Sex and age ratios should not be used in those ways because they are not telling something about the population's dynamics.

Caughley (1974) showed by computer simulation that a sudden rise or fall in mortality rate that affects all age classes equally has no effect on the age distribution and hence on the age ratio, however measured. It remains constant at its level before the demographic change. Other manipulations, particularly of the age-specific fecundity schedule, act upon the age distribution through their influence on r, but they can lead to either an increase or a decrease according to their nature.

These points are far from academic. Age ratios are used in conservation and wildlife management for purposes that they are ill-designed to serve. A recent example (Bodmer et al 1994) in an otherwise excellent paper had overharvesting diagnosed from both the shape of the age distribution itself, as converted from age at death of a collection of skulls, and a comparison of that derived age distribution with those produced by the same method that came from an area known to be heavily hunted. Age distributions yield extremely limited information on populations and take much time and money to construct.

Using sex and size (age) classes of harvested animals to monitor harvesting has two disadvantages in addition to the above. First, assuming that the age and sex ratios of the harvested population will represent the population at large supposes equal catchability. Second, it assumes that there is no selectivity in the market and that representation of sex and age classes in the initial catch is maintained until the harvest is monitored (Box 8.3).

An alternative approach to determining if harvesting is depressing a population is to compare harvesting effort and yield as we do for the humpback whale, *Megaptera novaeangliae*, and tuna fish (Section 11.3). Regressing catch-per-unit effort against accumulated catch is common in fisheries for both vertebrates and invertebrates where estimating

Box 8.3.   The diagnosis of the effect of harvesting on *Tupinambis* spp. lizards in South America

Problem
Tegu lizards are found in most of South America east of the Andes, and these large and omnivorous lizards seemingly thrive even when forests are cleared. They have long been hunted for their meat, but the current level of national markets for leather and meat is unknown. In the late 1970s, international trade in reptile skins escalated, and in the 1980s an average of 1.25 million tegu skins was annually exported.

Diagnosis
In Paraguay more than half the hunters interviewed believed that lizards had declined, but this was not reported by hunters in Argentina. There is no other information on population trends, and apparently the lizards are difficult to count. Since 1988 in Argentina, the lizards are hunted under quotas based on the needs of the tanning industry and not the demography of the lizard populations. As there are both professional and opportunistic hunters, often with varied motives, there is no estimate of either total harvest or harvesting effort. In essence, there is no real basis to estimating the effect of the harvest, and its deleterious effects are only surmised from the scale of the export offtake.

Management
The lizards are sexually dimorphic, and the sexes can be recognized from the skins. Skins are traded in three size categories, with the tanneries preferring not to purchase the smallest skins. Representation of the sexes was not equal among lizards taken by the hunters or skins held at the tannery. This could be explained by differences in vulnerability of males and females to capture, geographical differences, or middlemen holding onto larger skins until later in the season. Monitoring the harvest as a check on trends in population size, then, has its problems.

(Modified from Norman DR. Man and Tegu lizards in eastern Paraguay. Biol. Conserv. 1987;41:39–56; Fitzgerald LA, Chani JM, Donadio OE. *Tupinambis* lizards in Argentina: implementing management of a traditionally exploited resource. In: Robinson JG, Redford KH, eds. Neotropical wildlife use and conservation. Chicago: University of Chicago Press, 1991:303–316.)

population size is impractical. There are a number of underlying assumptions to the method, and the effect of their violations varies between species (Jamieson 1993). If the species being harvested is patchy in its distribution either as static or mobile aggregations or colonies, the catch-per-unit effort may be initially insensitive to a decline.

8.1.6   Habitat changes

Habitat change and loss are often advanced as the dominant cause of extinctions. We are in no doubt of their importance but emphasize that "habitat" as such is a flustered concept that needs to be tightened up. To diagnose what is happening under the idea of habitat change, we have to reduce the notion of habitat to variables whose levels can be measured and manipulated. We define habitat as the suite of resources (food, shelter) and environmental conditions (abiotic variables—temperature—and biotic variables—competitors and predators) that determine the presence, survival, and reproduction of a population.

The *habitat matrix* is the combination of land system and ecosystem that contains patches on which the species is commonly found, and those patches will be distinguished by such attributes as species mix in the herb layer, height and spacing of trees, rotting logs, cracks in the soil, and so forth. This is the species' *core habitat*. The clues that an ecologist uses to identify core habitat may have nothing to do with the environmental requirements of the species; they might simply co-occur with them. Often the habitat matrix is obvious, but the core habitat is less so. Yet that is the level where changes have to be recognized and treated.

Habitat modification

If woodlands are modified by a change of fire regime or if large areas of it are felled, this would clearly have an effect on the wildlife density. But here is the rub: *we may not know why*. We know in broad terms what constitutes that species' matrix habitat, but we often do not know what elements of that habitat are essential to it. To find out, we would need to know more about the species' core habitat. Rather than evoking "habitat modification" to explain decline of a species, it is better to find out what element is being changed to the species' detriment. Failure to appreciate this has handicapped the conservation of species such as the northern spotted owl. Despite all the recent studies on the owl's ecology, Simberloff (1987) concluded that "the owl requires old-growth forest and enormous tracts of it, though there is not yet consensus on *why* the owl needs old growth." Hence our emphasis is on the specific resources that an animal needs from its habitat.

Diagnosing which resources are in short supply as a consequence of habitat modification is standard ecological fare, and its techniques thus can be gleaned from the appropriate literature. Manly et al (1993) explain the design and analyses for field research for resource selection. To sort out factors and their interactions brings manipulative experiments into their own (Underwood 1986b). Nonetheless, as case histories such as the large blue butterfly (Section 4.4) tell us, diagnosis can be tricky when the factors are not conspicious, and then it may take a fair bit of ecological acumen.

The importance of abiotic factors is easily overlooked but is often critical for smaller vertebrates and invertebrates. Numbers of salamanders were significantly less in clear-cut areas than mature forests in the southeast United States (Petranka et al 1994). The reason is the loss of their moist microhabitat as clearcuts had unshaded floors, less leaf litter, and warmer, drier soils.

A more difficult problem to diagnose is when the habitat was modified some time previously and we are left with a few populations, often one only, of an endangered species. There are two dangers of misinterpretation in the absence of diagnosis. First, the decline might have nothing to do with habitat, the habitat change being simply associated with the decline, not the cause of it. Second, the habitat remnant might be viewed as preferred habitat when in fact it is not. Unsurprisingly, remnant populations of an endangered species often end up not in the habitat most favorable to it but in the habitat least favorable to its agent of decline. Historical reports and distribution of subfossils will reveal its former distribution. The woodhen of Lord Howe Island became stranded in mountain cloud forest when its preferred habitat was lowland palm forest (Section 4.2). The Hawaiian goose (Section 4.11), dark-rumped petrel (Section 4.12), and Bermuda petrel (Section 4.3) used to occur in lowlands but are now restricted largely to highlands or small islands.

The takahe, a large rail, presently lives as remnant populations in alpine tussock grassland in New Zealand. This was assumed to be its preferred habitat (Mills et al 1984, 1988), and there it struggled to maintain its numbers despite expensive management (Craig 1994). The first step toward a diagnosis was interpretation of its subfossil record. Prehistorically, the takahe preferred lowland forest, and it was restricted to the alpine zone only after the Polynesians arrived (Beauchamp and Worthy 1988, Caughley 1989). Takahes translocated to islands with lowland grassland and forest thrived, which is further support for the hypothesis that the alpine grasslands are a suboptimal habitat and food shortage resulted. During feeding trials, takahes preferred pasture to tussock grass (Craig 1994).

The ideal experiment to test food shortage as a hypothesis is by measuring the effect of supplementary feeding on the population's rate of increase, the experiment using experimental controls and replication. The Hawaiian goose (Section 4.11) has been largely restricted to uplands and deprived of seasonal foraging in the lowlands. Its poor recruitment may reflect substandard nutrition. However, we do not know any trial to manipulate food supply which was not confounded by concomitant changes in the access of predators to nesting birds.

Food shortage after habitat modification was suspected to limit black robins (Section 4.7), red kites (Section 4.13), Seychelles magpie-robins (Section 4.5), and pink pigeons (Section 4.10), but the hypothesis was confounded by other factors and made more difficult to test as populations were small and could not be subdivided for replication and experimental control. "Before" can be used as a pseudocontrol on "after." The evidence is then circumstantial but at least provides a weak inference which must be acted upon with caution, but it is better than a guess. Such trials should be run for a period longer than adequate for a controlled experiment to confirm that the effect is not simply a reflection of year-to-year environmental variation.

Habitat fragmentation

Habitat fragmentation is the man-made change imposed on natural habitat heterogeneity and has received considerable attention (for example, Saunders et al 1991, Hobbs et al 1992 for semiarid woodland; Lovejoy et al 1984, 1986, Diamond et al 1987 for tropical forest; Askins 1993 for temperate forest, shrubland, and grassland). Usually the habitat is changed patch by patch into another land use, and initially those patches of modified habitat are embedded in the original habitat. Even at the early stages of fragmenting a habitat, relatively small areas of the altered habitat, especially if they are long and thin patches such as roads, can have large effects on the ecology of the interior. Then as the changes progress, bits of the original habitat become marooned as isolated patches within the converted habitat. Thus we can be dealing with a continuum of variation in size, shape, and dispersion when we talk about fragmented habitat.

Most attention is at the level where the matrix habitat ends up as patches, and consequently the species of interest ends up as populations separated one from the other. The dynamics of these individual populations will follow the behavior of single populations (Sections 5.1–5.6). These populations will also interact with each other according to the dynamics of metapopulations through the dispersal of individuals between the populations occupying different fragments (Section 5.6). A fragmented habitat may impose an entirely different population structure from the original one in continuous habitat, although this remains to be demonstrated. Habitat fragments are also prime candidates for the construction of reserve systems (Chapter 10), and the three topics—habitat fragmentation, metapopulation structure and dynamics, and reserve selection—are closely linked.

Fragmentation is simple to measure if the habitat is definable from aerial photography. At least two sets of aerial photographs are compared with each other, and with the present state on the ground, to deduce the past (or at least since the first aerial photographs in the 1950s). Digitized satellite imagery can produce a "difference image" by subtracting one from the other, thereby highlighting the changes over the period (Graetz et al 1992). Figure 8.5 shows an example of fragmentation of dry *Eucalyptus* woodland in southwestern Western Australia, which is the habitat matrix of the numbat (Section 4.14).

The fragment may simply not provide food, cover, or breeding sites for even a small population particularly if the environmental variation is high. Persistence of the bay checkerspot butterfly, *Euphydryas editha bayensis*, on fragments of its Californian grassland habitat during droughts correlated with topographic variation of the site and not its size (Murphy and Weiss 1988). This serves as a cautionary note: knowing the frequency and amplitude of environmental variation is part of the diagnosis for habitat change—weather records can be a starting place here.

The habitat fragment may be too small or its resources may have changed as a result of fragmentation. The first step is to examine the distribution and abundance of species at risk according to the sizes of habitat patch. Sampling effort has to be equal among patches rather than proportionate to patch size to avoid a confounding effect if larger patches have more individuals and species. The next set of steps are from the standard toolbox of ecological studies to diagnose whether there are enough resources to support a population (Box 8.4).

Effects may occur while the modified habitat is still patches embedded in the matrix of original habitat. Pitting the intact canopy with clearings even before a forest is partially cleared changed the microclimate for the world's only nesting population of Abbott's booby, *Sula abbotti*, on Christmas Island (Box 8.5, Figure 8.6). Diagnosis came from experiments in the field and laboratory.

The shape and size of a fragment control the "edge" between it and its surroundings, and that may change the habitat itself as the proportion of edge to center changes. Declines of woodland birds in the northeastern United States has been explained by woodland fragmentation (reviewed by Askins 1993). Forest fragmentation increased the proportion of edge, and experiments using artificial nests or surrogate eggs demonstrated that predation rates and cowbird parasitism are higher along edges. Flaws in experimental design or definitions of terms, however, limit generalizations about the relationship between the edges, predation, or parasitism and the decline of the birds (Paton 1994).

Increasing the amount of edge is a traditional objective of wildlife management to favor species that use one habitat for food and another for cover. Similar techniques to measure the effect of interspersing habitats to boost populations can equally well be applied to the other side of the coin: the effect of fragmented (interspersed) habitat on

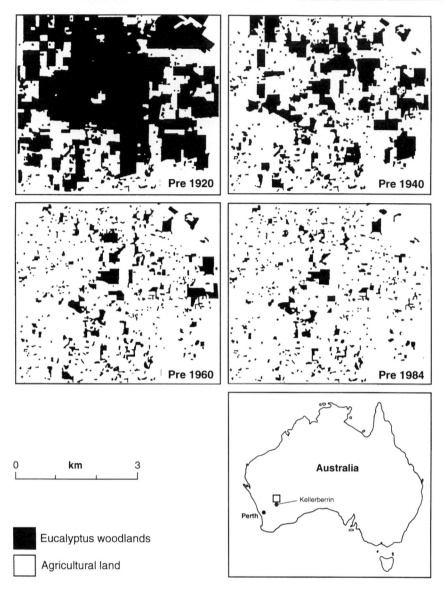

Figure 8.5. The history of fragmentation of part of Western Australia where the matrix habitat of many species is converted in a patchwork manner into wheat farms. (Modified from Arnold GW, Weeldenburg JR. The distributions and characteristics of remnant vegetation in parts of the Kellerberrin, Tammin, Trayning, and Wyalkatchem Shires of Western Australia. CSIRO Division of Wildlife and Ecology Technical Memo 33, 1991.)

reducing population size. A clear approach to testing the effects of edge that takes into account the degree of interspersion between two habitats and edge characteristics is Kremsater and Bunnell's (1992) study of black-tailed deer, *Odocoileus hemionus columbianus*. Their sampling

Box 8.4.   Diagnosis of the effect of habitat fragmentation on New England cottontail rabbits, *Sylvilagus transitionalis*

Problem
The New England cottontail colonized central and northern New Hampshire in the mid-1800s, but its range started to contract southward in the 1950s and now it occurs only as a few remnant populations.

Diagnosis
Step 1.   Assemble evidence for and against possible causes.
Suggestions that New England cottontail competed unfavorably with the eastern cottontail, *S. floridanus*, are unlikely to be true as its decline preceded the colonization of the eastern cottontail, which has only reached southern New Hampshire. Winter survival determines population persistence and depends on food and cover. Habitats with those attributes have a dense shrub understory typical of early successional stage of second-growth woodlands. Woodlands in New Hampshire have matured beyond their suitability for cottontails, and the remaining second growth has become fragmented by urbanization and other land uses.

Step 2.   Test predictions that habitat fragmentation has driven the decline.
The hypothesis that habitat fragmentation had driven the decline has two predictions:

1.   The changes and fragmentation of the habitat should coincide in time and area with the contraction of the cottontail's distribution.
2.   Survival of the cottontails would be lower in smaller patches of habitat.

Results
1.   Between 1880 and 1930, an average 13,000 ha/year of farmland was abandoned and became secondary forest. Secondary growth is only suitable as cottontail habitat for 10–25 years and the area of immature forest declined by 85% between 1960 and 1983. Increasing urbanization is fragmenting the remaining habitat (but our source papers did not report the rate and extent of fragmentation).
2.   Cottontail winter survival was 35% in habitat patches <2.5 ha and 69% in patches >0.25 ha. Patches >5 ha were always occupied, but smaller patches had only a 60% occupancy.

Step 3.   Test predictions of why cottontail's survival is low in small patches of habitat.
The next stage of the diagnosis was to determine why the rabbits fared poorly in the small patches. They might not find enough food, and then while searching for food outside the patches they could be vulnerable to predators. The predictions for rabbits on small patches (≤2.5 ha vs. ≥5.0 ha) were that

1.   Their physical condition (body weight and size) would be less.

(Continued)

Box 8.4.   (Continued)

2. Their use (based on fecal pellet counts) of habitats (classified by stem density) would be varied and they would have a greater distance to reach cover.
3. Their diet would be more varied and of lower quality, based on sampling of browsed and unbrowsed twig diameters and use of bark.
4. Mortality would be higher as measured by survival of radio-collared animals.

Statistical tests were nonparametric and the level of significance was $\alpha = 0.05$.

Results
1. Cottontails were smaller and in poorer physical condition on smaller patches.
2. Use of habitats with sparse understory density and distance from cover was greater on small patches.
3. Diets were not more varied, but cottontails did use lower-quality browse (larger twigs and more bark) on smaller patches.
4. Survival was lower and predation losses higher on small patches.

Conclusions
Cottontails dispersing to small patches of woodland were at risk of dying. More fragmentation of early successional woodlands into patches ≤5.0 ha will further the cottontail's decline.

(Modified from Barbour MS, Litvaitis JA. Niche dimensions of New England cottontails in relation to habitat patch size. Oecologia 1993;95:321–327; Litvaitis JA. Response of early successional vertebrates to historic changes in land use. Conserv. Biol. 1993;7:866–873.)

Box 8.5.   The diagnosis of the effect of forest clearing on Abbott's booby

Problem
Subfossils occur across the Pacific, but now the forests of Christmas Island are the only area left for this tree-nesting bird. It is long-lived and usually has only one independent offspring every 5 years. On Christmas Island, mining for potash picked up momentum in the early 1970s and by 1980 had caused the clearing of 25% of the forested nesting area. The population declined from 2,300 breeding pairs in 1967 to 1,900 pairs in 1983.

Diagnosis
Step 1.   Assemble evidence for and against possible causes of the decline.
The decline of breeding success and number of breeding pairs coincided with clearing the forests and evidence for other causes is lacking. Observations on breeding biology in 1967 found that the birds, specialized for a gliding flight, depend on an unencumbered approach into the prevailing wind to reach their nests and room for a free-falling takeoff. Beneath the nest an unbroken canopy reduces the risk of chicks or adults becoming grounded, as they are virtually unable to launch themselves from the ground.

(Continued)

Box 8.5. (Continued)

The mining created large clearings and disturbance over a third of the nesting habitat but the birds remained there because of the favorable prevailing winds (Figure 8.6). Breeding success and adult survival had decreased, which indicated persistence of the effects of the clearings. This implied that it was not simply the loss of nesting trees and their protective canopy that was the problem. Observations pointed to the effects of a change in the local winds.

Step 2. Test predictions of why the cleared patches of habitat reduce adult survival and fledging success.
The hypothesis that turbulence caused by winds crossing the clearings would decrease breeding success and increase adult mortality has three predictions.

1. If nesting pairs are avoiding the clearings, the density of nests in the vicinity of the clearings will be decreased.
2. If changes in winds around the clearings affect nesting success, it will be reduced downwind of the clearings.
3. If the clearings are affecting adult mortality, it will be increased downwind.

Data on breeding success, nest location, and adult mortality were collected after forest clearing from randomly selected nests within three sets of treatment blocks interspersed over the nesting area. Blocks were either within 305 m upwind or downwind of a clearing edge and distant (>305 m) beyond the clearing edge.

Results
1. Density of nests around the clearing decreased: the percentage of nests within 305 m of clearings was 73% preclearing compared to 42% postclearing.
2. Breeding success (independent fledglings/eggs) within 305 m upwind of the clearings was significantly higher. Downwind birds laid three eggs to every two of upwind birds, probably through replacing eggs lost to nest damage from turbulent winds in the canopy. Reduced success was from fledglings falling through the canopy during wind gusts. Grounded birds either starve or succumb to land crabs.
3. Adult mortality between 1983 and 1988 based on disappearance of the adult breeders was 3.2% upwind and 5.4% downwind.

The next stage was to determine the problems caused by the forest clearings. The prediction was that the clearings were causing increased wind turbulence in the canopy. Measurements from a model forest in a wind tunnel revealed that intermittent wind vortices were formed by the clearing and caused a distinct downwind zone with turbulent gusts (Figure 8.6). Their intensity was greatest in clearings 10 or more canopy heights in diameter.

Recovery treatments
The wind-tunnel experiments suggested that increasing the roughness of the clearing's surface by reforestation would reduce the turbulence generated, and

(Continued)

Box 8.5.   (Continued)

tree planting in the clearings started. Clearing was initially restricted to areas downwind of existing clearing and then stopped when, in 1989, the national park was extended to include the nesting areas.

(Modified from Nelson JB, Powell D. The breeding ecology of Abbott's booby *Sula abbotti*. Emu 1986;86:33–46; Reville BJ, Tranter JD, Yorkston HD. Impact of forest clearing on the endangered seabird *Sula abbotti*. Biol. Conserv. 1990;51:23–38.)

---

design used radiotelemetry (tested for independence; see Section 1.2.2) to compare deer locations with the distribution of random locations relative to habitat edges.

Separating effects of changes in the habitat of fragments compared to overall habitat loss can be tricky, as the two aspects may not be independent (Box 8.6). Furthermore the dispersion of the fragments relative to each other is another effect to account for. Conner and Rudolph (1991) suggested that even though enough foraging habitat is left, it might be inaccessible if the pattern of fragmentation forced the foraging woodpeckers to cross the territories of other groups. This was not experimentally demonstrated, but it does convey the subtlety that has to be addressed.

8.1.7 Impact of introduced species

Introduced animals are implicated in 40% of historic extinctions. An alien species can prey upon, compete with, or disturb an indigenous species. Mammals, birds, and reptiles dominate our examples, but some fish introductions have also been disastrous: notable is the loss of some 200 species of cichlids, *Haplochominidae*, following the irruption of Nile perch, *Lates*, introduced to Lake Victoria in the 1980s (Goldschmidt et al 1993). Less well known are the effects of alien invertebrates such as predatory snails, *Euglandina*, which were introduced to many Pacific islands. Hawaii has lost 600 of its 1,061 snail species, but the most detailed account comes from one of the Society Islands (Box 8.7).

The presence of an introduced animal is not the signal for control; documentation of its effect comes first, and that is not always obvious. Cats released on Dassen Island, off the west coast of South Africa, had about equal proportions of endemic seabirds and introduced rabbits in their scats. Scrutiny of prey remains, however, revealed that the cats scavenged seabird carcasses and killed rabbits (Apps 1984). Those findings indicate there is little to be gained by control of the cats—the cats may be beneficial if the effects of the rabbits were diagnosed to be detrimental.

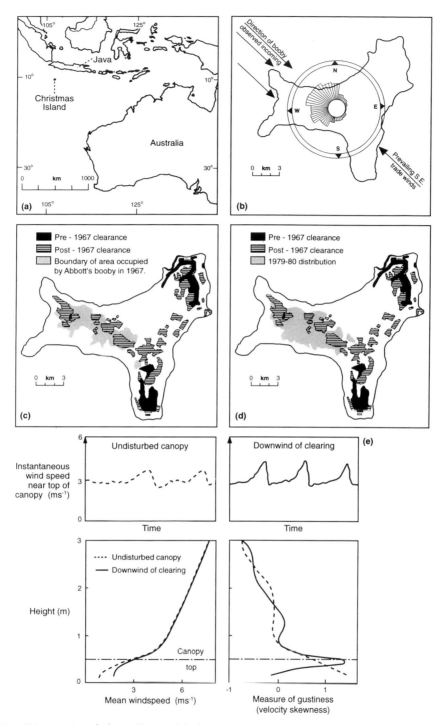

Figure 8.6. Diagnosis of the effect of habitat fragmentation on nesting success in Abbott's booby: (a) location of Christmas Island; (b) direction of the prevailing winds and the upwind approaches of boobies to their nests; (c) 1967 nesting distribution in relation to subsequent forest clearing; (d) 1970–1980 nesting distribution in relation to forest clearing showing no redistribution of nesting after clearing; (e) wind tunnel measurements showing an increase in gusts, although mean wind speed is unaffected. All downwind measurements were taken at a distance of four canopy heights from the clearing edge. (Modified from Nelson JB, Powell D. The breeding ecology of Abbott's booby *Sula abbotti*. Emu 1986;86:33–46; Brett D. Sea birds in the trees. Ecos 1989;61:4–8; Raupach MR, Bradley EF, Ghadiri H. Wind tunnel investigation into the aerodynamic effect of forest clearing on the nesting of Abbott's booby or Christmas Island. Canberra: CSIRO, 1987.)

Box 8.6.   The diagnosis of the effect of forest clearing on the red-cockaded woodpecker

Problem
Red-cockaded woodpeckers were widely distributed in the southeastern United States. They became restricted to about 30 isolated populations when the open pine savannas maintained by fire became converted for agriculture and forestry.

Diagnosis of factors driving the decline
Step 1.   Assemble evidence for and against possible causes.
The contraction of the woodpecker's distribution coincided with habitat loss. The woodpeckers live in colonies: a breeding pair with 0–4 others, usually young males who help raising the nestlings. Most young females and a few males disperse to replace a lost breeder in another group. The limitation to the formation of new breeding groups is the number of suitable nest trees. The woodpecker excavates its nest chamber in the heartwood of old pines and forages for insects on the pine's trunks and branches. Preliminary results were that forest removal within 400 m of a woodpecker colony caused it to decrease in size.

Step 2.   Test predictions of the effects of habitat fragmentation.
The experiment sought to differentiate as to whether habitat loss and fragmentation caused colonies to decline because of loss of foraging or impediment of dispersal and replenishment of groups. Habitat loss (percentage of forest removal) and measures of fragmentation (an angular measure of gaps in forest viewed from the center of each colony; and numbers of cut patches) were measured within 400-m and 800-m radii of a colony's center.

Predictions
1.   Habitat fragmentation would be greater around single male colonies than breeding pairs if female dispersal is impeded by fragmentation.
2.   Habitat loss will be greater around breeding pairs than breeding pairs with helpers if habitat loss reduces foraging opportunities.
3.   Colony size will diminish with increasing forestry loss in both low- and high-density populations if the effect is lack of foraging and not dispersal.

Results
1.   Habitat fragmentation was significantly greater within 400 m and 800 m of small colonies than large (>2 woodpecker) ones.
2.   Habitat loss (% forest removed) was significantly greater within 400 m and 800 m of small colonies than large (>2 woodpecker) ones. Breeding pairs with helpers had significantly less forest removed within 400 m and 800 m than breeding pairs.
3.   In the high-density population habitat loss was not significantly different around a breeding pair compared to a breeding pair with helpers.

(Continued)

Box 8.6.   (Continued)

Conclusions
The first two predictions were both supported, which suggests that it is not possible to separate the effects of fragmentation on dispersal and habitat loss on foraging. The nature of the interaction between the rate of removal and fragmentation limits drawing inferences about their effects. Forest removal had more effect on reducing colony size and loss of colonies in low-density woodpecker areas compared to high-density ones.

(Modified from Conner RN, Rudolph DG. Forest habitat loss, fragmentation, and red-cockaded woodpecker populations. Wilson Bull. 1991;103:446–457.)

There is always a temptation to control all alien species on the grounds that at least one of them is probably causing the problem. That scattergun approach almost invariably fails. Control techniques are highly species specific, and so a multispecies campaign simply dilutes the effort that should be concentrated on the species responsible.

The first step is to determine that the decline began soon after the introduction. If the decline began before that, the alien did not initiate it, although it may have exacerbated it. Timing can also be used to exonerate as well as implicate suspected culprits. The black rat was a popular suspect for the range contraction of the woodhen (Section 4.2). However, the bird was in decline by the mid-1800s, whereas the exact date of the black rat's arrival was pinpointed because they jumped ship—the SS *Makambo*, grounded on a reef at 8 P.M., June 14, 1918 (K. McFadyen, personal communication, 1994). The rats presumably arrived in the cargo bins that were jettisoned, and the ship was refloated after 9 days. Rats did not cause the woodhen's decline, but the disappearance of five species of bird and two species of large land snails coincided with the rat's arrival (Table 8.2). Other introductions whose timing is known are in the Caribbean (Figure 8.7). Mongooses spread throughout the Caribbean mostly after deliberate introductions (Hoagland et al 1989). The parasitic shiny cowbird, *Molothrus bonariensis*, spread itself, but even so the dates of its spread across the Caribbean are known as are the consequences (Section 9.4.3).

Coincidence between an alien's arrival and decline of an endemic species is not often as precisely known as the above examples. Alterna-

Box 8.7.   Effect of introduced snails on endemic Pacific snails *Partula* spp.

The introduced predator

Euglandina rosea is a carnivorous snail from the southeastern United States that has now been introduced to the Society Islands, Hawaiian Islands, Mascarene Islands, Seychelles, Vanuatu (New Hebrides), and New Caledonia.

The reason for introduction

Populations of the African giant snail, *Acatina fulica*, exploded to pest proportions after it was introduced for food on Pacific Islands. *Euglandina* was then brought in to control the African giant snail. The giant snail had already declined in some areas before the introduction of *Euglandina*—possibly a disease reduced its numbers.

The consequences

Euglandina ignored the African giant snail and selectively devoured endemic *Partula* snails, which were smaller and had more heavily calcified shells. The tree snails occur in a bewildering variety of pinks and browns, striped and plain because hybrids between species are fertile. *Partula* snails are found on most of the islands of the southwestern tropical Pacific Ocean. The tight coincidence between the *Euglandina* introduction and the decline of the endemic snails is known from the Tahitian island of Moorea (12 km²), where the fortuitous timing of a study of *Partula* genetics revealed the *Euglandina* blitzkrieg. On nearby Huahine Island, the carnivorous snail was not introduced and the two species of *Partula* tree snail are still abundant.

The Moorea Island extinctions

1967—African giant snail introduced to Tahiti and rapidly spread to all islands
1977—*Euglandina* predatory snail introduced and spread at 1.2 km²/year on Moorea
1980—*Euglandina* invaded over 4 km²
1982—Six species of *Partula* taken into captivity
1983—Three species of *Partula* extinct, populations of others shrinking
1986—All seven species of *Partula* on Moorea extinct in the wild

(Modified from Hadfield MG, Miller SE, Carwile AH. The decimation of endemic Hawaiian tree snails by alien predators. Am. Zool. 1980;33:610–622; Clarke B, Murray J, Johnson MS. The extinction of endemic species by a program of biological control. Pacific Sci. 1984;38:97–104; Tonge S, Bloxam Q. A review of the captive-breeding programme for Polynesian tree snails *Partula* spp. Int. Zoo Yearb. 1991;30:51-59.)

Table 8.2. The dates and presumed causes of the extinctions and near extinctions of birds on Lord Howe Island

Species	Year of extinction	Cause of extinction

White gallinule
    Nortornis alba
	Not certain but prior
	    to 1844
		Killed for food

Lord Howe pigeon
    Columba vitiensis godmanae
	Last seen between
	    1853 and 1870
		Killed for food

Lord Howe parakeet
    Cyanoramphus novaezelandiae subflavescens
	About 1870	Killed as an
		    agricultural pest

Lord Howe boobook owl
    Ninox novaeseelandiae albaria
	Not certain, but
	    probably after 1919
		Not known, but
		    possibly interbred
		    with introduced
		    N. novaeseelandiae
		    boobook

Vinous-tinted thrush *Turdus xanthopus vinitinctus*	Between 1919 and 1938	Exterminated by rats
Lord Howe warbler *Gerygone insularis*	Between 1919 and 1938	Exterminated by rats
Lord Howe fantail *Rhipidura cervina*	Between 1919 and 1938	Exterminated by rats
Robust silvereye *Zosterops strenua*	Between 1919 and 1938	Exterminated by rats
Lord Howe starling *Aplonis fuscus hullianus*	Between 1919 and 1938	Exterminated by rats

Lord Howe currawong
    Strepera graculina crissalis
	About 30 to 50
	    individuals survive
		Cause of decline not
		    known

Woodhen
    Tricholimnas sylvestris
	About 120 to 140
	    individuals survive
		Pigs and predation
		    by man

Modified from Recher HF, Clark SS. A biological survey of Lord Howe Island with recommendations for the conservation of the island's wildlife. Biol. Conserv. 1974;6:263–273.

tively, absence or scarcity of a species in trouble can be compared with the introduced animal's distribution. Whitaker (1973) compared lizards on islands in the same group according to whether Polynesian rats were present (Table 8.3). A similar approach comparing distributions of several predators may reveal which one is reducing a species. Table 8.4 identifies tenrecs, *Tenrec ecaudatatus*, as the more significant predator of geckos, although previously rats had been assumed to be the problem.

Within an island or area, range overlap between the alien and the indigenous species is a further clue to what is going on. The range of the indigene should contract to areas where, for one reason or another,

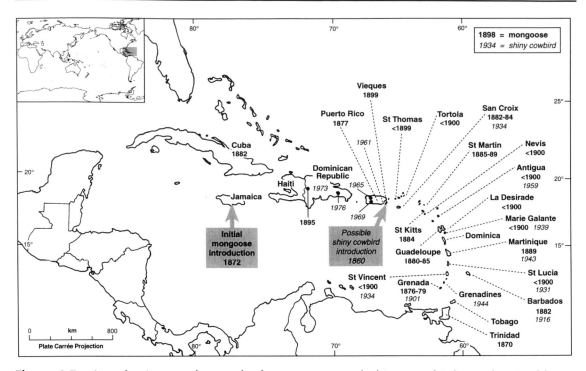

Figure 8.7. Introductions and spread of mongooses and shiny cowbirds in the Caribbean. (Modified from Hoagland DB, Horst GR, Kilpatrick CW. Biogeography and population biology of the mongoose in the West Indies. Biogeog West Indies 1989;1989:611–634; ICBP. Report of the 1986 University of East Anglia Martinique Oriole expedition. Cambridge: Birdlife International, 1987.)

the alien is uncommon or absent. Mapping respective distributions and estimating density of the alien where the indigene now is, compared with where it used to be, will sharpen, if not test, the hypothesis that the alien is the causal agent. Miller and Mullette (1985), for example, showed that the black rat was more abundant in the present contracted range of the woodhen than it was in areas sampled throughout the woodhen's former range (Box 8.8). Mapping respective distributions of the feral pig and woodhen "showed that the two did not overlap, but (Little Slope apart) fitted together very neatly. In several places the distributions were separated by only a low rock face" (Miller and Mullette 1985). Pigs became a prime suspect, which was confirmed by their experimental removal (Section 4.2).

Overlap in time and space between predator and prey leads to a hypothesis. To test that hypothesis comes back to knowing enough of the ecology, such as the predator's climbing ability and nest accessibility and the telltale signs of predation. The appearance of shells separated rat and carnivorous snail predation on Hawaiian tree snails (Hadfield et al 1993), and likewise the appearance of egg frag-

Table 8.3. The distribution and abundance of lizards and Polynesian rats on the Hen and Chickens Islands, New Zealand

	Area (ha)	Rats	Lizards							
			Hd	*Hp*	*Lsm*	*Lm*	*Lsu*	*Lo*	*Sp*	*S sp.*
Hen	484	A	F	R	F	F	F	—	F	—
Marotiri	138	A	F	—	F	F	F	—	F	—
Whatupuke	90	A	F	R	F	—	F	—	R	—
Coppermine	72	A	F	R	F	F	F	—	—	—
Mauitaha	22	A	F	—	F	F	—	—	F	—
Araara	1.8	?	F	—	F	F	—	—	—	—
Muriwhenua	4	—	A	A	F	—	A	A	—	—
Pupuha	1.0	—	—	A	A	—	—	A	—	—
"Middle Stack"	0.3	—	—	A	F	—	A	A	—	—
"Keyhole Rk"	0.3	—	—	A	A	—	—	—	—	—
"Gunsight Rk"	0.2	—	—	A	A	A	—	—	—	—
"Starfish Rk"	0.1	—	—	?	F	—	—	—	—	—
Sail Rk	2.1	—	—	A	F	—	—	—	—	A

Hd, Hoplodactylus duvauceli; Hp, H. paucificus; Lsm, Leiolopisma smithi; Lm, L. moco; Lsu, L. suteri; Lo, L. oliveri; Sp, Sphenomorphus pseudormalis; S sp., Sphenomorphus sp; R, rare; F, frequent; A, abundant. Modified from Whitaker AH. Lizard populations on islands with and without Polynesian rats, *Rattus exultans* (Peale). Proc. N.Z. Ecol. Soc. 1973;20:121–130.

Table 8.4. Gecko abundance and the occurrence of introduced predators on the Seychelles Islands

Island	Geckos	Rats	Cats	Tenrecs
Cousin	Abundant	—	—	—
Cousine	Abundant	—	—	—
Aride	Abundant	—	—	—
Chauve Souris	Abundant	—	—	—
Frigate	Frequent	—	●	—
La Digue	Frequent	●	●	—
Félicité	Frequent	●	?	—
Big Sister	Frequent	●	?	—
St. Anne	Frequent	●	●	—
Anonyme	Frequent	●	?	—
Conception	Frequent	●	?	—
Thérèse	Frequent	●	?	—
Praslin	Rare	●	●	●
Mahé	Rare	●	●	●
Silhouette	Extinct/rare	●	●	●

Modified from Gardner AS. The biogeography of the lizards of the Seychelles Islands. J. Biogeog. 1986;13:237–253.

Box 8.8.   Use of contingency table χ^2 to test for a difference between density of rats in the woodhen's former range against rat density in the range of the relict population (the reference area). Each sample is of 150 trap nights

Problem
The woodhens had been marooned on top of Mount Gower for years.

Diagnosis of problem driving the decline
If black rats were the problem, then the woodhens would have survived where there were fewer rats. Our null hypothesis is that the abundance of rats will not differ between sites without woodhens and the sites with woodhens. The abundance of rats was indexed by the results of 150 trap nights per site.

Site	Year	Traps catching	Traps empty	Total (χ^2)	χ^2 of totals	Heterogeneity (χ^2)
Reference	1978	68	82			
area	1979	25	125			
Little	1978	28	122			
Slope	1979	14	136	28.08	24.86*	3.22
Big Slope	1978	31	119			
	1979	14	136	24.21	21.68*	2.53
Mulley's	1978	31	119			
Drive	1979	12	139	25.95	21.68 (nt)	4.27*
Transit	1978	34	116			
Nichols	1979	13	137	21.51	19.71*	1.80
North Bay	1978	4	146			
	1979	3	147	93.92	88.75 (nt)	5.17*
Stevens'	1978	3	147			
Reserve	1979	3	147	97.03	91.56 (nt)	5.47*

*$p < 0.05$.
nt = not tested as years combined because of heterogeneity. Tested by separate years.

The table shows frequencies of capture on the reference area occupied by the woodhen and on sites where it used to occur. The first task is to test for heterogeneity between years in the comparison of each site with the reference area. Using Little Slope for purposes of demonstration, we first test the hypothesis that the two yearly samples from Little Slope tell the same story with respect to the reference area.

(Continued)

Box 8.8.   (Continued)

1978	Catching	Empty	Total	
Reference area	68	82	150	$\chi^2 = 24.51$
Little Slope	28	122	150	$(df = 1)$
Total	96	204	300	

The χ^2 of such a 2×2 contingency table is calculated as

$$\chi^2 = \frac{n(f_{11}f_{12} - f_{12}f_{21})^2}{(C_1)(C_2)(R_1)(R_2)}$$

$$= \frac{300(68 \times 122 - 82 \times 28)^2}{96 \times 204 \times 150 \times 150}$$

$$= 24.51$$

1979	Catching	Empty	Total	
Reference area	25	125	150	$\chi^2 = 3.57$
Little Slope	14	136	150	$(df = 1)$
Total	39	261	300	

1978 + 1979	Catching	Empty	Total	
Reference area	93	207	300	$\chi^2 = 24.86$
Little Slope	42	258	300	$(df = 1)$
Total	135	465	600	

Total $\chi^2 = 28.08$ $(df = 2)$

χ^2 of totals $= 24.86$ $(df = 1)$

Heterogeneity $\chi^2 = 3.22$ $(df = 1)$

The heterogeneity χ^2 of 3.22 tests for a difference between years in the trapping success compared between the two sites. It would have to exceed $\chi^2_{0.05,1} = 3.841$ to be significant, and it does not. We conclude that there is no heterogeneity between years and so can test for a difference in density between sites on the data from the 2 years pooled. Its χ^2, 24.86 with one degree of freedom, is far in excess of 3.841, and so we conclude that the density of rats on the reference area is significantly higher than on Little Slope.

Equivalent analyses for the other sites compared with the reference area reveal that three of the comparisons (asterisked) are heterogeneous between years. They must therefore be compared with the reference area one year at a time, as with the separate yearly analyses above for Little Slope, and those comparisons indicated

(Continued)

Box 8.8.   (Continued)

in each case that density differed from (and was lower than) density on the reference area in each year.

Conclusion
We conclude that rats were more abundant on the reference area occupied by the woodhens than on the sites without woodhens. Rats therefore are unlikely to be the cause of contraction of the woodhen's range.

(Modified from Miller B, Mullette KJ. Rehabilitation of an endangered Australian bird: the Lord Howe Island woodhen *Tricholimnas sylvestris* (Sclater). Biol. Conserv. 1985;34:55–95.)

Table 8.5.   Diagnosis of brown tree snake predation on endemic birds on Guam by live-bird-baited trapping

	Guam Island				Cocos Island
	Site 1	Site 2	Site 3	Site 4	
Occurrence of predators					
Tree snake	●	●	●	●	—
Monitor	●	●	●	●	—
Rat	●	●	●	●	●
Habitat					
Second-growth forest	●	●	●	—	—
Mixed forest	—	—	—	—	●
Savanna	—	—	—	●	—
Status of native birds	Extinct	Rare	Declining	Stable	Stable
Trapping: quail bait taken/number of traps					
Tree snake	20/20	23/32	9/16	2/24	—
Monitor	0/20	6/32	1/16	0/24	—
Rat	0/20	1/32	0/16	13/24	0/15

Modified from Savidge JA. Extinction of an island forest avifauna by an introduced snake. Ecology 1987;68:660–668.

ments separated rat from monkey predation on nesting pink pigeons (Section 4.10).

An alternative approach to identify a predator or to index the rate of predation is trapping with a surrogate for bait. The brown tree snake on Guam was identified as the most significant predator of birds and lizards by trapping with local geckos or introduced quail, *Corturnix corturnix*, as bait (Table 8.5). Subsequent trials revealed that even manure from the quails or domestic hens was an adequate bait to avoid using live birds (Fritts et al 1989).

Ultimately, identification of an alien species as causing an indigenous species' vulnerability must rest upon the result of an experiment in which the alien is removed or excluded from part of that range by a physical barrier (Section 9.4.2). In Western Australia, rock wallabies, *Petrogale lateralis*, have declined and foxes were suspected as predators. Kinnear et al (1988) excluded foxes by poisoning around rock outcrops where rock wallabies live. Their design was a treatment site replicated twice and an unreplicated control site (two unpoisoned sites were monitored but data were collected in different years). Treatment and control sites were not interspersed: the two treatment sites were in the northeastern study area. The control site did not receive the same procedures (except the actual baits) as the treatment sites. The authors did not analyze the effect of fox control on either abundance or rate of increase of the wallabies but described a proportionate increase in wallaby numbers. Only weak inferences can be drawn from this design.

Hone (1994) undertook a retroactive analysis. A two-way analysis of variance (fox control and years as factors) revealed no significant effect of fox control on numbers of rock wallabies. The effect of years and interaction of fox control and years was likewise insignificant. The conclusion would have to be that fox control had no effect. Hone went on to use another approach to test the data. Normally, if the statistical test is chosen as part of the experimental design, a test would not be replaced if found to be unsatisfactory (the experiment would have to be rerun with improvements). Hone was demonstrating the sensitivity of conclusions to the type of statistical tests. He analyzed the abundance at each site relative to initial abundance and found significant effects for fox control and years on rock wallabies' abundance.

The dynamics of predator–prey systems leave little room for complacency even once a diagnosis has been made and treatment implemented. Vigilance is still needed to detect changes in predator–prey relationships. Addition of an alternative prey can allow prey switching and support an increase in the number of predators. Cats and ground-nesting red-fronted parakeets, *Cyanorhamphus novaezelandiae*, coexisted for 80 years on Macquarie Island (Figure 4.2) until rabbits were introduced in 1879. Cats preyed on rabbits and increased, and the parakeets disappeared (Taylor 1979). This reconstruction of the sequence of events is based on coincidence in timing of the events. Another point to watch for is that introduced predators can replace each other. Rats raided 88% of artificial nests baited with quail eggs on Guam, but 28 months after the brown tree snake's introduction, it was responsible for 100% of the predation on the artificial nests (Savidge 1987).

Introduced herbivores can devastate vegetation. Rabbits have been released on at least 700 islands around the world and have often denuded the vegetation (Flux 1993). What initially may begin as a competitive relationship between an endemic and an introduced herbivore can become the total loss of endemic's food and that can set off other extinctions. Laysan Island (370 ha), Hawaii, is an oft-quoted example where rabbits affected vegetation and endemic birds (Box 8.9). The rabbits ate all the vegetation: a rail, honeycreeper, and a warbler became extinct and a duck and finch almost so. The disappearance of several noctuid moths (Opler 1976) is the likely key to the birds' disappearance. When the rabbits were grazing down grass tussocks, they outcompeted the caterpillars; this has to be assumed from the coincidence between the rabbits' arrival, loss of the birds, and the importance of moths as food for the birds. The birds also lost their nesting sites in the vegetation, but the significance of that in their decline is unknown.

The Laysan Island story is not a tight diagnosis as it happened decades ago and no one was there. A more detailed accounting of effects after introducing a herbivore comes from Cousin Island, one of the Seychelles (Figure 4.1). Black-naped hares, *Lepus nigricollis nigricollis*, were introduced about 100 years ago. A diagnosis of their grazing affects was to study their food habits in relation to habitat types using standard ecological procedures (Kirk and Racey 1992). The hare may prevent regeneration of *Casuarina equisetifolia* by eating seedlings, and *Casuarina* is a nesting tree for seabirds and landbirds forage in its foliage. A cyclone in 1982 destroyed almost all the mature trees, and without the protection of exclosures, hares would prevent regeneration. One recommendation was that the trees should not be replaced except by a salt-resistant species, as seasonally *Casuarina* loses its leaves from salt spray and landbirds including endangered species then lose a foraging opportunity. Such a recommendation would be contrary to another objective—to maximize seabirds.

Still other effects of introduced animals to watch for are genetic—hybridization and subspecies loss. Introduced feral animals may come to overlap the distribution of their wild progenitors. Genetic mixing between the European wildcat, *Felis sylvestris*, and domestic cats was detected (Hubbard et al 1992), but whether there is interbreeding between dogs and the rare High Arctic wolf, *Canis lupus arctos* (Miller 1978), is unconfirmed as the requisite genetic test is lacking.

8.1.8 Chains of extinction

This category comprises secondary extinctions, the extinction of one species caused by the extinction or decline of another species on which the first depends. A circumstantial case can be made to explain the disappearance of *Harpagornis moorei* in New Zealand. It was an immense forest eagle (males to 10 kg, females to 13 kg) that preyed on large ground birds, of which New Zealand had many (Holdaway 1989). It died out at roughly the same time as the moas became extinct around A.D. 1400 (Section 2.4.3).

Box 8.9.   Bird extinctions and declines following the introduction of rabbits to Laysan Island, Hawaii

The rabbits

The island (370 ha) is covered by shrub, bunchgrass, and low creeping plants. Large colonies of seabirds attracted guano miners (1891–1904), who introduced rabbits. By 1911, the rabbits were damaging the vegetation. In 1912, 5,000 rabbits were killed, but further control was not until 1923. By then vegetation was virtually destroyed.

Effect on the endemic landbirds

1.   Laysan duck, *Anas laysanensis*. The miners subsisted partly on the Laysan duck, apparently starting a decline. In 1909, feather hunters took many ducks for food, and in 1912 only seven ducks were counted. Despite the collection of 6 of the 20 counted in 1923, ducks numbered 400–600 by 1957. The ducks used clumps of grass to nest in and fed on cutworm larvae (*Agrotis* spp.). The rabbits denuded the plant cover, which probably led to the extinction of three species of moths (Opler 1976). The ducks also fed on flies around the brine lake and numerous carcasses of seabirds. The effect of removal of the rabbits on the ducks is unknown as it coincided with protection from large-scale hunting.

2.   Laysan finch, *Telespyza cantans*, has a heavy finchlike bill strong enough to break the eggs of terns and petrels. They scavenge seabirds and take plants and insects, and such a catholic diet may explain their survival. In 1912, the finches were everywhere on the island, but in 1923 only a few dozen were seen hopping among rocks. By 1938, their numbers had recovered to 1,000.

3.   Laysan rail was a small flightless rail which fed on insects, tern, and petrel eggs, plant seeds and leaves, and scavenged seabird carcasses. The rail nested in the grass tussocks and the chick was described as "... a black velvet marble rolling along the ground." By 1923, only two rails remained of the some 2,000 estimated in 1911. More than 100 rails were taken from Laysan to the Midway Islands in 1911 (Section 9.5.2). Between 1923 and 1936, the rails on Laysan had gone. What happened is unclear as the rails were the most omnivorous of the landbirds.

4.   Laysan millerbird, *Acrocephalus familaris familaris*, became extinct by 1923, although recorded as abundant in 1913. This small member of the warbler family fed exclusively on *Agrotis* spp. moths and caterpillars and nested in the grass tussocks. Presumably the rabbits ate out the tussocks, and the millerbird's food and nesting cover disappeared.

5.   Laysan apapane was a small scarlet honeycreeper that favored the tall grasses and bushes where it fed on honey from blossoms and small insects. In 1912, they were recorded as few, as the rabbits destroyed the plants, and in 1923, after a sandstorm, the last three honeycreepers could not be found.

(Modified from Berger AJ. Hawaiian wildlife. Honolulu: University Press of Hawaii, 1981; Warner RE. Recent history and ecology of the Laysan duck. Condor 1963;65:3–23.)

The black-footed ferret (Section 4.6) is a more recent, if not secure, example. Its decline is associated, almost certainly causally, with decline of the prairie dog, which was its food and provider of its shelter. The Mauritius kestrel's, *Falco punctatus*, decline coincided with clearing of evergreen forests. Ecological studies revealed that the kestrel's principal prey, geckos, *Phelsuma*, disappeared with the forests (Box 8.10).

Coincidence in timing and knowledge of ecology also revealed possibly why the Atlantic eelgrass limpet, *Lottia alveus alveus*, disappeared when disease eradicated its food (Carlton 1993). The limpet was abundant along the east coast of North America until, between 1930 and 1933, a slime mold killed 90% of the eelgrass, *Zostera marina*, upon which the limpet had fed. Neither the limpet nor the slime mold survives brackish water, but eelgrass could and it survived as remnant populations in estuaries. Tellingly, another mollusk which specialized in feeding on the eelgrass did not become extinct, and it, too, was a species that could tolerate brackish waters.

A final example of a chain of extinction is the white-letter hairstreak butterfly, *Strymonidae w-album*, which declined in Britain with the loss of most colonies in the 1970s. The decline coincided with Dutch elm disease killing most of the elms, *Ulmus*—the larval food plant. The check on the diagnosis was that the loss of colonies dropped when the surviving trees suckered and regrew (Warren 1993).

Extravagant claims are occasionally made for the prevalence of chains of extinction. Lovejoy (1974), for example, claimed that ". . . a species cannot become extinct without taking a galaxy of other species to extinction with it." But everything is not directly connected to everything else. Species do differ in the role that they play in community structure and function, and more than one species may have similar roles (Walker 1992). Another view of community structure emphasizes the importance of individual species in upholding community structure—the concept of the keystone species. This is a loosely defined and applied idea that the loss of one species will cause others to collapse (Mills et al 1993). Detection relies on knowing the natural history of the suspected keystone species and then experimentally verifying the hypotheses so generated.

Another chain of extinction is the loss of parasites when their host species become extinct. This has received little attention (Stork and Lyal 1993). Many species of mammal and bird have species-specific parasites, and those are lost if their host becomes extinct.

8.1.9   *Environmental contaminants*

Environmental contaminants, particularly organochlorines, are implicated or suspected as a causal agent of the declines, but diagnosis to demonstrate causality is often lacking. The peregrine falcon, *Falco peregrinus*; Mauritius kestrel; osprey, *Pandion haliaetus*; Bermuda petrel;

Box 8.10. The diagnosis of the causes of the decline of the Mauritius kestrel

Problem
Mauritius kestrels may have numbered between 164 and 328 pairs in 1753, assuming pairs occupied 5–10 km^2 territories and some 1,644 km^2 of evergreen forest existed on Mauritius (1,843 km^2). The rate of decline is unknown, but a compilation of sightings suggests that the former widespread distribution contracted during the 1800s. By the early 1960s kestrels were restricted to a small area in southwest Mauritius. The population was guessed to be some 50 pairs from the 1930s to the 1950s when it declined steeply to about 10 pairs. By the first census in 1973, the seven to eight birds left included only two pairs.

The kestrel's habitat was subtropical evergreen forests. Clearing of lowland forests began in the 1730s and continued in surges until, by the 1880s, only 3.6% of virgin forest remained, along with larger areas of degraded forest.

There was an intensive campaign of spraying organochlorines to eradicate malaria-carrying mosquitoes, *Anopheles* spp., between 1948 and 1965, which then continued at a reduced level for another 8 years. Other possible problems for the kestrels were introduced rats and monkeys, which are potential nest predators, and hunting by humans.

Diagnosis of factors driving the decline
The diagnosis of the forest clearing as the initial factor driving the decline of the falcon is based on the coincidence between the contraction of its range and forest clearing. The kestrel became restricted to the remaining patches of upland forest. The specific effect of habitat change was the loss of habitat for the kestrel's food species rather than loss of nest sites as fieldwork showed that the kestrel nested in holes in the many cliffs as well as trees. The field studies also showed the kestrel hunted in or below the canopy, taking day geckos, *Phelsuma* spp., from the trunks and branches. The number of geckos is related to the presence of native trees and shrubs. Competition for the geckos also may have come from the introduced and numerous red-whiskered bulbul, but this was not demonstrated, Indirect evidence of lack of food for the kestrels was the low number of pairs that attempted to nest (53%) compared to other small diurnal raptors.

Thinning of eggshells as a result of pesticide residue accumulation did not occur in eggs collected from before the 1940s compared to eggs taken 1981–1983. The captive population in the late 1970s did show eggshell thinning, and high levels of insecticides and some birds died with clinical signs of poisoning. Spraying for malaria-carrying mosquitoes was still going on in that area in the 1970s. This is suggestive that the sharp decline in the 1950s may have been from environmental contamination.

Monkey and rat predation may have occurred opportunistically, but evidence is sparse. People killing the kestrels as suspected chicken eaters may have previously been a factor, but after publicity the birds have been largely left alone.

(Continued)

Box 8.10.   (Continued)

Recovery treatments
The augmentation of the wild population as a recovery treatment had a shaky start. The wild population slowly increased to about 30 birds by the mid-1980s despite the removal of five adults and three nestlings into the captive population, which only produced one chick and died out by 1980. A second attempt with eggs and two nestlings between 1981 and 1986 was successful, and by 1992, 235 kestrels had been released (Cade and Jones 1993) and the number of pairs of kestrels had reached 42.

(Modified from Jones CG. The larger land-birds of Mauritius. In: Diamond AW, ed. Studies of Mascarine Island birds. Cambridge: Cambridge University Press, 1987:208–300; Cade TJ, Jones CJ. Progress in restoration of the Mauritius kestrel. Conserv. Biol. 1993;7:169–175.)

California condor, *Gymnogyps californianus*; eastern brown pelican, *Pelecanus occidentalis carolinensis*; and the St. Lawrence Seaway population of the beluga whale, *Delphinapterus leucas* (Figure 4.4), are all examples where contaminants are implicated in population declines. For at least North American raptors, the biochemical mechanisms by which eggshells are thinned and productivity reduced are now understood.

Finding a symptom, even one well known such as egg breakage, is not enough. The agent itself still must be identified. In the Netherlands, more great tits deserted their nests and the percentage of eggs with thin shells increased from 10% in 1983–1984 to 40% in 1987–1988 (Graveland et al 1994). Elsewhere, thin eggshells have been associated with organochlorine contamination, but DDT had been banned previously and raptor populations had recovered, which hinted that the problem lay elsewhere. Great tits nesting in areas with poor soils laid a higher proportion of fragile eggs, which led to a hypothesis that calcium deficiency was responsible. It was tested by feeding snail and chicken eggshell fragments to nesting tits, and those with the calcium supplement had significantly more hatchlings. The tits had been feeding on snails, but the snails had declined. Acid rain increases the solubility of soil calcium in poor soils, which had affected the snails and left the tits short of calcium during egg laying.

An alternative approach is to measure the levels of chemicals in the eggs with thin shells. Any level of organochlorine above trace in eggs should be viewed with disquiet, particularly if associated with excessive breakage of eggs in the nests. It can be linked to a population decline by the standard techniques of experimental design, comparing the rate of increase for populations returning low levels of the contaminant with others returning high levels. That correlation is a hypothesis, not a conclusion. A probit analysis or logistic analysis can then be run on fledging success of captive birds subjected in batches to differing levels of organochlorines in their food.

The case for diagnosis of contaminants proposed to explain other declines and extinctions is sometimes only based on coincidence. The extinction or almost extinction of a caddis fly, several mayflies, and 18 freshwater bivalve mollusks is attributed to contaminants, but that diagnosis is only secured by their disappearance from large rivers with industrial development (Polhemus 1993, Bogan 1993).

Environmental contaminants are not always industrial products. In 1968 several species of seabirds suffered heavy mortality on the Northumberland coast of England (Coulson et al 1968). About 80% of breeding shags, *Phalorcrocorax aristotelis*, died on the Farne Islands (Figure 8.8). Pesticides were immediately suspected, but exonerated. The cause was "paralytic shellfish poisoning" produced by toxins synthesized by dinoflagellates mainly of the genus *Gonyaulax*. The filter-feeding mechanism of mussels concentrates the dinoflagellates and hence the toxin. The toxin also reached seabirds along another pathway: from dinoflagellates to zooplankton filter feeders (e.g., crustacea) to fish and thence to the seabirds.

Environmental contaminants can set the stage for outbreaks of disease (see the following section), and that can be difficult, if not almost impossible, to detect without experiments under controlled conditions. Experimental exposure to organochlorines, oil, and selenium reduced immune function and disease resistance in mallard ducks, *Anas platyrhynchus* (Whiteley and Yuill 1991), and seals (Swart et al 1994).

Sometimes disciplined data on the effect of environmental contaminants are difficult to obtain in the field. The problem is then deciding whether to take action as soon as an environmental contaminant is identified or whether to wait until its effect on a species has been determined beyond doubt. Zimmerman (1975) provided a readable analysis of this dilemma. The debate triggered by the collapse of the peregrine falcon populations in the early 1960s was about whether to act on circumstantial evidence of low breeding success and eggshell thinning or wait for experimental verification before banning organochlorines. The decision was to wait, and is told in the voluminous story of the decline of peregrines, edited by Hickey (1969), and their recovery, edited by Cade et al (1988).

8.1.10   Disease

We found relatively little information on the role of pathogens or parasites as a factor causing declines of species in trouble, but that may simply reflect a tendency to underrate disease as a factor in population dynamics. Conversely, diseases are successful in eradicating populations but under the guise of pest control. An introduced disease

Figure 8.8.   Map of Europe.

(transspecies infection) or a disease as an added factor for a species already in trouble are situations where we know a disease can become the ultimate factor in an extinction. The last known wild black-footed ferrets collapsed as canine distemper virus killed most individuals except the few who were hastened into captivity on the assumption that they, too, would otherwise succumb (Section 4.6). The epidemiology of distemper in the wild is unknown, but it is probably a natural feature of the ferret's ecology and is maintained by other species acting as reservoirs.

The unlucky heath hen, *Tymphanuchus cupido cupido*, of North America was buffeted by a transspecies infection. Domestic turkeys, *Meleagris gallopavo*, were introduced to the island where the last population of heath hens was already struggling with a series of problems.

With the turkeys came a poultry disease called blackhead, and it spread to the heath hens, killing many but not all (Simberloff 1986).

The Indian wild ass, *Equus hemionus khur*, an endangered subspecies, was faced with an enzootic and an epizootic disease at the same time that human activities were encroaching on its habitat (Gee 1963). Definitive diagnoses are lacking beyond the coincidences in timing. The enzootic disease was surra, *Trypanosoma evansi*, a protozoan disease spread by horseflies, *Tabanidae*, and asses died from surra in the late 1950s. The epizootic disease was African horse sickness, a viral disease spread by mosquitoes, *Culicidae*. In 1959 it spread from Africa and reached India in 1960 where it killed asses. Wild ass numbers dropped from 3,000 to 5,000 in 1946 down to 368 in 1969 (possibly a low estimate) and 720 in 1975 (Ali 1986).

Pathogen-host dynamics predict that pathogens maintained by a reservoir host may pose a greater threat than single host-pathogen systems. The latter can markedly reduce the number of individuals constituting a species, but the reduction, although often spectacular, tends to be transient. Recently, an epidemic of phocine distemper virus swept through European waters and killed at least 18,000 harbor seals, *Phoca vitulina*, in Europe (Thompson and Hall 1993). Populations varied in their susceptibility for several reasons such as the timing of the outbreak relative to use of breeding colonies, which would have encouraged spread of disease between individuals.

Regional differences in levels of pollution may add to geographical variation in the severity of disease as pollution depresses the immune system in seals, for example (Swart et al 1994). Acid rain has been postulated as a factor stressing amphibians, decreasing their immune systems and leaving them vulnerable to fatal bacterial illness. Carey (1993) described this possible chain of events as a hypothesis and offered supporting observations and the steps to be tested.

These examples point to some of the steps to identifying the causative agent. A disease putting a species at risk is not necessarily obvious. Even deadly diseases do not usually leave conspicuous numbers of dead and dying animals: sick animals may hide and scavengers quickly remove evidence. Secondary pathological signs may confound diagnosis, and it is not always easy to identify an agent, especially if it is a transspecies disease or a new strain. Knowledge of recent introduction of other animals or epidemics in neighboring areas may help. Stress or pollutants may be a hint of reduced resistance to disease. Other diagnostic steps are to collect information of the suspected disease in dying and the general population, especially the incidence rather than prevalence of parasites. The pathogens to worry about are those that decrease fecundity and those that are maintained by a reservoir host. McCallum (1994) concluded that the most certain diagnosis followed experiments and offered pointers for their design.

8.1.11   *Inbreeding depression*

We have not found any examples of where inbreeding depression was demonstrated to be a factor in the decline of a species in the wild. Nonetheless we have included it because it is sometimes advanced as an explanation. The rationale is either assumed or based on a measured low genetic variation (Section 6.3.2), but to drive a decline the low genetic variation has to reduce fitness, and that is not easy to measure (Sections 6.3.7 and 6.3.9).

A more circuitous approach to diagnosing inbreeding depression is to rule out other factors. The cheetah is an example where ecological observations of predation on cubs overturned an assumption that high juvenile mortality in the wild had a genetic basis (Caro and Laurenson 1994). We caution that those ecological findings support the role of predation, but they do not conclusively dismiss an explanation of reduced fitness. There is a counterargument that the cub's susceptibility to predation was genetical. Entire litters were lost which would argue for a possible effect of relatedness, except that lions raided the lairs and thus entire litters would be expected to disappear. We conclude that predation is the simplest explanation consistent with the facts.

The wolves on Isle Royale (215,740 ha) have been a small population for decades. A recent slide in their numbers from 50 to only 14 by 1990 was caused by poor reproduction—only four pups survived in 6 years. The competing explanations for that were inbreeding depression, a canine parvovirus epidemic, or lack of food (Wayne et al 1991). Genetic variability was measured and was low. Wolves on the mainland had died from parvovirus, and the low genetic variability and small population size would increase their susceptibility to disease. Pup survival then increased in 1993 to eight pups when the moose population was at an all time high, which increases the likelihood that the previous decline was a lack of food (Peterson 1994).

8.2   PLANNING THE DIAGNOSTIC STUDY

Before taking measures to reverse a population decline, we must determine the cause. The selection of hypotheses is critical. Equally critical is the decision on the order in which hypotheses are tested and on whether hypotheses must be tested singly or whether some can be tested concomitantly. Before any hypothesis is tested we must prepare a detailed plan of campaign which will separate the independent effects of each putative agent of decline.

Where the effect of two or more agents are not separated in an experiment, they are said to be confounded. It is a common mistake, and surprisingly easy to make even by experienced biologists. (We have no intention of confessing how often we have fallen into this trap.) Black and Banko (1994) may have done this in their experiment to assess the effect of introduced predators on the Hawaiian goose (Section 4.11). They placed captive geese in ten predator-proof enclosures but "commercial poultry ration and fresh drinking water supplemented the naturally-occurring vegetation in the pens." Thus any increase in recruitment rate of the geese in the pens, relative to that of the free-ranging birds of the experimental control, might be owing to reduced predation, or increased food, or a combination of the two. The effects of predation and food supply are confounded.

Box 8.11 provides as an example a plan for determining why caribou are declining on Banks Island in the Canadian High Arctic (Figure 4.4). Only two plausible hypotheses are identified, food shortage and predation, but the first may act through three possible mechanisms and the second through two. These must be teased apart because the problem cannot be treated until the mechanism is identified.

A scheme of possible causes and their testing helps us to avoid the seduction of settling for causes that are the most amenable to testing. Either the effects are conspicuous—overharvesting, predation, plowed grasslands—or easy to measure—holes in trees. Not all agents are so conspicuous which in no way lessens their importance, but they may tax our skills and ingenuity as ecologists.

8.3 SUMMARY

Determining the factors driving a decline needs an analytical approach founded on scientific method. The initial step is to use trend in numbers, a contraction in range, or an obvious pending threat to determine that there is a problem. The next step is to acquire a basic understanding of the species' ecology to devise hypotheses for which factors are pushing it toward extinction. Testing the hypothesis through the rigor and discipline of scientific method leads to the surest diagnosis.

The strongest diagnosis for overharvesting is to compare annual offtake to the maximum sustainable yield, but that is rarely possible. Instead we have to rely on indirect methods using estimates of annual offtake, not always the easiest statistics to reliably gather. Coincidence in timing of a decline and a factor such as an introduced species are helpful but lead to weak inferences. Stronger inferences come from using the standard toolkit of ecological methods to reveal whether a species is killed by introduced predators or loses out in a competitive relationship. Habitat modifications, whether uniform or leading to fragmentation of the original habitat, have to be reduced to which resources (food or shelter) have become limiting. Loss of trees for

---

Box 8.11.   Hypotheses to explore the cause of the decline of caribou on Banks Island, Northwest Territories, Canada

---

Hypotheses to account for the decline
Either **A**, food shortage, or **B**, increased predation

If (**A**) then mechanisms may be: **A1**, increase in weather events, such as freezing rain, that affect availability of food; **A2**, competition for food with muskoxen, *Ovibos moschatus*, which are increasing; or **A3**, caribou themselves reducing the supply of food.

If (**B**) then mechanisms may be: **B1**, wolf predation, or **B2**, human predation.

The food-shortage hypotheses (**A**) may be tested against the predation hypotheses (**B**) by checking body condition. Hypotheses **A** predict poor body condition and low fecundity during a population decline; hypotheses **B** predict good condition and high fecundity during a decline.

If this test identifies the **A** hypotheses as the more likely, then **A1** is separated from **A2** and **A3** by its predicting a positive rate of increase in some years. **A2** and **A3** predict negative rates of increase in all years.

A2 (competition with another species) is separated from **A3** (competition between caribou) by checking for concomitant decline of caribou where muskoxen are not present in the same climatic zone.

(Modified from Caughley G, Sinclair ARE. Wildlife management and ecology. Cambridge, MA: Blackwell Science, 1994.)

---

nesting sites or shelter are recurring themes in diagnosis, because they lend themselves to identification and measurement.

Loss of one species may start a slide for other species, but these chains of extinction are not inevitable. Hypothesis testing of relationships is as necessary to document the links in a chain of extinction as with any purported cause of a decline. Detecting reduced survival or productivity caused by environmental contaminants or disease is rarely straightforward and as the two factors may work together—pollution increasing susceptibility to disease—an experimental approach is often the only answer. The diagnosis of inbreeding depression is another factor not necessarily easy to diagnose except by virtue of ruling it out rather than demonstrating it.

Few factors act in isolation; we can think of overharvesting and some instances of environmental contamination or introduced predators. More usual are factors acting in concert either simultaneously or sequentially. Then the order in which research hypotheses are posed becomes critical to determining limiting factors.

CHAPTER **9**

Treatment of declines

The last step in our sequence for endangered species recovery (Section 8.1) is to reverse a decline through removing or neutralizing its agents. We describe treatments that supplement resources or modify biological variables such as predators, parasites, or competitors. If there are too few individuals left at risk of further decline, then they can be supplemented from protected stock. A treatment applied as an experiment has the power to confirm the diagnosis and to evaluate success, which is measured by an increase in the species in trouble and not by treatment level.

We categorized treatments according to what effects they are dealing with—limited food, shelter, or breeding sites; mortality and populations too small likely to persist. Examples illustrate how researchers have treated those effects, but for most examples experimental design is conspicuously absent, yet the treatments appeared to work. To this, we have two comments. Not many of us expose failures to the scrutiny of our peers, so published examples are biased toward successes. More importantly, we do not know for sure, without an experimental design, that the treatment and success were causally related. That the treatment appeared to work in the absence of an experimental design is not a rationale for excluding one in future treatments.

Treatments can be, despite difficulties for some species, embarrassingly successful. History of introductions and some repatriations make this clear, but even removal of a predator can release a population to outstrip its resources. The Seychelles warbler, *Acrocephalus sechellensis*, is an example that we refer to in the following sections. Planning treatment needs to include projections about population trends and as an essential adjunct, monitoring. If the population becomes capped by its resources, likely on a small island or in a reserve, then can its individuals colonize other areas, or can they be translocated to new areas, or will their numbers erupt and decline?

We should not lose sight of the complexity and subtlety of ecological relationships in our haste to help a species in trouble. These following words were written during sometimes heated discussions on culling animals in national parks, but they are just as apt for other issues in conservation biology (Caughley and Walker 1983).

The world is not simple and things are not always as they seem. In many instances the obvious management response to a problem is not appropriate, it may even produce an effect directly opposite to that intended. This reflects the inherent difficulty in establishing cause and effect relationships in complex poorly understood systems.

9.1 DEPENDENCE ON DIAGNOSES

Other than by coincidence the effectiveness of treating a decline depends on the accuracy with which its causes have been diagnosed

(Chapter 8). More than one factor may be driving a decline which increases the need for an experimental investigation. The case histories give examples of where a shaky diagnosis led to ineffective treatment. Often a decline is not detected until it is well under way and only a small population remains. Even under those circumstances a shotgun approach with a suite of recovery treatments is seldom, if ever, the answer. Diagnosis will reveal the magnitude of how different factors are affecting the decline and that, in turn, will guide priorities for the treatments.

We do not have a heading for habitat modification, but instead treatments are listed as supplementing specific resources. Too often a decline is ascribed to habitat change and the treatment advocated is setting up of a reserve containing the "right habitat." As we emphasized in Sections 8.1.6 and 10.3, diagnoses of which specific resources were lost in the habitat changes are essential for successful treatment.

Reducing habitat to a model of the specific resources required has the inestimable advantage that it may supply sufficient understanding to offer alternative resources. We are not just thinking of artificial nest boxes but how to make a modified habitat acceptable. We cannot rely on serendipities, such as the survival of Mauritius kestrels released outside the native forests once their numbers had outstripped the capacity of the remnant patch of forest to support more territories. It had been assumed that kestrels needed to live in native forest to feed on tree-living geckos (Box 8.10). Captive-raised kestrels adjusted to patches of degraded forest interspersed with exotic trees (Cade and Jones 1993) to which the *Phelsuma* geckos had adapted. As well, the kestrels learned to hunt in suburbs and farmlands, taking small mammals, birds, and a lizard, *Calotes versicolor*, introduced from India.

9.2   TREATMENTS AND EXPERIMENTAL DESIGN

We need to know which treatments are successful, either to maintain them or to switch to another treatment if the first is not succeeding. For this we must be able to unambiguously define their success or failure and that requires that the treatments be experimental (Section 1.2). Furthermore, we need to know what worked and what did not work so that we can transfer successful treatments to similar problems. The treatment's results test whether the factors were correctly diagnosed.

Two arguments commonly raised against testing the efficiency of treatments experimentally (although seldom stated explicitly) are, first, that there is not enough time and, second, that the population contains too few individuals to allow statistical analyses. If there is time to apply a treatment, there is time to apply it according to the standard protocol of experimental design: this means treatments compared with controls, and the treatments (of which the control is one) suitably replicated (Section 1.2). Statistical analysis is essential. The common cry, "We did not use statistics because numbers were too low," misses the point. It is only when numbers are high that statistics may not be needed. Statistics is the mathematics of low numbers. It answers the question: "Am I justified in accepting the conclusion that the data suggest given the dearth of data that I have to work with?"

Box 9.1 offers the type of experimental design for determining the effect of one or other two factors on a declining species, and their interaction if present. That design would allow us to determine whether excluding rats and supplementing nest sites have a significant effect on density or rate of increase and by how much. But converting that statistical result back into a management recommendation is not necessarily straightforward (Section 1.2).

Experimental designs, such as we illustrate in Box 9.1, are relatively clear, although that has not made their employment in conservation biology commonplace yet. When it comes to other treatments such as reduction in harvests, an experimental approach is possible and desirable. Even if the situation looks complex there are techniques, and some of these for large-scale experiments are to be found in Walters and Holling (1990). Hairston (1989) uses examples from the published literature on experiments in different habitats that have sought to address general ecological questions such as interactions between species. Those experiments can be used to furnish ideas on design and pitfalls to be sidestepped.

9.3 SUPPLEMENTATION OF RESOURCES

A resource may be limited by habitat modification or lost through a chain of extinctions or through a competitive relationship with another, usually introduced species. The treatment is then to supplement the resource and compare resource supplementation with the rate of population increase. Mostly we are dealing with consumable resources that are reduced by the user on either a preemptive or consumptive basis. Preemptive use prevents other animals using it (woodpeckers and nest holes). Consumptive use reduces the resources available to other animals (feeding stations for vultures). Whether the use is preemptive or consumptive helps us select which simple model to predict the effect of resource supplementation on the population's persistence time (Sections 6.1 and 6.2). The predictions can then be incorporated into a hypothesis to describe the treatment's effect.

Habitat changes are not inevitably detrimental to a species, and a relationship between a habitat and a species is not invariably obligatory. Given the accelerating rate of habitat changes in the world, we

Box 9.1.   The logic of an experimental design to test the effect of two factors (predation and supplementation of a resource) on the response variable, fledging success

---

Problem
Rats were inadvertently introduced by loggers, who selectively removed 50% of the cavity-bearing trees on a small island in 1990. There are now only 50 pairs of the imaginary parrot, *Exparrotus insurmountus*, a tree-cavity nester. The fledging rate is only 10%, much lower than the mean of 30% reported from ecologically similar species elsewhere. The population has declined sharply since 1990.

Diagnosis of factors driving the decline
The arrival of rats and the initiation of logging both coincided with the decline of the parrots, which means that the two factors are confounded if we seek the effect of each on fledging success by simple observation.

Restrictions on the experiment
1.   The rate of decline is too rapid to allow testing the two factors in consecutive years, and anyway there is the possibility of interaction between them. We may lose the whole population before the results are in.
2.   The difficulty of removing rats from the experimental treatment blocks and the prevention of leakage of rats between blocks demand that for demonstration of the effect of rat predation, the rats have to be excluded from the parrot nests rather than be exterminated.

Step 1
a.   Determine that rats are predators of eggs or fledglings (various techniques are artificially baited nests, hair tubes, traps, and detection of sign).
b.   Determine that parrots will accept nest boxes with a guard to exclude rats.

Step 2
Design an experiment to test for the effects of excluding rats and supplementing nest sites.

Null hypotheses
1.   There is no significant difference between fledging success in treatments excluding rats and those in which rats are not excluded.
2.   There is no significant difference between fledging success in treatments where nest sites are supplemented and those where nest sites are not supplemented.
3.   There is no interaction between the effects of excluding rats and the supplementation of nest sites.

Design
There are two factors: rat exclusion (factor A) and supplementation of nesting sites (factor B), each with two levels of each factor (0 and 1 for each). As each level of each factor is evaluated in the presence of each level of the other factor, there are

(Continued)

Box 9.1. (Continued)

four treatments. Each treatment has two replicates. The response variable is the successful fledging as measured by the number of fledgling parrots per treatment.

	Factor A (rat exclusion)	Factor B (supplemented nests)
First treatment	Factor A level 0 (rat exclusion)	Factor B level 0 (no nest supplementation)
Second treatment	Factor A level 1 (no rat exclusion)	Factor B level 0 (no nest supplementation)
Third treatment	Factor A level 1 (no rat exclusion)	Factor B level 1 (nest supplementation)
Fourth treatment	Factor A level 0 (rat exclusion)	Factor B level 1 (nest supplementation)

Data analysis
1. Two-way analysis of variance is applied to determine if the main effects have a significant effect.
   a. Determine if there is significant interaction. If there is, graph the response variable against the variable level to identify the type of interaction. Essentially if there is an interaction, the results are ambiguous as you cannot state the effect of either factor without knowing the level of the other variable.
   b. If there is no interaction, determine whether one or the other or both of the main factors have an effect—an unambiguous answer.
2. For a factor returning a significant effect in the absence of interaction, determine its magnitude by comparing the mean number of fledglings from blocks with rat-excluded nests with means from blocks from which rats were not excluded. Then repeat this for the fledging success from blocks with and without supplemented nests. The magnitude of difference which would signal that management would be cost effective should be set before the experiment.

need to find if a modified habitat can still include the resources that a threatened species needs for survival. Again it comes back to using ecology to determine how flexible the species is in its requirements. For example, Kirmse (1994) categorized raptors as to how genetically or behaviorally plastic is their choice of nest sites. Armed with this information, we would be alert to whether the raptors could nest in a modified environment and what sort of coaching they might need to change their behavior.

9.3.1 Management of breeding sites

Management of breeding sites covers, for example, artificial nest sites for birds, breeding ponds for amphibians, and providing appropriate plants for butterflies to lay their eggs. Protecting beaches for turtles to come ashore and lay their eggs is another example: leatherback turtles, *Dermochelys coriacea*, come ashore on the Andaman and Nicobar Islands (Figure 9.1). Feral dogs injure turtles and take eggs, and turtles may return to the ocean without laying if disturbed. A year after nests were protected with a palisade of stakes, more turtles came ashore (Misra 1993), but whether that was the protection or annual variability in the return of breeding females is unclear.

Deforestation often selectively removes large trees with either natural cavities or suitable for cavity excavation. Providing artificial nest sites as a treatment lends itself to experimental verification that shortage of nest sites was a limiting factor and that artificial nests do not have unwanted effects instead of boosting population size such as redistribution or increased predation or parasitism (Snyder 1978). The green-rumped parrotlet, *Forpus passerinus*, adapted from nesting in trees to hollows in fence posts in Venezuela (Figure 4.15). Lines of fence posts and high density of parrotlets (16.3 nests per km^2) were an ideal situation for an experimental design. Nest boxes were used, but were they simply drawing birds in from elsewhere or increasing the number of breeding pairs? As the authors said themselves, "In the absence of carefully constructed experiments, it is difficult to estimate directly a change in the number of birds breeding . . ." (Beissinger and Bucher 1992).

A successful experiment showed that artificial nest cavities (Box 9.2) significantly increased numbers of colonies of red-cockaded wood-peckers (Walters et al 1992). By comparison, another social cavity-nesting species, the Puerto Rican parrot, did not increase despite the provision of artificial sites (Section 4.9), suggesting that something else was amiss.

9.3.2 Management to supplement food

Food shortages can be either absolute (loss of prey or forage plants) or relative (loss of foraging opportunities), and an animal can be short of food (quantity) or a particular nutrient (quality). Provision of entire carcasses at feeding stations for Cape griffons, *Gyps coprotheres*, in South Africa offered quantity but, initially, not quality. The carcasses did not have the broken bones typical of having been fed on by large mammals. When the vultures could not supply their chicks with enough bone splinters to meet their calcium requirements, the chicks developed osteodystrophy, manifested as deformed wings. Once splin-

Box 9.2.   Experimental provision of nesting cavities to increase the number of colonies of the red-cockaded woodpecker

Problem
Agriculture and forestry reduced the forests and the woodpeckers declined.

Diagnosis of factors driving the decline
Step 1.   Assemble evidence for and against possible causes.
Even where there is habitat, woodpeckers rarely either reoccupy abandoned or colonize new territories. Woodpeckers are cooperative breeders, and each group defends a foraging territory and a cluster of nesting trees. It takes the woodpeckers months or even years to excavate nest cavities in live trees, which might be why the woodpecker is slow to colonize.

Step 2.   Test predictions that colonization depends on availability of nest cavities.
a.   The null hypothesis was that there should be no difference in the occupation between sites provided with artificial cavities and control sites on previously unoccupied sites.
b.   The null hypothesis was that there should be no difference in the occupation between sites provided with artificial cavities and control sites on abandoned sites.

The acceptability of the artificial cavities was first tested and woodpeckers adopted them. The experimental design was paired-control treatment plots replicated 20 times. On unoccupied and abandoned areas, 40 were chosen and paired for similar habitats. At each of the 40, five live pines were chosen and, in the 20 experimental plots, equipped with an artificial cavity.

Results
The woodpeckers were significantly more likely to occupy experimental than control sites in previously unoccupied or abandoned areas.

Conclusions
Cavity construction outside existing territories is effective in prompting the woodpeckers to colonize new areas or reoccupy areas with inadequate cavities.

(Modified from Walters JR, Copeyon CK, Carter JH III. Test of ecological basis of cooperative breeding in red-cockaded woodpeckers. Auk 1992;109:90–97.)

tered bones were provided, the incidence of osteodystrophy fell, from 17% in the early 1970s to 4% in 1983 (Armstrong 1993).

Habitat management is usually the longer-term answer to food shortage. Researchers decided that supplementing black robins' food was too difficult except briefly during breeding (Section 4.7). Instead, emphasis was on planting trees to encourage leaf litter accumulation where the robins could find insects. Similar logic applied to supplementing food availability for the Seychelles magpie-robin (Section 4.5).

Supplementing food, or any resource, through habitat management requires a vein of logic that factors in the species' ecology. The sequence can be something like this example: Bourn and Thomas (1993) first demonstrated that choice egg-laying sites were a limiting factor for the brown argus butterfly, *Arica agestis*, a species scarce in England. Then they found out why females select particular plants. Finally they determined site management necessary to grow those preferred plants which have fresher, greener leaves, as measured by higher organic nitrogen content and a thicker mesophyll.

9.3.3 Management to supplement shelter

Shelters run the gamut from a refuge to escape predators, a place to sleep, estivate, or hibernate, to a place with its own, and preferable, climate. Sometimes we have to rely on ensuring another species is there to provide the shelter; mention has been made of the prairie dog as the architect and engineer of the black-footed ferret's shelter (Section 4.6). More often habitat modification has reduced or altered a suite of plants or some attribute such as leaf litter that offered a suitable microclimate. Some shelters are conspicious and tend to be featured in both diagnosis and treatment. Timber removal depletes the availability of tree hollows, which leaves birds and arboreal mammals homeless. Research must first determine exactly what the species need for shelters and then how those needs are affected by the availability and turnover time of trees with hollows. The next step is to see if forestry practices can accommodate those needs. We have taken Leadbeater's possum, *Gymnobelideus leadbeateri*, as an example (Box 9.3), and in this case the treatment is anticipatory. We include it to illustrate an approach to determining a species' shelter requirements.

Many species of bat roost in caves where they may be vulnerable to people disturbing them. However, blocking cave entrances has, in some cases, proved disastrous for a cave's attributes as a shelter. A population of Indiana bats, *Myotis sodalis*, plummeted after the entrance to their roosting cave was modified (Box 9.4). In this case the diagnosis was sound (disturbance of bats in their roosts), but the treatment did not take into account the microclimatic needs of roosting bats. The problem is not insignificant: on public lands in the western United

Box 9.3. The type of forest needed to provide trees with hollows for Leadbeater's possum

Leadbeater's possum is an attractive Australian marsupial rediscovered in mountain ash, *Eucalyptus*, forests in 1961. The possum shelters in tree cavities and emerges at dusk to search for tree crickets and gum from *Acacia* trees. The crickets shelter under flaking bark, which is only plentiful on older or dead trees. The availability of hollows limits the number of possums.

Habitat requirements of Leadbeater's possum
The possum's use of trees with hollows in a mountain ash forest (1125 trees on 152 sites) was observed: characteristics of the occupied trees and their sites were used to predict occupancy rates. An important point that emerged was that it was not just the presence of trees with hollows but also the amount of decorticating bark on the neighboring trees and the availability of *Acacia* trees that determined occupancy rates.

Availability of habitat requirements
Trees do not develop hollows until they are at least 120 years old, but another 20 years of age makes them more attractive to the possums. Hollow trees eventually collapse at an annual rate of 4%. The form of the tree determines its tendency to collapse.

Implications for forestry practices
The cutting rotation is 80–120 years, which is less than the age when trees start to develop hollows, so the clear-felling will lead to a loss of shelter trees for Leadbeater's possum. Current forestry practices aim to leave 10 "habitat" trees/ha, but research predicts that this will be inadequate given the rate of loss and distribution of the hollow trees for Leadbeater's possum. However, the predictive models from the detailed understanding of the habitat requirements, especially the site characteristics, allow a specific set of recommendations to be made which, if implemented, would prevent a collapse of Leadbeater's possum populations.

(Modified from Lindenmayer DB, Cunningham RB, Tanton MT, Smith AP. The conservation of arboreal marsupials in the montane ash forests of the Central Highlands of Victoria, south-east Australia. I, II. Biol. Conserv. 1990a,b;54:111–131, 133–145.)

Box 9.4.   The recovery of an Indiana bat population after modification of its roosting cave

Problem
The Indiana bat, *Myotis sodalis*, declined by 28% between 1954–1966 and 1980–1981 and then by 36% during the next decade. The species is vulnerable as half the population now winters in only two caves. The species continued to decline despite legal protection from hunting and disturbance. Measures such as blocking entrances to caves to protect roosting and hibernating bats were undertaken.

Diagnosis of inappropriate treatment
When the two entrances to a cave in Kentucky (Hundred Dome Cave) were blocked in the 1960s to prevent people from disturbing the bats, the bat population fell from 100,000 to 50 by 1991. But, as Richter et al (1993) state, "the inference [of disturbance as causing the decline] was based on historical events and qualitative observations rather than designed experiments and quantitative data."
    Richter et al (1993) compared fat metabolism of hibernating bats wintering in two caves in 1976–1977. The airflow in one cave was restricted by a blocked entrance. The cave with restricted airflow was warmer, and hibernating bats in that cave were more active. The greater activity caused them to lose 42% of their body mass through fat metabolism. The bat's body mass fell below the level critical for survival. No bodies were found possibly because the starving bats moved deeper into the cave. After the barrier to the cave entrance was removed in 1977, the number of bats increased from 1,000 to 3,000 to 13,000 by 1991.

(Modified from Richter AR, Humphrey SR, Cope JB, Brack V Jr. Modified cave entrances: thermal effect on body mass and resulting decline of endangered Indiana bats (*Myotis sodalis*). Conserv. Biol. 1993;7:407–415.)

States, there are 100,000 abandoned mines, and, while not all harbor bats and need bat-friendly closure, all six of the United States' endangered bats roost in mines (Dukto 1994).

9.4   CONTROL OF DIRECT POPULATION LOSSES

Harvesting, predation, disease, and environmental contamination all directly remove individuals from a population and can do so to the extent of driving a decline. The treatment seems obvious—reduce or remove the cause—but, as our examples show, that is not always as easy as it sounds.

9.4.1 Legislation and education

Legislation to reduce harvesting is the standard treatment for a decline caused by overharvesting. Species ecology, harvesting method and type (legal or illegal; commercial or subsistence), and economic incentives to harvest have to be factored into an effective treatment. Once a decision is made to reduce the offtake, the next step is to decide whether effort or the offtake itself is to be regulated. The safest approach, especially if monitoring is infrequent relative to environmental variation, is to regulate effort through control of seasons, numbers of hunters, nets, or hooks (Caughley 1987).

Prohibiting harvesting has mixed success. When the product is valuable and its value increases with rarity, more than prohibition is needed to uncouple the positive feedback loop between value and demand (Section 11.5). The steps are to lower its value and desirability, and to ensure that the penalties for possession outweigh benefits. To crimp availability, individual nest sites or even individual animals have been guarded—the red kite (Section 4.13), Spix's macaw, and black rhinos are examples. Substitute products may reduce desirability, such as water buffalo, *Bubalus bubalus*, horn instead of rhino horn (Leader-Williams 1992).

The high value of rhino horns meant that legislation to prohibit hunting and antipoaching patrols were less than successful. An expensive and as yet unproven deterrent is capturing and releasing rhinos to remove their horns and the incentive to poach them. Between 1991 and 1993, 212 black rhinos and 120 white rhinos, *Ceratotherium simum*, were dehorned in Zimbabwe (Milliken and du Toit 1994). Dehorned rhinos were still poached partly because of problems in maintaining effective antipoaching. But horn regrowth is also sufficiently rapid that unless dehorning is annual, the regrowth is attractive to poachers (Berger 1993). Apart from doubts about the effectiveness of dehorning in removing the incentive to poach, the ecological consequences are uncertain. Rhino cows use their horns to deter predators and calf survival of dehorned rhinos may be reduced. A comparison of the cow's horn size and whether her calf had been maimed by predators was insignificant, but the small sample size left the probability of a Type 2 error high (Berger 1994). Even if dehorning cows increased calf mortality, the significance of that compared to other reasons why calves die is unknown and, without that, evaluating dehorning is incomplete.

More common than outright prohibition is regulation to harvest sustainably, so that neither harvesters nor the species are unduly penalized. Much is written about sustained-yield harvesting. Theoretically it is quite simple (Caughley 1977), but in practice it has proven elusive (for example, Shaw 1991). At least as an initial step, quotas can be set using whatever information is at hand and then their effect monitored and the quotas adjusted accordingly. This is the approach taken by Surinam (Figure 4.15) to hold the export of its parrots to sustainable levels, compared to other countries that set their export quotas on the basis of average export levels (Thomsen and Brautigam 1991).

Regulation of international trade has to be supported by national legislation (Sections 12.1 and 12.2). It has taken a combination of international and national legislation to reduce commercial trade of fruit bats, *Pteropus mariannus*, in the Pacific group of Western Caroline Islands (Figure 1.1) and bat numbers recovered (Box 9.5, Figure 12.1).

A recovery treatment that limits hunting without balancing education and effective enforcement is usually doomed to failure. Public support and understanding pays dividends when it anticipates and accompanies legislation. In the eastern Caribbean, the recovery of endangered parrots is beginning once people understood what was happening to the parrots (Box 9.6). Part of the parrot's problem was capture as pets which there, as elsewhere, is big business. In 1986, the value of live trapping of parrots from South America was $6.4 (U.S.) million gross income to the trappers and $308.2 (U.S.) million to the retailers (Thomsen and Brautigam 1991). Incentives to illegally trade are clear from the amount of money involved, but solutions are not simple.

Local involvement was instrumental in reducing the illegal take of wildlife in the Luangwa Valley, Zambia (Figure 4.1). Interviews of local people made plain that subsistence poaching supplied much-needed meat and cash (Balakrishnan and Ndhlovu 1992). After local people became involved in enforcement, Lewis et al (1990) alluded to a tenfold decrease in the density of poached carcasses of black rhinos and African elephants and an increase in apprehended poachers. On the other hand, Leader-Williams and Milner-Gulland (1993), also investigating elephants and rhinos killed in the Luangwa Valley, suggested that such schemes employing local people will not reliably deter poaching organized by groups based outside the area. Those authors distinguished between subsistence poaching and organized poaching, when discussing the balance between detection and penalties to reduce the illegal take.

The need for public support becomes even greater, and the effect of protective legislation even less, when the declining species is a threat to people. Sankhala (1977) suggested that half the losses of tigers, *Panthera tigris*, in India was attributable to a deep-seated fear well founded on the fact that tigers kill people (Section 10.3). Misconceptions about and fear of the banded iguana, *Brachylophus fasciatus*, in Fiji were why people killed them despite their protected status. Familiarizing school children with iguanas helped to counter the misapprehensions (Gibbons and Watkins 1982).

Reducing offtake is sometimes possible by a shift from wild- to

Box 9.5. Legislation, commercial trade, and the recovery of fruit bats on Yap State, Western Caroline Islands

Problem
The decline of fruit bats on Guam led to their import from other Pacific islands, including Yap State (Figure 12.1) in the 1970s. Traditionally, bats had been harvested for local use by catching them with nets thrown from platforms erected near feeding trees. Realization of the cash return from a market for the bats and the opening up of the country by roads increased the harvest. The export of bats increased from a few in 1974 to 865 in 1975. By the mid-1970s, bats were being shot in their roosts and were becoming scarce. In 1981, only 1,000 to 2,000 bats were left.

Legislative treatment
Legislation to impose a closed season was passed in 1975, but was ineffective. More bats were exported during the closed season than the open season. Some local hunters gave up the hunt by 1979, but the trade driven by foreign businessmen continued. The Yap legislature then passed a regulation to protect bats year-round and to prevent the private use of guns. The trade to Guam ceased in 1982, but a trickle of bats continued to reach the Northern Marianas. The exception was a 1-month season in 1988 which allowed export of 1180 bats in 1989. Initially, the legislation was effective as the bats increased from 1,000 to 2,000 in 1981 to approximately 5,000 in 1986.

(Modified from Wiles GJ. Recent trends in the fruit bat trade on Guam. In: Wilson DE, Graham GL, eds. Pacific Island flying foxes: proceedings of an international conservation conference. Biological Report No. 90. Washington DC: U.S. Department of the Interior, 1992:53–67; Falanruw MC, Manmaw CJ. Protection of flying foxes on Yap Islands. In: Wilson DE, Graham GL. Pacific Island flying foxes: proceedings of an international conservation conference. Biological Report No. 90. Washington DC: U.S. Department of the Interior, 1992:150–154.)

captive-raised animals to supply the market. On one hand, captive raising may signal a reprieve for the wild populations, and this apparently was the case for some crocodiles (Luxmore 1992). On the other hand, apart from the regulatory problems of identifying products from captive- versus wild-raised, supplying a market may encourage it and continue the pressure on the free-ranging population. Captive raising did not decrease the pressure on musk deer, *Moschus moschiferus moschiferus* (Green 1987). The wild product may be perceived as superior, and economic logic (Swanson 1992) predicts that captive breeding

---

Box 9.6. Education and legislation as a treatment for recovery of parrot species threatened by human take, Lesser Antilles, Caribbean

---

Problem
Only four of 14 parrot species on the Lesser Antilles are extant, and they are endangered through habitat loss, hunting, and hurricanes.

Diagnosis of factors driving the decline
The diagnoses were based on observation and the coincidence of the declines with forest clearing, hurricanes, and known take for food or pets.

Treatment
Acts in 1901 and 1914 and the enactment of forest reserves had failed to reverse the declines of parrots. In the early 1980s the predicament of the parrots was advertised through question-naires, posters, and school visits. Subsequent surveys revealed more knowledge and less illegal hunting and trapping. Parrots increased in number, although we found no data measuring causes of the increases. In response to public interest the governments of St. Vincent and Grenadines set aside reserves for parrots, boosted fines for illegal possession, and ratified CITES.

The four species have increased since the 1980s after decades of decline:

St. Lucia Amazon *A. versicolor*	150 ± 25 (1980)	300–350 (1990)
St. Vincent Amazon *A. guildingii*	421 ± 52 (1982)	440–500 (1988)
Red-necked Amazon *A. arausiaca* (Dominica, West Indies)	150–225 (1980)	500 (1992)
Imperial Amazon *A. imperialis* (Dominica, West Indies)	40–60 (1981)	80 (1990)

(Modified from Butler PJ. Parrots, pressures, people, and pride. In: Beissinger ER, Snyder NFR. New world parrots in crisis. Solutions from conservation biology. Washington DC: Smithsonian Institution, 1992:25–46; Collar NJ, et al. Threatened birds of the Americas. London: ICBP, Washington DC: Smithsonian Institution Press, 1992.)

may be a disincentive for the conservation of wild stock and especially their habitat. Some idea of the controversy attached to captive raising of endangered species such as crocodiles and the difficulties can be gauged from the tangles that the Convention on International Trade in Endangered Species of Wild Fauna and Flora (CITES) has become ensnared in on this topic (Fitzgerald 1989, Luxmoore 1992). A case-by-case approach is likely to be more helpful than inflexible generalities.

Success of reserves where hunting is prohibited depends on enforcement and attitudes of people living outside the reserves (Section 10.2). Demonstrating its success as a treatment to reverse a decline caused by harvesting is difficult if environmental variation is not factored in, as this example shows. The centuries-old harvest of seabird eggs in the Caribbean declined in the 1970s to about a tenth of that in the 1920s, which was believed to reflect a decline in breeding birds and not harvesting effort (Haynes 1987). A quota system failed and an island sanctuary took its place. After 5 years, numbers of brown noddy, *Anous stolidus*, chicks and sooty terns, *Sterna fuscata*, more than doubled in the sanctuary (Haynes 1987). But any inference that this increase resulted from no harvesting is weak. Environmental variation could have caused the population increase, and a balanced design using replicated sites with and without harvesting would have coped with that.

9.4.2   *Eradication and control of predators*

A voluminous literature exists on eradication and control of introduced species, and Hone (1994) is a thoughtful summary. The effects of introduced predators are relatively easily diagnosed, and control or eradication is a common treatment. The usual measure of successful treatment is an increase in the population of prey and not numbers of predators killed or kilograms of poison. Eradication of cats from Marion Island (Figure 4.9, Box 9.7) was measured by an increase in great-winged petrel, *Petrodroma macroptera*, breeding—the stated objective for killing the cats. We are usually, but not necessarily, considering introduced predators. Changes in habitats may lead to natural predators adding to the troubles of an endangered species (Paton 1994, Nour et al 1993).

A proviso for predator removal is that merely because a species is judged undesirable is no reason to allow lapses in humane treatment. With the development of quick-kill traps, the use of leg-hold traps or any technique that imposes stress or pain is questionable. Having said that, we realize its difficulties in practice as what is accepted as suffering is a cultural value and not necessarily one easily transferred between cultures. Whatever the method, clearly precautions may be needed to ensure that nontarget animals will not be at risk. Merton's (1987) description of the precautions taken to protect endemic reptiles during poison baiting to eradicate rabbits from Round Island, Mauritius, serves as an example of the checks required. And while we

---

Box 9.7.   Recovery of breeding in petrels after eradication of feral cats on Marion Island, southern Indian Ocean

---

Problem
Marion Island (290 km^2) has 12 petrel species. Five cats were introduced in 1948–1949 to control introduced house mice, *Mus domesticus*, and increased at 23% per year to reach 3,045 cats in 1977. The cats preyed on eight species of burrowing petrels (adults, chicks, and eggs). As the petrel populations shrank, the cats shifted their attention more toward house mice. The great-winged petrel, *Pterodroma macroptera*, was especially vulnerable to the cats because it breeds in winter, has a long breeding season, and uses larger burrows than other petrels.

Diagnosis of the cause of the decline
In 1975 cats killed around 48,000 great-winged petrels, which became relatively rare compared to numbers on the neighboring (and cat-free) Prince Edward Island.

Prince Edward Island	No cats	33% nests ($n = 30$) with chicks in 1979
Marion Island	Cats present	1% nests ($n = 109$) with chicks in 1979

An introduced disease reduced cats, but then the survivors built up the numbers. Outside a catproof exclosure, no petrel nests had chicks, compared to 50% of nests with chicks inside.

Recovery treatment
1977—Feline panleukopenia (FLP) introduced
1982—Cats reduced to 620; subsequently FLP antibodies decreased in the cats
1986–1990—952 cats removed by shooting and trapping
1990—Petrel recruitment increased from 100% chick mortality in 1979–1984 to 0% chick mortality in 1990
1991—Cat eradication believed complete
1992—Reports of house mice increasing

(Modified from van Aarde RJ. The diet and feeding behaviour of feral cats, *Felis catus* at Marion Island. S. Afr. J. Wildlife Res. 1980;10;123–128; Bloomer JP, Bester MN. Control of feral cats on sub-Antarctic Marion Island, Indian Ocean. Biol. Conserv. 1992;60:211–219; Cooper J, Fourie A. Improved breeding success of great-winged petrels *Pterodroma macroptera* following control of feral cats *Felis catus* at subantarctic Marion Island. Bird Conserv. Int. 1991;1:171–175; Howell PG. An evaluation of the biological control of the feral eat *Felis catus* (Linnaeus 1758). Acta Zool. Fenn. 1984; 172:111–113; von Rensbury PJJ, Skinner JD, van Aarde RD. Effects of feline panleucopaenia on the population characteristics of feral cats on Marion Island. J. Appl. Ecol. 1987;24:63–73.)

---

might be squeamish about admitting this, generous allowance has to be made for human error.

Unwanted consequences of predator control stalk the unwary. Control of one predator is risky without considering that it may release other predators to assume the limiting role for a species in trouble. This is not justification for controlling all predators at the onset, but it does argue for systematic follow-up to determine if another predator becomes a factor. Increases in feral cats apparently followed fox control implemented to reduce predation on numbats, and the cats maintained predation pressure on the numbats (Section 4.14). In western Queensland (Figure 1.1), feral cats erupted in the vicinity of a greater bilby colony, *Macrotis lagotis*—an endangered marsupial and vulnerable to cat predation. Pettigrew (1993) postulated this sequence: 3 years of rains stimulated a flush of grasses; long-haired rats, *Rattus villosissimus*, escalated, which supported an influx of cats. This coincided with control of dingoes, which had been holding the cats in check. Only coincidence links the events, but even so it does convey the ecological complexities that are the reality of any treatment, not just predator control.

Eradication is difficult because it requires a huge commitment to maintain the effort as the numbers of predators decrease. Reinfestation is frequent. In a small area (current attempts to remove rats have capped at about 300 km²) and where subsequent reintroduction of the predator is unlikely or can be controlled, eradication is worth attempting (Box 9.7). Fencing can partition an area into smaller units to allow more efficient removal. Other pointers to success come from eradicating muskrats, *Ondatra zibethicus*, over a large area in southern England and Scotland. The muskrats had escaped from fur farms and spread over some 1,800 km² in only 3 years, and then it took 5 years of trapping to remove them (Sheail 1988). The campaign had included research on muskrat ecology, and the biologists heeded the experiences of other European eradication schemes. The third factor in favor of success was public and political support.

New Zealand wildlife managers have, by 1993, eradicated rats from about 30 islands. Once the rats are removed, the islands become reserves for repatriation of insects, reptiles, and birds. Use of potent and palatable anticoagulant poisons in the last 10 years has increased success in clearing rats from larger islands, but it is not a task to undertake lightly. We mention the level of effort that Taylor and Thomas (1993) expended to eradicate Norway rats from a forested and rugged island 170 ha in area as an illustration: 743 bait stations maintained for 22 days with subsequent monitoring for reinfestation every few months for 2 years. Eight people racked up 564 operator days. The reason for all the effort was to create an reserve: endangered weevils, *Hadramphus stilbocarpia* and *Anaotus fairburni*, and an endemic wattle-bird, the saddleback, *Philesternus carunculatus carunculatus*, were repatriated and Fiordland skinks made their own way to the island.

An alternative to eradication of a predator is control: simply holding down their numbers to an acceptable low level by killing them, removing them from a buffer area, or excluding them by a barrier. Fencing has been tried on a grand scale—Australia's 5,309-km dingo fence is to protect sheep (Holden 1993), an irony as stock grazing is implicated in the decline of arid-land marsupials and rodents. On a smaller scale, exclusion of predators by barriers (fences, moats) can be effective. Lokemoen and Woodward (1993) found that the fledging success of ducks was 54% on fenced (electrified mesh) peninsulas after the predators had been removed, compared to 17% success on the control (unfenced) peninsulas.

Control may be synchronized with a vulnerable time for the species in trouble. For example, the baiting and trapping of black rats around colonies of dark-rumped petrels is only during the breeding season (Section 4.12). A factor in deciding the approach to control is alternative prey. If the endangered species is seasonal in its exposure to the predator, and the predator has few alternative prey, then the time to control is when the predator is food stressed.

In New Zealand, poison baiting has been tested to create a predator-free buffer zone, to prevent black rats from boarding ships which then inadvertently carry them to rat-free islands (Hickson et al 1986). The experimental design did not examine what influences the rate at which rats board ships. The experiment was short-term (16 days baiting and 4 days postbaiting monitoring). Black rats rapidly move into vacated territories, however, and their numbers can recover within 3 months. If even a few survivors are enough to pose a threat by boarding vessels, then experiments are needed to test bait scheduling and intensity to hold the rats below that level or whether eradication is necessary from the buffer zone.

Eradication or exclusion is not the only approach to reducing predation. Conditioned taste aversion is a behavior learned in a single trial. When an animal feeds on a food with a distinct taste and is sick a few hours later, it develops an aversion to that taste. Nicolaus and Nellis (1987) used laboratory and field experiments to teach mongooses to avoid eggs. Mongooses are difficult to exclude from an area, such as a turtle-egg-laying beach, by fencing. The effect of trapping is temporary, as other mongooses quickly replace the trapped ones. After eating hens' eggs poisoned with carbachol, mongooses subsequently avoided eggs, compared to the control site where they readily ate them. The aversion only lasted 14 days, but that could possibly be modified by using different substances as the identifying flavors. Similar laboratory, then field, experiments revealed that carbachol also successfully prevented crows, *Corvus brachyrhynchus*, from eating eggs (Nicolaus et al 1989). Conditioned aversion has potential, therefore, for reducing egg predation by mammals and birds. It could serve as an effective treatment in urban areas where pet cats or dogs are an actual or perceived problem, and control methods that kill fuel public outcry. It has the disadvantage that the predator is not removed, which could mean that it then shifts to another food source, which could in turn create a problem.

9.4.3 *Control of parasites*

Brood parasitism is implicated in the decline of songbirds in the Americas. Controlling brown-headed cowbirds, *Molothrus ater*, and Kirtland's warblers is revealing about diagnosis and treatment (Zimmerman 1975). The warbler's numbers dropped from 502 territorial males in 1961 to 167 in 1974. Cowbird parasitism, known since 1924, had increased to 70% of all warbler broods by the 1960s. In 1971, trapping cowbirds began, whereupon brood parasitism dropped to 5%. But the warblers remained capped at about 200 territorial males during the 1970s and 1980s (Mayfield 1978, Trauger and Bocetti 1993) and first-year survival remained less than 20%. The reasons were unknown, and cowbird control was questioned (Wiley et al 1991). In 1991, the number increased to 347 pairs after more cowbirds were trapped in 1989 (Collar et al 1992), but it was not stated whether trapping increased or there were more cowbirds. At the same time, however, considerable efforts had been expended to restore the warbler's habitat (Trauger and Bocetti 1993). The cowbird control is confounded by the second treatment, habitat restoration.

A different species of cowbird is spreading through the Caribbean and causing problems. The shiny cowbird is endemic in South America and the Caribbean Islands of Trinidad and Tobago. In the last 100 years the cowbird has spread along the island chains, reaching Puerto Rico in the late 1950s and the Dominican Republic in the 1970s (Figure 8.7). The species already patchily distributed and vulnerable to the brood parasitism included the yellow-hooded blackbird, *Agelaius xanthomus*, and parasitism was reasoned to be the factor driving the blackbird's decline. Accordingly, shiny cowbird control began on Puerto Rico in 1980 (Box 9.8). Treatment success (increase in population size) is unrecorded despite the significant improvement in fledging success in 1980.

9.4.4 *Control of competitors*

Competition is a notoriously intractable subject, and here we use the term assuming a competitive relationship when a herbivore or omnivore is introduced and a habitat resource is reduced, such as through feeding or trampling. Regardless of the control technique, knowing the competitor's ecology and behavior is essential. A simplistic example is using an animal's social behavior. On San Clemente Island (148 km²) 29,000 feral goats were killed between 1972 and 1989 to protect the habitat for seven endemics. The last of the now-scattered and wary herds was quickly located by radio-collared "Judas" goats (Keegan et al 1994).

Rabbits had been released on Round Island (151 ha), Mauritius, be-

---

Box 9.8. Control of shiny cowbirds to reduce brood parasitism in the yellow-shouldered blackbird, Puerto Rico

---

Problem

The yellow-shouldered blackbird was abundant and widespread on Puerto Rico but is now restricted to Isla Mona and two mangrove forests. Its decline, starting in the 1940s, coincided with arrival of the shiny cowbird. Shiny cowbirds have not reached Isla Mona, and the blackbirds are stable in number (400–900). The population on the island of Costa Rica was 1800 in 1975–1976 and 270 in 1985 and 1986.

Diagnosis of factors driving the decline

First, the decline in the blackbirds coincided with the arrival of the shiny cowbird. Second, field observations found that, between 1973 and 1982, 95% of the blackbird nests were parasitized. Third (almost), there were blackbirds but no cowbirds on Isla Mona, but this population does not seem to have been used as a control. Fourth, the treatment (removal of the cowbirds) significantly increased the proportion of nests fledging chicks.

Treatment

In 1980, cowbirds were caught in decoy traps at Boquerón Commonwealth Forest. Their fate was according to the following experimental design.

Treatment	Proportion of blackbird nests parasitized	Proportion of blackbird nests fledging 1 + chick
1. Control (banded and released)	11/12 (92%)	5/12 (42%)
2. Cowbirds killed	5/11 (45%)	9/11 (82%)
3. Female cowbirds killed	3/10 (30%)	8/10 (80%)
4. Male cowbirds killed	6/9 (67%)	5/9 (56%)

(Modified from Wiley JW, Post W, Cruz A. Conservation of the yellow-shouldered blackbird *Agelaius xanthomus*, an endangered West Indian species. Biol. Conserv. 1991;55:119–138; Heisterberg JF, Nunez-Garcia F. Controlling shiny cowbirds in Puerto Rico. In: Crabb AC, Marsh RE, eds. Proceedings of the Thirteenth Vertebrate Pest Conference, University at California, Davis, 1988:295–300; Collar NJ, et al. Threatened birds of the Americas. London: ICBP, Washington DC: Smithsonian Institution Press, 1992.)

---

fore 1810 and goats between 1844 and 1865, which was the same time as the endemic Mauritian tortoises, *Geochelone* spp., became extinct, apparently from grazing competition by introduced herbivores. Rabbit removal in 1986 was the most ambitious attempt so far to clear introduced herbivores from a large island (Merton 1987). The vegetation swiftly regrew once the rabbits were gone. Increases in six species of endemic reptiles followed not so much from the loss of a competitor but probably the plant regrowth was cover for them and their insect prey (North et al 1994).

Flux (1993) reviews attempts to control rabbits. His conclusion was unexpected: the introduction of hares (*Lepus timidus* and *L. europaeus*) was significantly more effective than cats or myxomatosis, even though it was unclear how the apparently competitive relationship worked. Poisoning, gassing, and trapping have cleared rabbits from islands, but failure to record unsuccessful attempts precludes analysis of relative efficiency (Flux 1993). Introducing hares to replace rabbits, however, might just change the problem rather than treat it.

Introduced animals, be they carnivores or herbivores, often erupt and decline, and this is evident from introductions to the sub-Antarctic islands (Leader-Williams 1988). Those cycles suggest that, for effectiveness, control should be initiated as soon after an introduction as possible or during the decline phase. Most introductions are, however, not recent, and then the costs to the local fauna and flora have to be weighed against letting the introduced animal erupt and start to decline before control is initiated, and that comes down to knowing something about their ecology. Leader-Williams et al (1989) argued that the introduced reindeer on South Georgia Islands (Figure 4.15) indirectly helped endemic white-chinned petrels, *Procellaria aequinoctialis*, whose eggs and chicks were killed by Norway rats. The reindeer's overgrazing of tussock grass, *Parodiochloa flabellata*, reduced the food and cover for rats, which were then exposed to attack by skuas, *Catharacta*.

9.5 SUPPLEMENTATION OF POPULATIONS

We can add individuals to a population and we can increase the number of populations. We might want to boost a population to leave behind the dangers of being small, or we may be establishing a species back in the wild. Individuals from the wild, or captive individuals raised on site (in situ) or at a more distant location, often a zoo (ex situ), may be used. Supplementing population size is a well-mined topic for reviews, and coverage by Olney et al (1994) and Gipps (1991) is excellent.

Whether and how to supplement a population depends on its size and a realistic appraisal of the risks it faces from demographic and environmental stochasticity and reduced genetic variance (Chapter 6). The population size will determine which risks pose the greater threat, and that begs the question of, at which size does one risk take precedence over the others? The answers depend on the individual species and its environment, and simulations of its dynamics will, at least, lead to predictions of persistence that can be tested (Chapter 7).

Very small populations face greater risks from demographic stochasticity (Section 6.1). Augmentation of numbers to boost population size is the desired treatment. For somewhat larger "small" populations, environmental stochasticity is a threat (Section 6.2). The emphasis is usually on adding more breeding animals to increase the mean rate of increase, but decreasing its environmentally induced variance is an option. For example, food and shelter removed by a hurricane can be supplemented.

Supplementing population size will stem loss of genetic variation. The effective population size is partly determined by the less common sex, but an even sex ratio reduces the potential rate of increase (Section 6.1.2). With an even sex ratio in the wild, where choice of mates cannot be enforced—not that it always works well in a captive setting—not all pairs may be compatible. In species with flexibility in the number of offspring, management steps to maximize offspring survival will maximize intergenerational exchange of heterogeneity. We stated in Section 6.3.3 that heterogeneity is determined by the size of the progeny generation.

We can spread the risk of extinction from environmental calamities by translocating individuals to start a another population, and this has become known as "metapopulation management" (Craig 1994). It is about reducing the vulnerability of a single population from catastrophes and increasing genetic variation—that of a metapopulation sums to more than any one constituent population (Section 5.6). But common sense is warranted as those advantages can be offset by two disadvantages. The genetic advantage is a long-term one as it takes time for the constituent populations to diverge from their parent population. Initially the small populations themselves are vulnerable, and perhaps more significantly their establishment will have reduced the source population. This was one of the successful arguments against estab-lishing a third captive population of Puerto Rican parrots (Section 4.9).

The decision to use wild stock depends on their availability. The single population may be large enough itself: the endangered Bahamian hutia, *Geocapromys ingrahami*, numbered 12,000 when it was rediscovered on East Plana Cay (Figure 4.8). To start a second population, in 1969, 11 hutias were moved to an island in the Exuma National Sea and Land Park (Figure 4.10), and by 1985 they had increased to an estimated 1,200 (Campbell et al 1991). On small area where even a small population may quickly become capped, "surplus" individuals can be repatriated to start another population. This happened with the Seychelles warbler, which was successfully repatriated to two other islands, and that was an experimental test of food as limiting factor on the source population (Komdeur et al 1991, Komdeur 1990, 1992, 1994).

9.5.1 Captive breeding and raising

Stanley Price (1991) noted the lack of guidelines for when to take animals into captivity, compared to all the criteria developed for releases. Removal into captivity should occur before the population is so small that capture of individuals will accelerate a decline (except the rescue of a doomed population facing an emergency). Once a population is down to a handful of individuals, removal of some for captive breeding can be difficult, as was the case for the echo parakeet, *Psittacula echo*. This small green parakeet feeds in the canopy of native forests on Mauritius. As the forests were cleared, the parakeets disappeared, and by the early 1970s only 30–60 remained. In 1974–1975, four birds were caught but died. In 1983 when the population was 7–11 birds, two attempts to catch them failed (Jones 1987).

Information on raising and releasing a captive species is difficult to gather from a species reduced to a few individuals. The case of the Puerto Rican toad (Section 4.18; Johnson 1994) makes the point that field studies may underpin successful captive breeding and release. And when a species is on the brink of extinction is not the time to be training people in the necessary techniques and skills to raise it in captivity.

Low numbers are not an automatic criterion for taking animals into a captive-breeding program. The black robin was not taken into captivity because of the difficulty of raising small insectivorous birds (Section 4.7). The decision depends on a realistic appraisal of potential success, whether the agent of decline is known and treatable, and what the effect of removals on the source population will be. Debate over conservation of the two smaller Asian rhino species centered on whether captive breeding was advisable or even possible (Santiapillai and Mackinnon 1993, Ramono et al 1993, Leader-Williams 1993). Both the Javan, *Rhinoceros sondaicus*, and Sumatran rhinos, *Dinoceros sumatrensis*, occur as small populations in reserves facing the twin perils of habitat loss and poaching (Box 10.5). The immediate cause of decline is known and treatable—increased protection and its costs were less than those to establish and maintain zoo populations. The captive-breeding record was not encouraging. A third of the 27 Sumatran rhinos captured since 1985 died and none bred (Santiapillai and MacKinnon 1993). Removal of animals from the wild would reduce already small populations.

Captivity is a useful short-term rescue treatment. If a wild population suddenly faces catastrophe, rescue for captive breeding may be the only option, as was assumed to be necessary for the black-footed ferret (Section 4.6). Captive breeding can be used to buy time while the agents of a decline are being dealt with. The Guam rail, *Rallus owstoni*, and the Guam Micronesian kingfisher, *Halcyon cinnamomina cinnamomina*, were two of five native forest birds threatened by the introduced brown tree snake (Box 8.1). In the early 1980s, the last known individuals of the rail and kingfishers were caught to start captive-breeding populations. Plans to take two other forest birds had to be abandoned as it was too late; they were extinct (U.S. Fish and Wildlife 1990). Guam is not yet free of snakes, but the kingfisher and rail have been translocated to Rota, a neighboring island which is snake-free.

The other side of the coin is when emphasis on captive breeding is at the expense of dealing with the problems of wild populations and their habitat. The Bali starling, *Leucopsar rothschilde,* is a beautiful bird, sought by the live-bird trade. Habitat loss and trapping had reduced numbers to 13 in the wild by 1990, and those birds only occur in the Bali Barat National Park (Figure 9.1). Some 700 birds were in captivity, and establishment of a breeding center in Bali is under way. Increased protection against poaching did more to increase the wild population than the release of captive birds, and further releases are limited by availability of suitable habitat (van Balen and Gepak 1994). MacKinnon and MacKinnon (1991) were not shy in pointing out that an argument for establishing a captive-breeding population of kouprey, *Bos sauveli,* was the greater ease of raising funds for captive breeding against a greater difficulty of funding reserves in Indochina. The benefits of publicity have to be balanced by a realistic appraisal of whether captivity is necessary.

Figure 9.1. Map of Southeast Asia.

Captive breeding

Captive breeding catches the headlines both in conservation biology and with the public. Nonetheless three problems limit captive breeding as a recovery treatment. First, behavioral and genetic changes adaptive to captivity may leave an animal released in the wild incapable of coping with finding food or a mate or avoiding predators (Jiménez et al 1994, Marshall and Black 1992). Second, the space available in zoos is often limited. Captive breeding of the pink pigeon and the Jamaican hutia faltered when space ran out (Sections 4.10 and 4.17). The proportion of endangered species held in zoos is low, and some species are represented by only a few individuals (Figure 9.2). Space is not equally available and larger mammals dominate, and even within taxa endangered species may be underrepresented. North American zoos with 100 or more reptiles and amphibians have only about 25% of their 3,000 enclosures for snakes listed in the American Association of Zoological Parks and Aquariums Species Survival Plans (Quinn and Quinn 1993).

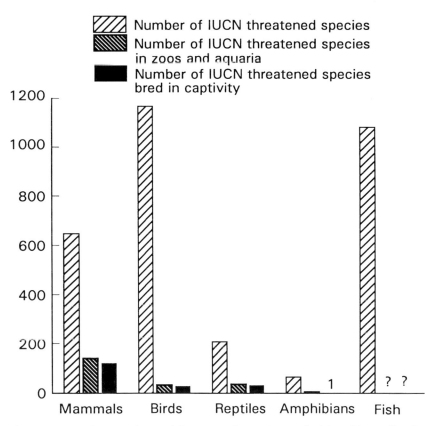

Figure 9.2. The numbers of threatened vertebrates held and breeding in captivity. (Modified from Rahbeck C. Captive breeding—a useful tool in the preservation of biodiversity. Biodiv. Conserv. 1993;2:426–437.)

Captive breeding has some contradictory objectives. As many individuals as possible are needed for releases, but that conflicts with genetic management such as equalizing founding family sizes. Equalizing family sizes to allow each animal to contribute the same number of offspring halves the amount of space needed to maintain the captive population at a given level of genetic variance (Ralls and Meadows 1993). The captive breeding of the California condor ran into this quandary, and the number of individuals took precedence over equal parity in representation of the 14 founders (Wallace and Toone 1992).

The third limitation is expense and time. It takes time to build up the knowledge needed for success, and endangered species may be difficult to breed. This was the case for endangered species of parrots, which, as Derrickson and Snyder (1992) wryly note, are harder to breed than common species used as surrogates. Captive breeding is usually expensive. One of the few cost breakdowns published is for the golden lion tamarin, *Leontopithecus rosalia*, and we give this figure as a guide to the magnitude of costs (Box 9.9).

Successes as measured by the establishment of self-sustaining populations in the wild are few, although it is too soon yet to evaluate success for species released only a few years ago. Mere return of captive-bred individuals is an insufficient measure of success, for example, the Hawaiian geese (Section 4.11). Kleiman et al (1991) discuss criteria used to rate success of the golden lion tamarin's repatriation.

The choice of whether to go for captive breeding on site or at a more distant location depends on the anticipated problems and their solutions (Table 9.1). A deciding factor may be the ease of building up numbers quickly. If this is anticipated, then that argues for a local site. If difficulty is expected, then a more specialized site with greater technical backing may be more expedient. The appropriate diet may be more easily procured locally, but this is not without its own problems. Local mice and lizards fed to captive Mauritius kestrels on the island were contaminated with organochlorines. The chemicals were implicated in the kestrels' failure to breed and in deaths (Jones 1987). On the other hand, feeding entire carcasses to cheetahs at the breeding center in South Africa was better than commercial cat diet fed at North American zoos (Setchell et al 1987). Commercial food contained enough estrogens from soybeans to cause liver disease in 60% of adults and suppress breeding. Only 9–12% of females produced cubs, compared with 60–80% at the South African center.

Captive raising

Two techniques are common under this heading of captive raising. *Cross-fostering* is taking eggs from wild birds and using a related species, either in captivity or the wild, to incubate them. The choice of a foster parent with a comparable life history and the timing of the

Box 9.9. Costs of captive breeding and release of the golden lion tamarin

Captive breeding
1958–1969—180 tamarin caught in the wild
1960–1975—294 tamarin bred in captivity

In 1975, the outlook for the tamarin in captivity was poor: few females bred and survival was poor. Research and cooperation turned things around and numbers increased.

Our estimate of the cost of captive breeding is cobbled from the available information.

1. Estimated annual costs (1989 U.S. dollars) for maintenance of 550 captive tamarin in 100 zoos was $1,675 per animal; 550 is the recommended minimum target, and as the current (1991) captive breeding population is 224, we have arbitrarily halved the cost per animal.
2. Estimating the overall costs for the total captive-breeding program 1972–1984 is not possible, so we have used the midpoint captive population between 1986 (368) and 1991 (224) to estimate costs for the period when the tamarin were released, which started in 1984:

296 tamarin at $800/year = $236,800 (U.S.)

1984–1991 (7 years) = $1,657,600 (U.S.)

Release program
Releases started in 1984: 1984–1989, 71 captive born were released, 27 survived, and 21 of 26 born survived. The total cost of the 1983–1989 field program, including salaries, is $1,083,005 (U.S.).

(Modified from Kleiman DG, Beck BB, Dietz JM, Dietz LA. Costs of reintroduction and criteria for success: accounting and accountability in the golden lion tamarin conservation program. Symp. Zool. Soc. London 1991;62:125–144.)

fostering to avoid imprinting on the foster parent are keys to success. Black robins imprinted on their foster parents, but that was subsequently minimized by using foster parents only to incubate robin eggs (Section 4.7). Cross-fostering of black stilts, *Himantopus novaezelandiae*, to the related pied stilts, *Himantopus himantopus*, was unsuccessful when fostered birds followed their migrating foster parents. The net result was that increased fledging success did not boost the overall number of black stilts (Reed et al 1993).

Table 9.1.   Advantages and disadvantages
of captive breeding on or off site

Factors to be considered	On site	Off site
1. Duration of travel for captured or individuals to be released	Short	Long
2. Acclimatization stress	None	Probable
3. Similar food to wild available	Likely	Unlikely
4. Specialized food available (special enriched formulations, etc.)	Unlikely	Likely
5. Quarantine restrictions	No	Likely
6. Exposure to new diseases	Unlikely	Likely
7. Introduction of diseases to wild population after captive exposure	Unlikely	Likely
8. Availability of veterinary medical assistance and other expert facilities	Unlikely	Likely
9. Expense of construction and staffing	Lower	Higher
10. Potential for education and publicity	Variable	High

The safest approach is an experimental one. In preparation for captive raising of the endangered piping plover, *Charadrius melodus*, Powell and Cuthbert (1993) used two surrogates, the killdeer, *Charadrius vociferus*, and the spotted sandpiper, *Actitis macularia*. Captive-raised and foster-parent–raised chicks responded to predators and to their foster parents and showed no evidence of behavioral problems such as imprinting.

Head-starting has become the term for gathering either turtle eggs or hatchlings and raising them in captivity. It originated in the observation that most turtle hatchlings die and this natural feature of turtle population dynamics is often amplified by human or introduced predators taking many clutches of eggs. Head-starting became a common technique with lots of public appeal but its value is now questioned (Frazer 1992, Grossblatt 1990). There is little point in the captive raising of turtles as a treatment for an incorrectly diagnosed or untreated decline. By-catch of turtles during shrimp fishing is the most serious factor limiting Kemp's ridley sea turtle, *Lepidochelys kempi,* in the United States (Box 9.10).

Captive raising has trade-offs: increasing a population rapidly by raising a larger number of "uncultured" offspring may compare unfavorably to a slower rate of increase by letting parents raise their young in the appropriate culture for the particular environment. Captive-raised Hawaiian goslings fared better when raised by their parents than only with other goslings (Section 4.11). Alternatively, lack of pa-

> **Box 9.10.**   Diagnosis of factors driving the decline of Kemp's ridley sea turtle, *Lepidochelys kempi*, in the United States
>
> **Problem**
> Kemp's ridley sea turtle is the most endangered sea turtle, and unlike other sea turtles it has only one nesting beach, Rancho Nuevo, Mexico. In 1947, 40,000 female turtles came ashore, but, despite 15 years of protection, only 350–650 females returned in 1988. Predation of the eggs and hatchlings on the beaches is negligible. The eggs are removed to coyote-proof enclosures, and armed guards prevent poaching. Trend in population size is more sensitive to changes in the survival of large juveniles and subadults based on analyses of the loggerhead turtle.
>
> **Diagnosis of factors driving the decline**
> Information on causes of mortality is from strandings which are monitored by volunteers and incidental catches in fishing gear. The numbers of stranded turtles increased coincidentally with the opening dates and locations of the shrimp fisheries. The greatest declines of loggerheads coincided with high-intensity shrimp fishing and, conversely, the turtle populations increased where shrimp fishing was low or absent. The proportion of dead or comatose turtles in shrimp trawls increases with the duration of the trawl. Direct estimation of mortality based on observers on the shrimp vessels had sampling problems and tended to underestimate mortality. Examination of stranded turtles for net marks, boat propeller wounds, and plastic ingestion and records of turtle deaths during dredging and strandings coincident with oil-rig removal supported the diagnosis that shrimp trawling is the dominant cause of mortality.
>
> **Treatment**
> The diagnosis of the effect of trawling led to the development of devices to exclude the turtles from bottom trawl nets. By 1989, regulations were implemented and, if successful, a decrease in strandings can be expected.
>
> (Modified from Grossblatt N. Decline of the sea turtles: causes and prevention. Washington DC: National Academy Press, 1990.)

rental guidance may be useful if an animal needs new behaviors to adapt to a modified habitat. Mauritius kestrels raised by kestrels did not disperse but could not establish territories as the area was saturated with breeding pairs. Kestrels, raised in captivity and released, adapted to an atypical habitat (Cade and Jones 1993).

9.5.2   *Translocation*

There seem to be as many definitions of the terms relating to the intentional release of individuals as there are reviews. We use terms as given in *Webster's Dictionary*. *Translocate* means to remove from one place to another (i.e., without prejudice as to what the purpose of the move is). *Establish* means to render stable or to set up. *Introduce* means to bring in a new or exotic plant or animal. *Repatriate* means to return to the native land. *Augment* means to make more numerous.

Wildlife has been moved around for a variety of reasons for decades, and its rising popularity as one of the tools of conservation biology has prompted a number of reviews. Moore and Smith (1991) and Stanley Price (1989) concentrate on large mammals, Black (1991) on guidelines for birds, Dodd and Seigel (1991) on reptiles and amphibians. Thomas (1989) picks up the cause for butterflies. Compilations specific to conservation biology are Wilson and Stanley Price (1994), Beck et al (1994, in press), and Kleiman et al (1994). Woodford and Rossiter (1994) and Viggers et al (1993) decribe diseases, both those that are carried by translocated animals and those that they may face. Many translocations are relatively recent and as such it is too soon to rate their success, or failure. However, in the United States, repatriations were started decades ago and their scale may come as a surprise; between 1890 and 1971, more than 32,000 white-tailed deer were repatriated in the southeast United States (Leberg et al 1994). Measuring genetic variation should, in theory, reveal whether repatriation supplemented deer numbers or if it was local stocks that recovered. However, the two studies that looked at this question disagreed on their results as their sampling design and techniques differed (Ellsworth et al 1994, Leberg et al 1994).

An analysis of the reasons for failure is just as informative as reviewing success. Short et al (1992) garnered hitherto unpublished details on six unsuccessful translocations of wallabys. They concluded that clear objectives and, regardless of the actual objective, an experimental design to determine the outcome are needed. And, although it should be part of the experimental design, a commitment to monitor success or failure is essential.

Some release plans, notably that for the black-footed ferret, have been precise in their objectives which, for example, specified the level of expected survival of released ferrets (Oakleaf et al 1992, 1993). Viability analyses to estimate persistence can be useful in establishing objectives, which can then be tested as hypotheses.

Small population size is an effect and not a cause. It is of no value to translocate animals to supplement a population or to establish one without treating the cause of its decline, and this point is emphasized in the IUCN guidelines for repatriations. This has not always happened: two translocations of the Laysan rail, *Porzana palmeri*, were to islands with the same problem as the source island—rabbits had destroyed the

Table 9.2.   Translocations of endemic landbirds within the Leeward Islands[a], Hawaii

Source	Destination	Date	Number	Fate
Laysan rail				
1. Laysan	Eastern[b]	1887	?	?
2. Laysan	Eastern	1891	Pair	Successful until introduction of rats, 1940
3. Eastern	Sand[b]	1910	?	Successful until introduction of rats, 1940
4. Laysan	Eastern	1913 ⎱ 100+		Successful until introduction of rats, 1940
5. Laysan	Lisianski	1913 ⎰		Died out 1916–1923; vegetation destroyed and rabbits starving in 1912
6. Midway	Pearl and Hermes	1929	7 pairs	Died out 1930; storm inundation and lack of food (?) vegetation; removal of rabbits
Laysan finch				
7. Laysan	Southeast[c]	1967	112	Thrived; estimated at 730 birds in 1973 and had spread to 3 close islets
8. Laysan	Midway	1891	?	Successful until introduction of rats, 1940
9. Laysan	Midway	1905	?	Successful until introduction of rats, 1940
Laysan duck				
10. Laysan	Southeast	1967	12	Wing-clipped but disappeared 1967
Nihoa finch				
11. Nihoa	East[d]	1967	10	Unsuccessful; storm after release
12. Nihoa	Tern[d]	1967	32	Successful

[a] Leeward Islands are Nihoa, Necker, Laysan, Lisianski, Pearl, and Hermes Reef Atoll (Southeast[c] is the largest of the eight islets), Midway Islands (includes Eastern[b] and Sand[b] islands), Kure Atoll, French Frigate Shoals (includes Tern[d] and East islands[d]).
Modified from Berger AJ. Hawaii wildlife. Honolulu: University Press of Hawaii, 1981.

vegetation (Table 9.2, Box 8.9). A sad example was when Richard Henry caught and moved 400 kakapo to an island in New Zealand's Fiordland in 1894. In 1900, he made the discovery that stoats had reached the island and the translocation failed. And that was perilously close to the end for this nocturnal parrot. A remnant population of males was found in Fiordland in 1960 (Section 6.1.1) and Stewart Island in 1975. On the other hand, researchers were careful to determine that islands destined to receive Seychelles warblers had a plentiful supply of insects. The source population was limited by a shortage of insects (Komdeur 1994).

Doubts about treating a decline's cause, whether it was possible or practicable, has led to using translocations as a rescue technique. New Zealand pioneered this by repatriating or introducing birds, reptiles, and insects to small islands which were clear or cleared of alien predators, as it was impossible to clear rats from the original habitat on larger islands. The best-known example is the saddleback, which was translocated to seven islands between 1964 and 1975 (Merton 1975). As black rats spread, the saddleback's distribution had contracted from across New Zealand's South Island to two islands by the 1960s. One was Big South Cape and black rats exploded there in 1964 and eradicated the greater short-tailed bat (Section 3.2.3), geckoes, skinks, and bush wrens (Section 3.1.3). The saddleback was rescued because techniques had been practiced in 1963 and the birds were easier to hold in an aviary during the captures. The bush wren and the Stewart Island snipe were less lucky: their diet was unknown. The last snipe died in the aviary, and the wrens died out after the translocation.

Even if treatment is possible it has to be effective; otherwise trans-locating animals is a risky venture. In Australia, 673 quokkas, *Setonix brachyurus,* a small rotund wallaby, were turned loose in a fenced enclosure over 12 years. Despite the fence and poisoning and shooting predators, the quokkas did not establish a population because suffi-cient predators survived (Short et al 1992).

If in doubt as to whether the cause of decline has been removed, a probe release can be informative. Captive-bred tree snails were re-leased to test whether the introduced predatory snails on Moorea, one of the Society Islands (Figure 2.2), had died out (Box 8.7). The answer was plain when a remnant population of the predators killed the tree snails (Meadows 1989).

Repatriation of a species back to its former habitat and community may not be simple. The released animals may themselves be changed by the differences arising from their sojourn in captivity, and it does not necessarily take long for the odds to shorten for survival of released captive-bred animals (Jiménez et al 1994). Translocations of swift foxes in 1990 to repatriate the species in Canada illustrates this. Wild-caught foxes from Wyoming and Colorado, where they are still quite common, were released along with captive-bred foxes (captive breeding had started in 1973). After 12 months, 47% of the wild-raised foxes had survived and 85% of those reproduced (Brechtel et al 1993). Com-parable figures for the captive-bred foxes were 11% survival and 25% of them bred. The figures for the captive-bred foxes were possibly biased downward as they were all young, 5–12 months old. Field studies are providing clues to how captivity can be modified so the foxes can sharpen their survival skills (Henry 1994).

The repatriated species' former community may have changed. First, the decline or even loss of the endangered species may in itself change the community. Second, the cause of its decline may have operated against other species. Third, plant and animal communities may have changed through their own dynamics. Removing the factor that forced a species' decline does not necessarily mean that the community will revert to its former state. Changes in marine communities after com-mercially fished stocks collapsed (Beverton 1990) tantalizingly hint that, after perturbations, the communities do not necessarily return to their earlier structure. Changes in grazing regimens on rangelands do not always return the ranges predictably to their ungrazed state but instead can be driven by single events to jump to different states (Walker 1993).

The swift fox disappeared from Canadian prairies in the early 1900s, driven to extinction by poisoning and hunting for its pelt. By the time captive-bred swift foxes were released in the 1980s, the prairie ecosystem had changed. Ranchers had reduced ground squirrels, *Spermophilus richardsonii*, and badgers whose burrows would have been bolt holes for the foxes. Extermination of grizzly bears and wolves that had killed bison reduced scavenging opportunities and, in the absence of wolves, coyotes had increased. With a decrease in escape cover and food, the released foxes were in trouble: 40% of known mortalities were kills by coyotes (Herrero et al 1991).

On the other hand, changes in a community may favor a translocation, at least initially. A few pairs of beavers, *Castor fiber*, were repatriated to a Swedish province in the 1920s. The rate of increase of colonies was rapid for the first 25 years, but then the colonies stabilized or declined (Hartman 1994). The normal predator of beavers was the wolf, but they had disappeared, which may have been why the colonies were able to increase so much.

Repatriation of an endangered species needs public understanding and support. As a step toward garnering that, Reading and Kellert (1993) interviewed 935 people on their concerns about the black-footed ferret repatriation (Section 4.6). Half the ranchers were opposed to ferret releases, and 97% favored controlling prairie dogs on which ferrets depend. Part of the resistance came from concerns about the U.S. Endangered Species Act (Section 12.2.1). The amount of publicity can become a trade-off between program recognition and disturbance. Disturbance from photographers during the first release of black-footed ferrets reduced their survival (Oakleaf et al 1993).

The size of the release depends mostly on the numbers available, but the species, its life history and breeding system, and the objectives of the release are all part of determining the size. Realization of the importance of breeding systems has perhaps lagged; we have an unimaginative tendency to think in pairs. Species such as the kakapo with its leks and the red-cockaded woodpeckers with their nonbreeding helpers suggest that release of appropriate social groupings is as important as total numbers.

It took the release of 706 swift foxes between 1983 and 1993 at four sites to end up with a conservative estimate of 150 foxes in 1993, and even then there is only cautious optimism that the swift fox is repatriated to Canada (Brechtel et al 1993). On a larger scale but with a smaller mammal, the Russians released about 20,000 steppe marmots at 300 release sites in 19 regions between 1977 and 1987, and populations were established in half of the regions (Bibikov 1991).

Small releases run the risk of sheer bad luck. Accidents struck down four of the eight California condors released since 1992 (Mestel 1993). A severe storm wiped out the translocation of 10 Laysan finches,

Loxiodes cantans, the night after they were released (Berger 1981). Przewalski's horse is only known from captive-bred animals (Section 3.2.3). Some of the 1,000 horses in captivity were released in southern France to establish a herd in preparation for release in Mongolia (Patel 1993). Three of four stallions released died from congenital defects and the fourth was felled by a heart attack after a lightning bolt (Patel 1993). That story has, however, a happier ending. By 1994, two herds were released in a Mongolian reserve after acclimatization in enclosures. On the other hand, there have been conspicuous successes with releases of a few individuals. The black robin takes the cake for one of the smallest and most successful releases (Section 4.7), but another example of a successful single-pair release was a pair of Laysan rails (Table 9.1).

If animals are turned loose without shelter or food, it is a "hard" release. A "soft" release offers animals shelter, food, or other resources and an "education." Release into predator-proof enclosures was successful support for the two endangered subspecies of mice in Florida (Figure 4.4). The Choctawatchee beach mouse, *Peromyscus polionotus allophrys,* and the Perdido Key beach mouse, *P. p. trissyllepsis,* are dune-dwelling mice whose habitat was lost to beachfront developments (Lowe et al 1990). The soft release of pink pigeons (Section 4.10) contrasts with no-frills release of woodhens (Section 4.2). Both approaches were tried for the repatriation of swift foxes to southern Canada (Herrero et al 1991). Soft release was holding pairs of foxes in enclosures on site for the winter and releasing them in the spring. Hard release was simply taking the foxes from the breeding station to a release site and letting them go. This became the technique of choice because survival of foxes, either soft or hard released, was similar but hard releases were cheaper (Brechtel et al 1993).

Releases are starting, especially with long-lived species, to incorporate planning to help the released animals know about their environment through cultural transfer. The rationale for leaving the last Spix's macaw in the wild was fear of losing the macaw's "local knowledge." The plan is to release a female for the male rather than take him into captivity (Collar et al 1992). A similar way to familiarize a released animal with its environment is to first release individuals, from ecologically similar species, as surrogates to act as "trainers" familiar with the area. Prior to the repatriation of the California condors, seven Andean condors, *Vultur gryphus,* were released in 1988–1989 to learn about the area, and they quickly found slopes with soaring updrafts to hone their flying skills (Wallace and Wiley 1991). As events transpired, the perils of power lines prevailed and, sadly for all the efforts and hopes, three of the six California condors released in 1992 were gone in a flash (Mestel 1993).

Monitoring is essential to track the fate of released animals. Its intensity and type depend on the original objectives for the release, but it is also a compromise between the amount of information needed to evaluate success, logistics, and effects of monitoring. Monitoring can merge into essentially individual protection: in 1989, the first Indian rhino, *Rhinoceros unicornis,* calf born to the rhinos translocated to Dudhwa National Park was guarded by a team of 10 people for 9 months (Sinha and Sawarkar 1993).

Radiotelemetry is an efficient technique for monitoring the fate of individuals, but it is not without shortcomings. Attention has to be paid to the frequency of monitoring to respect the data's independence (Section 1.2) and to ensure that the radiocollars do not affect the individual. Obvious as that seems, it does not always happen, if for no other reason than that some animals are more difficult to design collars for than others. The black-footed ferret's sinuous form sliding though close-fitting prairie dog burrows posed problems, which may be why radio-collared ferrets dispersed over longer distances and fared less well than ferrets unencumbered by a radiocollar (Oakleaf et al 1993).

Monitoring is necessary to determine how closely the objectives of the translocation have been met and when support can be withdrawn (supplementary feeding, predator exclosures, and control). In the black robin's case (Section 4.7), numbers had increased enough to allow evaluation of the breeding success among a group of robins left alone, a second group with minimal management (nests protected), and a third group with cross-fostering to tomtits. The unmanaged group fledged half as many chicks as the cross-fostering group, but the group with nest protection fared poorly owing to random problems and small sample size (eight pairs). Intensive management program (cross-fostering) was then ended and management shifted to monitoring breeding and banding chicks.

9.6   SUMMARY

Treatment is the next step after diagnosis in endangered species recovery. We have belabored the point that effective treatment depends on accurate diagnosis of whatever is limiting a population. Populations in trouble are often small, but that and urgency are not reasons not to use an experimental design when applying a treatment. We categorized treatments according to what effects they are dealing with—lack of resources, halting or reducing mortality, or supplementing populations too small to persist. Supplementing resources such as food and shelter was either direct (providing food, nest boxes) or indirect (habitat management).

Conspicious causes of mortality (hunting, predation, and parasites) tend to be diagnosed and treated first. Control of legal hunting usually needs support through explanation and education. Control of illegal hunting is often difficult, expensive, and ineffective. Predator control or eradication is most successful on small islands, and for larger areas control pulsed in time or space can give sufficient protection to an endangered species. In any treatment we should not lose sight of the complexity and subtlety of ecological relationships which can trigger unanticipated effects. As often as not it comes back to knowing the species' ecology.

If a population faces risks because it is too small, it can be augmented from captive-bred or captive-raised stock or other populations can be established. Spreading the risk of extinction through managing captive animals to create several populations has to be traded off against the different risks inherent in very small populations. Translocations, especially using captive-raised stock to establish a population or return a species to the wild, are often a long and complicated procedure. The success of translocations, especially if a long time has elapsed for the target community to have changed, is not predictable.

CHAPTER 10

Reserves in theory and practice

Most countries in the world have protected lands and have had for one purpose or another since time immemorial. Reserves protected sacred species or conversely protected the rights of hunters. In at least one instance, this led incidentally to saving a species—Père David's deer, *Elaphurus davidianus* (Box 10.1). The rationale for most reserves is that similar patches of habitat will eventually be converted to a land use incompatible with the continuing vigor of the species designated for protection. Loss or alteration of habitat is the driving factor in the decline of 76% of the threatened mammals and 60% of the birds listed by IUCN (Groombridge 1992). Reserves may also be established as sanctuaries to protect a species from harvesting rather than to protect its habitat.

Reserves as treatment to prevent or reverse declines of populations show a clear division between theory and practice. Most current writing about reserves is theoretical, focuses on reserve selection, and follows the premises of the small-population paradigm. This is partly the consequence of a shift in the approach to reserves (Simberloff 1988). Until the 1970s, the idea was to garner as much information as possible about a species' ecology, especially its habitat requirements, before setting up reserves. Then focus shifted to theory drawn largely from island biogeography, and debate on size, shape, and connections between reserves took the stage. Those theories quickly transmuted into guidelines for the selection, size, and shape and their uncritical pursuit (Simberloff 1988). Now the circle may be coming around again, as the emphasis is starting to edge back to discovering why some species persist while others are lost (Margules et al 1994a).

Number of reserves and thence application of guidelines on size, shape, and so on, will increase as habitat losses continue. IUCN's Action Plans, which cover 418 threatened species (Section 12.3), recommend protected area management and protected area establishment 410 and 116 times, respectively (Stuart 1991). Creating reserves is one of the measures taken or proposed for 39% and 70%, respectively, for the endangered species in 22 families of birds in the Americas (compiled from Collar et al 1992).

Before we can either establish new reserves or review existing ones, the first step is to define what the reserve is supposed to do—its objectives. Then we can, at least in theory, determine reserve size, number, shape, location (Section 10.1), and how the reserves can be connected (Section 10.2). Species have been saved and lost in reserves, and we describe some reasons why (Section 10.3). Setting aside land specifically as reserves is not the only answer to protecting habitat and populations. A variety of ways of regulating the use of land to maintain its compatibility with conservation has become lumped as off-reserve conservation (Section 10.4).

Box 10.1. Saving Père David's deer (the milu): a story of two reserves

---

The first reserve: the Imperial Hunting Park, Beijing
The wild is apparently only known from subfossils and may have been extinct in the wild for as long as 2,000–3,000 years. Its appearance is distinguished by its long tail and large spreading hooves. How and why it became isolated in a walled hunting park is not recorded, at least in European literature. The existence of the deer was unknown in Europe until the French missionary Armand David saw the deer and arranged to smuggle bones and a pair of hides from the park in 1865. A pair of deer reached an English park in 1898. Meanwhile in China most of the Père David's deer had escaped from the reserve but only to be eaten. The Boxer Rebellion was the final blow to the Chinese population of the deer, all of which were killed or, ironically, sent to European zoos.

By 1918 the only Père David's deer in the world were the 50 held in an English park. They were the offspring of the original pair and descendants of deer obtained from European zoos before the 1914–1918 war as a cooperative effort to save the species. The initial population was seven males, nine females, and two calves.

The second reserve: Dafeng Milu Reserve, near Shanghai
The Dafeng Forest Farm is within the subfossil range of Père David's deer and was one of the areas selected during 2 years of evaluating sites. It is a 1,000-ha forestry reserve and sparsely populated. The deer were gathered from different collections in England. The 13 males and 26 females were initially held for their first winter in a quarantine pen. In 1987, the deer were released into a 120-ha fenced area for eventual release into the whole reserve.

(Modified from Whitehead GK. Deer of the world. London: Constable, 1972; Thouless CR, Chongqui L, Loudon ASI. The milu or Père David's deer *Elaphurus davidianus* reintroduction project at Da Feng. Int. Zoo Yearb. 1988;27:223–230.)

10.1 OBJECTIVES OF RESERVES

Before designing a reserve network we must decide its function, which means define its objectives. The world currently has some 8,500 areas that IUCN lists as protected areas and they cover a modest 5.2% of the earth's surface. IUCN classified the objectives of protected areas (Table

10.1) to allow comparisons of reserves among countries, but this is only for areas larger than 1,000 ha. With the exception of one category (resource reserves), all categories have at least three conservation objectives, but some are mutually incompatible. For example, scientific reserves are to sample ecosystems in their natural state and conserve genetic resources. This raises the question of incongruity between conserving processes (ecosystems are dynamic) and states (genetic resources).

This is a basic dichotomy (Caughley and Sinclair 1994): whether to conserve states (species and species assemblages) or ecological processes. If states are to be conserved, the problem comes down to maximizing the number of species captured by the system, or the taxonomic distinctiveness of the assemblage so captured, a topic explored by May (1990) and Vane-Wright et al (1991). These are variations on the theme of "biological diversity" (or the term's common contraction, biodiversity), but that phrase is defined in numerous disparate ways to the extent that it is essentially meaningless. Its common shorthand appellation is "species diversity," which is another dubious phrase with a long history of changing definition. These days it may mean the number of species occupying an area, a statistic also called, and less ambiguously, "species richness."

The number of species per se is of less interest than the set of species. It does not necessarily follow that areas with many species are more desirable as reserves than areas with fewer species (Margules et al 1994b). The potential contribution of an area to a reserve system depends on which species the area contains. If an area has a few species but they do not occur elsewhere, then the contribution is high. This is the idea of complementarity in reserve selection (Pressey et al 1993). Reserves may also be dedicated to the protection of a single species, although that usually will have the effect of "carrying" some other species with it.

Objectives for reserves that deal with the conservation of processes are well encapsulated by Frankel and Soulé (1981), who maintain that ". . . the purpose of a nature reserve is to maintain, hopefully in perpetuity, a highly complex set of ecological, genetic, behavioral, evolutionary and physical processes and the co-evolved, compatible populations which participate in these processes." To recast that purpose as tight and measurable objectives requires understanding the ecological processes to be conserved. Community dynamics, especially the time span and tightness of relationships between consumers such as herbivores and producers such as plants, can cause nightmares for reserve administrators. This follows because ecological processes mean changes are inevitable and species will come and go. The problems are compounded as community responses to natural perturbations are not always predictable (Pimm 1991).

The distinction between whether the objective of the reserve is to conserve a state or a process is of practical importance as it establishes

Table 10.1. IUCN's classification of protected areas and their objectives

Conservation objective	Protected area designation (IUCN category number)							
	I. Scientific reserve	II. National park	III. Natural monument	IV. Managed nature reserve	V. Protected landscape	VI. Resource reserve	VII. Natural biotic	VIII. Multiple-use area
Sample ecosystems in natural state	●	●	●	●	○	○	●	—
Ecological diversity and environmental regulation	○	●	●	○	○	○	●	○
Conserve genetic resources	●	●	●	●	○	○	●	○
Education, research and environmental monitoring	●	○	●	●	○	○	○	○
Conserve watershed, flood control	○	●	○	○	○	○	○	○
Control erosion and sedimentation	○	○	○	○	○	○	○	○

Table 10.1. (Continued)

Conservation objective	Protected area designation (IUCN category number)							
	I. Scientific reserve	II. National park	III. Natural monument	IV. Managed nature reserve	V. Protected landscape	VI. Resource reserve	VII. Natural biotic	VIII. Multiple-use area
Indigenous use or habitation	—	—	—	—	●	○	●	○
Protein from wildlife	—	—	—	○	○	○	○	●
Timber, forage, or extractive commodities	—	—	—	—	○	○	○	●
Recreation and tourism	—	●	●	○	●	—	○	●
Protect cultural, historical, or archaeological heritage	—	○	○	○	●	○	●	○
Scenic beauty	○	●	○	○	●	—	—	○
Open options, management flexibility, multiple-use	—	—	—	—	○	●	—	●
Rural development	○	●	○	○	●	○	○	●

● = primary objectives; ○ = compatible objectives.
Modified with permission from Groombridge B. Global biodiversity: status of the Earth's living resources. London: Chapman and Hall, 1992.

the type of management, if any. Again, we return to Caughley and Sinclair (1994):

1. If the aim is to conserve specified animal and plant associations that may be modified or eliminated by wildfire, grazing, or predation, then intervene to reduce the intensity of wildfire, grazing, or predation.
2. If the aim is to give full rein to the processes of the system and to accept the resultant, often transient, states that those processes produce, then do not intervene.
3. A combination of both—if the aim is to allow the processes of the system to proceed unhindered unless they produce "unacceptable" states, then intervene only when unacceptable outcomes appear likely.

When setting reserve objectives, then, we have to know a species' needs and the ecological relationships to be conserved, bearing in mind that community dynamics with or without chance events, such as droughts, can shift successional patterns. Most of all, we have to remember that objectives cannot be established in isolation from the people whose way of life will be colored by the reserve. A telling tale of ignoring those realities is Parker's (1983) account of the Tsavo National Parks, Kenya (Box 10.2). Failure to perceive the history and needs of those who lived in the area before it was a park, changes in numbers of large herbivores and their habitat, and natural events such as droughts all contribute to the tale.

10.2   THE THEORY OF RESERVE DESIGN AND SELECTION

10.2.1   The influence of island biogeography and metapopulation theory

Answers to the questions of the size, shape, and numbers of reserves have been largely sought from island biogeography. MacArthur and Wilson (1963, 1967) and Mayr (1965) launched island biogeography when they combined the well-known species–area relationship with the effect of counteracting rates of colonization and extinction to explain why some islands hosted many species and others few. The equilibrium number of species on an island was envisaged as a result of the average number of species that colonized the island over a given period and the average number of established species that died out over the same period. The rate of colonization is a function of the distance from a source (usually a continent) of colonizers and how

Box 10.2. The history of people, elephants, and woodlands in the Tsavo National Parks, Kenya

The people

The Tsavo National Parks (21,800 km²) were gazetted in 1948 on the land used by five tribes. The people were nomadic hunters and pastoralists who had lived there for at least 500 years. They were evicted from the parks and settled around them. Their numbers increased by 185% between 1948 and 1979.

The elephants

Hunting for ivory may have depressed numbers in the early 1900s, but then elephants increased during the 1950s and 1960s. Expanding settlement around the parks displaced elephants from those lands into the parks. Thus numbers in the parks continued to rise and reached 50,000 despite recruitment falling as forage declined. In 1971, at least 9,000 elephants died in a drought. In 1979, the estimated population of elephants was 9,500 and declined further to perhaps 5,400 in 1987.

The people and the parks

People continued to hunt in Tsavo until an intensive antipoaching campaign in 1957–1958. They were so intimidated by it that a small ranger force was adequate to keep the parks free of hunters for the next 14 years. The 1971 drought and die-off left thousands of tusks available for the taking. People who had lost crops and stock in the drought were driven by necessity to enter the parks. They did so on a scale beyond the parks' force to prevent them. Having discovered the weakness, and when the ivory ran out, people hunted the surviving elephants or rhinos and a decade of intense poaching followed. Had the parks' authorities been sensitive to the predicament of people during the drought and hired them to salvage the ivory, they might not have exposed their limitations to prevent incursions and the poaching.

Parks' habitat

The increase in the number of elephants coincided with a decrease in the woodlands by the late 1950s and large areas became grassland. Seasonal fires were more prevalent and hindered regeneration of the woodland. The change to the grassland increased the vulnerability of the elephants and rhino to hunting. With the decline in elephants, the grasslands are reverting to thicket and woodland.

(Modified from Parker ISC. The Tsavo story: an ecological case history. In: Owen-Smith RN, ed. Management of large mammals in African conservation areas. Pretoria: Haum, 1983:37–49; Cumming DHM, Du Toit RF, Stuart SN. African elephants and rhinos. Status survey and conservation action plan. Gland, Switzerland: IUCN, 1990.)

large a target the island presents. The rate of extinction is determined principally by the average size of a population, which itself is determined by the size of the island. The discrepancy between the number of species held by two islands equidistant from a source of colonizers will be determined by their respective areas, conforming to the observed relationship that a tenfold difference in area is associated with a two-fold difference in the number of species (see May 1975 for a review of species–area relationships).

It was not a large conceptual leap to see the ecological equivalence of oceanic islands in patches of habitat set aside as reserves (see, particularly, Brown 1971, Diamond 1975, 1976, Whitcomb et al 1976, Margules et al 1982). The predictions of island biogeography were quickly adopted, and debate on whether one large or several small reserves took center stage. However, the final comment in Simberloff and Abele's (1976) account of island biogeography's application to reserves is worth repeating as its message has been overlooked: "In sum, the broad generalizations that have been reported are based on limited and insufficiently validated theory and on field studies of taxa which may be idiosyncratic." And criticisms of support for the theory of island biogeography have, if anything, increased since those comments were written (for a summary see Shrader-Frechette and McCoy 1993).

Metapopulation dynamics (Section 5.6) predicts that for a species which occurs in more than one reserve of a network, the chance of its dying out is a function of the number of reserves that it occupies, the chances of losing a population in a single reserve over a specified period, the rate of dispersal among reserves, and the correlation among the reserves in environmental fluctuations from year to year. Thus a reserve network acting as a metapopulation emphasizes connections and dispersal between the individual reserves (i.e., corridors). As well, metapopulation dynamics, in particular those of sink and source populations, argue for the importance of habitat quality, while island biogeography rests more on consideration of area.

10.2.2   How big?

The required size of a reserve is determined by what it is supposed to conserve. Population modeling suggests that a self-perpetuating population of 50   90 grizzly bears may need between 1,000 and 13,500 km² of suitable habitat (Shaffer 1983), whereas field observations indicate that for some 25 species of British butterflies a few hectares may be adequate (Section 10.3). We illustrate the use of population viability analysis (Chapter 7) to estimate the appropriate area for a reserve. If the structure of the model is sound and if the data fed into it are real, a viability analysis will give an estimate of the mean population size necessary to retain that species at designated levels of probability and time. The estimate, divided by the average density of the species in that environment, returns the minimum size of a reserve. Section 7.2.5 demonstrates such an analysis for red kangaroos. Although we can predict the area needed to maintain a population with estimated probabilities that it will survive, we lack experimental verification that this will indeed be the case in practice.

Some progress has been supplied by studying temperate plant and animal communities marooned on mountain tops by post-Pleistocene warming (Brown 1971, Patterson 1984, Belovsky 1987, Grayson and Livingston 1993). If such a patch is big enough, we can assume tentatively that it held most of the species occurring in the region 10,000 years ago and that any species missing has been lost from that patch over that interval. The rate of loss can then be regressed on patch size to determine the relationship between persistence and area.

10.2.3 *Single large or several small?*

Although we are generally repelled by acronyms, we do like the one covering this question: SLOSS. The problem is too determine whether one large reserve or a network of smaller reserves that sum to the same size will be more effective in retaining species. It is not the same question as whether a single large reserve or a set of smaller reserves will begin with the greater number of species.

The question of a single large or several reserves was initially answered from application of island biogeography theory, and Simberloff and Abele (1976) convey the gist of this. Akçakaya and Ginzberg (1991), summarized by Burgman et al (1993), investigated the SLOSS problem in the context of the mountain gorilla, *Gorilla beringei*, in Uganda (Figure 4.7). They stressed that the simplifying assumptions of their occupancy model and lack of detailed ecological information restrict their conclusions to the general effects of spatial factors on the risk of extinction. The answer to the SLOSS question is not simple but depends on

1. The difference between the extinction probabilities of a small and a large population
2. The number of populations
3. The correlation in year-to-year fluctuation of the environments of the populations
4. The probability of recolonization of a patch emptied by local extinction

In their conceptual example, three small patches led to a persistence time greater than for one patch summing to the same area when no recolonization was allowed and environmental fluctuations were uncorrelated between the patches. With environmental correlation added, the populations in the small patches fared better or worse than that in the large patch according to the strength of the correlation. Recolonization rates above zero favored the smaller populations in aggregate over the larger.

Järvinen (1982), Lahti and Ranta (1985), and Robinson and Quinn (1992) further discuss the question of a single large or several smaller reserves. The answer depends on the interaction between interdependent factors that will differ case by case. The dominant consideration for the mountain gorilla, for example, is likely to be the relative efficiency with which unauthorized offtake can be reduced while operating on a fixed budget spread across three reserves or concentrated on one.

Metapopulation dynamics emphasize habitat quality as a factor in reserve design, specifically when several small reserves are more effective than a single large one. Population persistence for some species depends on a high degree of dispersal between ephemeral habitat patches. A single large reserve encompassing patches of ephemeral habitat may be the necessary design. The more classic metapopulation comprises several habitat patches occupied by local populations going extinct and being recolonized. In that case, a reserve network with individual reserves, each with a habitat patch capable of supporting a local population, would be one answer—if the reserves were connected to allow dispersal. Another metapopulation structure is a source population in a favorable habitat supplying colonists to sink populations in a less favorable habitat. It might not take a high degree of environmental variability to knock out the peripheral populations, but the source population would be buffered by its high-quality habitat. Protecting a source population in a reserve would be the priority. These hypothetical models equate habitat patch with a reserve, but scales will rarely overlap so conveniently. Instead, in practice, we will have to depend on population structure to tell us the required dispersion of habitat patches. Then those patches have to be within an area determined by the species dispersal behavior.

10.2.4   What shape?

Diamond (1975) suggested that a reserve should be as nearly circular as other considerations permit, to minimize within-reserve dispersal distances: "If the reserve is too elongate or has dead-end peninsulas, dispersal to outlying parts of the reserve from more central parts may be sufficiently low to perpetuate local extinctions by island-like effects." Another reason for favoring a circle over any other shape is that it minimizes the ratio of circumference to area and thus minimizes the effect of external influences on the reserve.

The contrary argument is that long narrow reserves oriented at right angles to isoclines of rainfall, altitude, and so on, will contain a greater diversity of habitats and thus hold more species. This reflects a penchant among some park planners for believing that the more species a reserve contains the better the reserve. Thus reserve boundaries are often drawn to capture a maximum diversity of habitats. Quite apart from the problem Diamond (1975) draws attention to, the shaping of a reserve to maximize the number of species within it will lead to an excessive rate of extinction because of the inverse relationship between persistence time and population size. The more habitats contained within a reserve the smaller the area occupied by a given habitat and the fewer the numbers of individuals of a species that depends upon that habitat.

10.2.5   Where should the reserves be?

Placing a reserve is usually a pragmatic compromise between budget, the ability to secure land and enact legislation, and coping with the problems posed by surrounding landowners and tenants. Practically, as we will not be free as to where we can place a reserve, we have to make choices, initially depending on our objectives. Each objective requires a different algorithm and leads to a different set of reserves. The minimum steps are

Step 1:   State the objective of the reserve clearly and unambiguously.
Step 2:   Identify patches available for use as reserves within terms of the objective.
Step 3:   At least survey the species occupying each patch. An estimate of the abundance of each species is more helpful.
Step 4:   Formulate the starting rule (i.e., the patch or patches to be selected first and to which subsequently chosen patches will be added to make up the reserve system). These will usually be previously established reserves which may bear only a tenuous relationship to the objective of the system as now defined; no reserve system is perfect. Alternatively, it might be best to choose the patch with the rarest species of a suite of species the reserve system is designed to conserve.
Step 5:   Formulate the iteration rules of choice and sequence according to the objective defined in Step 1.

1. How much redundancy should be built in?
2. Should patches be chosen to minimize the distance between them and thus encourage interpatch dispersal and metapopulation dynamics (Section 5.6)?
3. Should patches be chosen to maximize the distance between them and hence minimize the correlation of environmental fluctuation among them?
4. Should patches be the smallest on offer to minimize purchase costs or as large as possible if the land does not need to be purchased?
5. Should patches be limited to a maximum size to minimize the cost of fencing and subsequent management?

Suppose the aim were to design a system of 10 reserves, each of 5 km², that would capture the maximum number of species of swallow-tail butterflies. There are about 100 species worldwide. Most of the reserves would be placed in areas of diverse habitat (thereby maximizing the number of species per reserve) in the Philippines and Madagascar, which between them have 34 endemics. The rest would be shared among other countries with a lesser but still significant level of endemism. However, the reserve network would be different if instead we aimed to maximize the average persistence of the species captured by the network. We would not then place the reserves in areas of diverse habitat because that would lower the mean number of individuals per species held within the network and thereby reduce their chances of persisting over the long term.

A number of algorithms have been published for choosing the minimum reserve network that will capture a given percentage of species in a region (Pressey et al 1993). One set gives priority to rare species (Margules and Nicholls 1987, Margules et al 1988, Pressey and Nicholls 1989b, 1991, Pressey et al 1990, Rebelo and Siegfried 1992) and another to areas containing many species (Kirkpatrick 1983, Kirkpatrick and Harwood 1983, Scott et al 1988), but nonetheless they tend to end up with much the same selection (Pressey and Nicholls 1989a). Box 10.3 gives one such algorithm (Nicholls and Margules 1993) that has the added advantage of clustering the reserves where possible, thereby enhancing the persistence of species within them. Figure 10.1 shows the result of a selection procedure to identify regions into which reserves should be placed to conserve all species of owls in the world.

An alternative approach to selecting reserves is to use environments rather than populations, species, or communities as the unit of choice. A sets of sites is selected to include a given proportion of each environment, the rationale being that if all environments are selected so will most of the species (Mackey et al 1989, Belbin 1993). Still another approach to selecting reserves that maximize species richness is to use higher taxa such as families as the sampling unit rather than species. Family richness is quicker and easier to score compared to the detailed surveys needed for species (Williams and Gaston 1994). The methods described above work on occupancy models: sites are either occupied or not by a given species. They take no account of population size nor do they address, except in passing, the problems of too many species for a given area, interreserve dispersal, or persistence within a reserve.

Guidelines for reserve selection need testing against how existing reserves hold a suite of species or ecological processes. Admittedly, experimental verifications are long-term projects and require considerable logistical support but are essential if the theories are to contribute to reserves. An example of what is required is an experiment to predict rates of extinction and colonization using three different sized fragments of eucalyptus forest in Australia replicated six times (Margules 1992, Margules et al 1994a).

Some national parks and reserves have been established since the first one or two decades of this century and yet there are few analyses of persistence and turnover of species or communities in existing reserves. Margules et al (1994b) took this approach when they evaluated

Box 10.3. An algorithm for choosing the minimum set of sites that captures a given proportion (say 10%) of the range occupied within a region by each of the species of a taxonomic group

Step 1.	State the objective ("to put together a reserve system that captures a given proportion of the range occupied within a region by each species of a taxonomic group—say the genus *Eucalyptus*").
Step 2.	This is an optional step that allows the inclusion of some sites before the selection process begins. Examples might be existing reserves, sites that fulfill international treaty obligations, or sites with known rare and endangered species.
Step 3.	Select all sites with unique occurrences of species.
Step 4.	Find the next rarest species and select the site that, when added to those already selected, will represent that species plus the greatest number of additional species at or above the required proportion of their area of distribution.
Step 5.	If there is a choice, select the site that is nearest in space to a site already selected.
Step 6.	If there is still a choice, select the site that also contributes the largest number of as yet inadequately represented species.
Step 7.	If there is still a choice, select the site that achieves the required level of representation of the rarest species remaining underrepresented.
Step 8.	If there is still a choice, select the site that contributes the most to achieving the required level of representation of the rarest group of species remaining underrepresented.
Step 9.	If there is still a choice, select the site that either contains the smallest percentage area needed to achieve the required level of representation of the species under consideration or which contributes the largest percentage of that species' range if no one site achieves adequate representation.
Step 10.	If there is still a choice, select the smallest site.
Step 11.	If there is still a choice, select the first suitable site on the list.
Step 12.	Go to Step 4.

(Modified from Nicholls AO, Margules CR. An upgraded reserve selection algorithm. Biol. Conserv. 1993;64:165–169.)

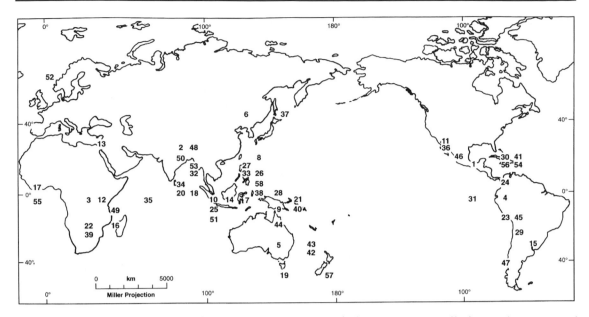

Figure 10.1. The sequence of 58 priority areas needed to represent all the owl species of the world. The starting rule is to pick the region with the greatest number of species and the iteration rule is to add in sequence to the next region that adds the most additional species. (Modified from Pressey RL, Humphries CJ, Margules CR, Vane-Wright RI, Williams PH. Beyond opportunism: key principles for systematic reserve selection. Trends Ecol Evol 1993;8:124–128.)

their procedure for identifying a minimum sets of sites which would capture species (in this case plants on limestone pavements in Britain) rated as nationally rare or uncommon. They compared a minimum set after 11 years and found that high species turnover (possibly confounded by sampling error) meant an additional 42% more sites would have coped with species loss and gain. They continued the exercise for minimum set of sites for more than one population of each species, and again the original site selection was inadequate. More sites are needed to capture the minimum set when the existing reserves are used as the starting point simply because the reserves were established for a variety of reasons.

It is too early to judge what the current vogue for applying island biogeography and metapopulation dynamics to reserve selection will contribute in practice. One of the only published examples of the systematic application of theories of reserve selection is the planned network of habitat conservation areas for northern spotted owls (Murphy and Noon 1992). However, the testing of its success is years away.

Application of the theories to establish networks of reserves for smaller-bodied species is easier because generation time is short, and the required patches small and thus more amenable to study. Murphy and his colleagues (summarized in Murphy 1988) describe the information required to protect the bay checkerspot butterfly whose grassland habitat had become reduced to fewer and smaller patches in California. They treated the local populations as belonging to a metapopulation and identified the habitat patch that supported the source population. This could supply dispersers and recolonizers have to be protected even at the expense of some sink populations. It was not the area of the patches that determined the persistence of populations but their topographic heterogeneity, specifically slope aspect that correlated with larval survival.

The marsh fritillary, *Eurodryas aurinia*, is a British butterfly characterized by large fluctuations in population size related to parasites, drought, and habitat suitability. Its colonies have high rates of extinction of which about a third occur in marginal habitat (Warren 1994). The species disperses over a larger distance than do most other colonial butterflies. Unlike the bay checkerspot butterfly, there was no obvious core population acting as a reservoir. An appropriate reserve network would be several patches within the species dispersal range, and Hanski and Thomas's (1994) model could predict the rate and pattern of occupancy.

10.2.6   *Corridors*

It seems axiomatic that a reserve network will operate most efficiently when some reserves are linked by corridors of natural habitat. The dispersal of individuals along corridors would help gene flow between reserves, increase effective size of the component populations, and encourage metapopulation dynamics whereby a declining population in one reserve might be rescued by dispersal from another. Diamond (1975) and Wilson and Willis (1975) advanced it as a logical extrapolation of the equilibrium theory of island biogeography. The proposition seems so commonsensical that it was swiftly assimilated as an article of faith by reserve planners.

Simberloff and Cox (1987) expressed the first doubts. While acknowledging certain advantages of corridors, they pointed out disadvantages. Corridors could help spread diseases and fire, and increase exposure to unauthorized hunting, predation, and competition with domestic animals. Translocation of individuals or even artificial insemination and embryo implantations between reserves might be more cost effective than waiting for individual animals to use corridors (Simberloff and Cox 1987). The need, however, for the translocation of individuals between reserves to stave off inbreeding depression in small populations is yet to be empirically determined.

Simberloff and Cox's (1987) paper elicited controversy, the flavor of which can be sampled by reading Noss (1987) and Simberloff et al (1992). The latter pointed out faulty sampling and misleading conclusions and cited four of five papers that failed to look for movements outside the corridors as a comparison to movements along corridors.

Nicholls and Margules (1991) pointed out that the hypothesis must be worded precisely, that it should be tested experimentally, and that previous attempts failed through flawed experimental design (other comments are in Section 1.2.1). They posed the simple hypothesis, "Recolonization will be greater on [habitat] remnants connected to source areas by corridors than on unconnected remnants" and showed that a minimum experimental design to test it requires four treatments (with replication): remnants in which the species under study had been driven to local extinction, remnants in which it was extant, and those two treatments combined with connection, or the lack of it, with a source area. One would not have to try this on an endangered species— an ecologically similar surrogate would fit the bill instead. Until such tight hypotheses are devised, and the appropriate experiments run, the efficacy of corridors linking reserves will remain theoretical speculation.

10.3   RESERVES IN PRACTICE

Most of the theory that we have described is about establishing new reserves or bolstering existing ones, and assumes that all the practical problems of land availability are surmountable. In this section we look at reserves already established for a variety of reasons and describe some conspicuous successes and not a few failures. A frequently cited success is the southern white rhino, *Ceratotherium simum simum*, which had been hunted to the point of its presumed extinction in 1892. For- tuitously, a dozen rhinos were found in the early 1900s and protected in a game reserve. By 1990, that dozen southern white rhinos had increased to 4,300 (Cumming et al 1990) and then to 6,000 by 1994 (S. Stuart, personal communication, 1994).

Reserves as a treatment for a species in trouble are rarely as simple as the success of the southern white rhino. In practice, reserves often have three problems. First, they do not maintain the specific resources needed to support a protected species. Merely setting aside tracts of land is rarely enough. Second, they may be host to conflicting land uses, whether legal or illegal. Third, their objectives may require ma- nipulative management that may change settings for other species and this often generates public controversy.

Setting aside reserves for a specific species only works if the reserve meets the needs of that species. The problem is that habitats naturally change, although the dynamics and causes are a source of argument among ecologists. The assumed model of a simple linear progression to climax vegetation is being replaced by a more complex model. Often when a reserve is declared, former land use practices halt, but the result may not be the original habitat. Alternatively, the land use practices may have been maintaining a particular community, and with their cessation the community changes: the large blue butterfly failed to survive in a reserve for this reason (Section 4.4). The consequences of reserves not maintaining the required habitat was the poor survival of 35 species of British butterflies (Thomas 1984a; Thomas 1990; Warren 1993, 1994). The rate of extirpation for colonies of the six butterfly species listed in the Red Data Book was similar on unprotected and protected sites (11.6% and 8.6% per decade).

The poor success of reserves in butterfly conservation is that the plants for egg laying and larval development of many species are tied to a successional habitat stage. The reserves are too small to have a series of habitat patches at different sequences of succession where natural or man-made perturbations set back the succession. Without that intervention, the habitat patches change and the butterfly population dwindles. The minimum areas required by most species are relatively small (Figure 10.2). The majority of species have closed populations, but whether they have a metapopulation structure depends on the dispersion of the habitat patches and dispersal of butterflies between them. A key point in the probability of the recolonization of a patch emptied by a local extinction is whether the changes that caused the local extinction are still extant.

We need ecological information before we can design reserves with the right set of conditions for a species. The large blue butterfly (Section 4.4) is one example, and another was assuming that establishing a reserve is sufficient to rescue a species in trouble, which was almost the final blow to the western swamp turtle, *Pseudemydura umbrina*, in Western Australia. In the 1950s the turtle was rediscovered, and once its last two refuges were set up as reserves all was assumed to be well. Unfortunately the reserves had marginal habitat, and one population disappeared in the early 1980s from the larger reserve (155 ha). In the smaller reserve (65 ha) only 20–30 turtles survived in 1987, the world's only population in the wild (some 20–30 turtles are in captivity). The turtle was threatened by the ephemeral nature of the swamp and fox predation. The solutions are to increase the reserve's size, rehabilitate the swamp, and fence it to prevent ingress by foxes (Kuchling et al 1992).

Problems arise if there are conflicts among the objectives for the reserve. The central Indian barasinga deer, *Cervus duvauceli branderi*, formerly widespread in central India (Figure 4.9), became restricted to a single reserve through loss of its habitat. Its decline continued until its problems were identified and management steps taken to reduce the conflict in the objectives (Box 10.4).

Reserves often are set up as sanctuaries with no legal hunting, but that only works with surveillance against illegal hunting. Project Tiger in India established 15 reserves for tigers, although not without problems for people (see later this section). Subsequently, by the 1990s the

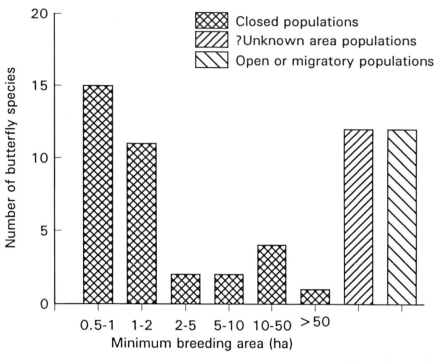

Figure 10.2. The minimum area required to support a viable colony of butterflies in the United Kingdom (Modified from Thomas JA. The conservation of butterflies in temperate countries: past effort and lessons for the future. In: Vane-Wright RI, Ackery PR, eds. The biology of butterflies. London: Academic Press, 1984a:333–353.)

early successes had been offset by increased poaching, although its extent is difficult to gauge. The poaching is mostly for traditional medicine, and tiger bone is particularly sought after. Tiger numbers have sharply declined: already three and possibly four of the subspecies are extinct and only 4,600–7,700 tigers are left (Jackson 1994).

A species in a reserve is a predictable location for those interested in finding it, be they tourists or poachers. Tourism is often a mixed blessing and, unless managed appropriately, can disturb wildlife and cause localized habitat damage. We say "appropriately" because management for one species and tourism can spill over as a problem for another. The baiting for tigers increased predation of the barasinga (Box 10.4), as mentioned. Demonstrating the effect of tourism is difficult in the absence of well-defined experiments. This was a conclusion from a study of tourists watching the nocturnal nesting of turtles in Costa Rica (Jacobson and Lopez 1994).

As reserves become the isolated enclaves of populations, their attraction to poachers increases, while in theory deterring poaching should be easier in predictable locations. The northern subspecies of the white rhino, *Ceratotherium simum cottoni*, will serve as an example: their num-

Box 10.4. The problems faced by central Indian barasinga in its last refuge, the Kanha National Park

Problem
The central Indian barasinga occurred in deciduous forests across India's central highlands. By the 1920s deer were declining as agriculture and cattle took over the grassy openings. Deer in Kanha National Park had protection from hunting when a sanctuary was established in 1933, but still the 3,000 deer estimated in 1938 had dwindled to 66 in 1970. By then the subspecies had disappeared from the rest of its range. The park was for the deer, agriculture, and tourism. Burning of grasslands started in 1902 and cattle grazing was prevalent. In the 1960s tigers were baited for viewing by tourists.

Diagnosis of factors driving the decline
In the late 1960s, fawn survival was low despite most all females being pregnant. Brucellosis had been suggested as the cause of the poor survival, but as other deer species in the park were increasing, diseased deer had not been found, and the proportion of females fawning was high, this diagnosis had little credibility.

The artificiality of the level of predation was not realized until 1969, and then the baiting stations for tigers were moved into the woodlands. The tiger kill of barasinga dropped from 16 in 1964 to only 3 between 1971 and 1973. The diagnosis of predation compared to habitat as a factor in the decline is complicated by the changes in land uses within the park at the same time the bait stations were moved.

Cattle grazing (25,000 head) on the grasslands was halted, eight villages were relocated outside the park, annual burning of the grasslands ceased, reconstruction of perennial water bodies, and enlargement of the park from 318 to 446 km² all played a role in improving the grasslands for the barasinga. The grass grew, providing cover during fawning, and the taller grass meant more food in the dry season. Coincidentally, fawn survival increased; the total numbers rose (283 in 1977) and the deer spread into the additional grasslands.

(Modified from Panwar HS. Decline and restoration success of the central Indian barasinga *Cervus duvauceli branderi*. In: Threatened deer. Morges, Switzerland: IUCN, 1978:143–158; Martin C. Interspecific relationships between Barasingha (*Cervus duvauceli*) and Axis deer (*Axis axis*) in Kanha National Park, India, and relevance to management. In: Wemmer CM, ed. Biology and management of the Cervidae. Washington DC: Smithsonian Institution Press, 1987:299–306.)

bers in Zaire's Garamba National Park reached 1,000–1,300 in 1960. Illegal hunting during an uprising in 1964 reduced them, but they recovered to 490 ± 270 by 1976. A further bout of illegal hunting reduced numbers to 13–20 by 1983. This last population of the northern white rhino is now increasing in the park in response to well-organized protection from illegal hunting, but the annual costs are high: $10,000 U.S. per rhino in 1991 (Smith and Smith 1993).

Great one-horned (Indian) rhino numbers fell and rose with the degree of protection of poaching within national parks (Box 10.5, Figure 11.10). The Indian, Javan, and Sumatran rhinos owe their existence mostly to protection within reserves from habitat loss and poaching. Only a few scattered remnant populations, none of which exceed 20 individuals, occur outside reserves. Again, the tale of the conservation of tigers and rhinos through reserves reinforces the point that a reserve is not enough without continual vigilance to stem illegal hunting. With sufficient protection, the reserves serve their purpose: Ujung Kulon National Park (78,359 ha) has protected Javan rhinos since 1921 and now holds the only population in the world except a handful of individuals in Vietnam (Figure 9.2).

Maintaining appropriate habitat or protection is, however, not the only requirement for a reserve to successfully conserve a species. In Section 4.14 we described how the numbat survived in two of the forest reserves. Clearing of the eucalyptus forest left only designated reserves (although not primarily for wildlife), but they were insufficient to save the numbat as fox predation had replaced habitat loss as the driving force in the numbat's decline. Once foxes were controlled, numbats increased in the reserves. Conserving species in a reserve is the same as anywhere: the problem has to be diagnosed and treated.

Managing habitat or introduced predators or hunting may not be enough if the reserve population of the endangered species is already too small to withstand natural catastrophes. The heath hen increased initially from 50 to 2,000 by 1915 in its 648-ha island reserve off the U.S. east coast (Simberloff 1986). But then, despite protection from hunting and habitat destruction, a sequence of environmental catastrophes drove the single population, isolated in its island reserve, to the level where demographic bad luck apparently finished the species.

Conflicts between the legal and illegal use of reserves is a recurrent and pervasive problem and is not restricted to hunting. Those problems are often the consequence of how the reserve was set up in regard to people living in the neighborhood or even who was evicted to make way for the reserve (Box 10.2). The uncertain fate of the hispid hare in Indian and Nepalese wildlife sanctuaries (Section 4.16) illustrates conflict between people's and wildlife's needs. The few remaining pockets of hispid hares are all within national parks, forest reserves, and wildlife sanctuaries where the only patches of suitable grasslands occur. Those reserves have the firewood, grass for thatch, and forage for livestock which are irresistible to people who lack those essentials. The

Box 10.5. Reserves and the status of the three species of rhino in Asia

Javan rhino

This species was distributed throughout lowland forests of southeastern Asia, but forest clearing and poaching reduced it to the world's most endangered large mammal. The males have one small horn and the females barely have a horn. A tiny population of 8–12 rhinos survives in one of Vietnam's last lowland forests (Lam Dong Province). Selective logging opens habitat for the rhino and poaching is a pressing threat. The aim is to protect the rhinos in a guarded sanctuary.

The only other population is the 57 rhinos left in the Ujung Kulon National Park, Java. In the early 1900s the rhino was already restricted to the national park, established first as a nature monument in 1921. Java is a densely settled island with 100 million people, and 92% of its forests have been cleared; yet with protection the Javan rhino hung on. In 1955, Ujung Kulon had 24–48 rhinos which were carefully protected. Nevertheless, poaching increased in the 1960s, and by 1967 only 25 rhinos survived. Protection was intensified and the rhinos doubled their numbers by 1980.

Sumatran rhino

This, the smallest of the rhinos, has a sparsely haired skin which has only a few platelike folds. The male's two horns are conspicuous compared to the female whose posterior horn may be almost invisible. The contraction in its range to Indonesia (Sumatra and Kalimantan) and Malaysia was apparent early in the 1900s, and the twin threats of forest clearing and poaching that drove its numbers down still hang over its future. The two-horned species has fared better than the one-horned Javan rhino probably because it is more of a highland species and mountain areas were cleared later. On Sumatra in 1993, less than 400 rhinos occur in three national parks with some individuals outside them (S. Stuart, personal communication, 1994). Poaching is serious: at least 10 rhinos were killed in 1990 from just one of the parks. In Malaysia, about 36–65 rhinos occur in two parks and a scatter of 14 other forests have less than 50 rhinos.

Great one-horned rhino

The Indian rhino is large with conspicuous skin folds lending it a plated appearance; its single horn is usually 24–30 cm. The rhino's habitat is marsh and open grasslands, and its conversion for agriculture and harvesting reduced the rhino to small populations in India and Nepal totaling some 1,700 rhinos in 1989.

In 1910, the shrinking numbers of rhinos led to their protection from hunting and a system of sanctuaries. In Assam, conversion of land for tea plantations left only 12 rhinos by 1908 in an area protected first as a sanctuary and then as a national park (Kaziranga). In 1955, competition with domestic cattle grazing and the risk of cattle-borne diseases were threats. Protection from poaching resulted in an increase to 1,129 rhinos in 1989. In the late 1980s, poaching increased from 12 (1972–1978) to 235 (1983–1989), but measures against illegal traders and poachers reduced the kill in 1991.

The other large population of Indian rhino survived in the isolated and thinly inhabited Rapti River Valley, Nepal. Since 1846 the rhino and its habitat was

(Continued)

Box 10.5.   (Continued)

recognized as "royal," and although the Ranas hunted rhinos a retroactive calculation was that the annual harvest was about 3%. The powers of the Ranas lapsed, and people moved into the Rapti valley where malaria had been eradicated. Rhino numbers collapsed from about 1,000 in 1950 to 81–108 in 1968. The habitat decreased by 70% and the rhino population by 88%. The decline since 1950 was at a constant rate, whereas poaching fluctuated. The supposition was that it was grazing by cattle that had a greater effect driving the decline of the rhino. The area became a sanctuary and then the Royal Chitwan National Park in 1973. The rhinos and the grasslands were heavily guarded and the population recovered to 375 in 1989. In the 1990s that protection may have wavered as poaching appears to be increasing. This population was the source for a translocation to the Royal Bardia National Park in 1986 and Dudhwa National Park in 1984–1985.

The other populations are all in reserves. Their numbers do not individually exceed 65, and in spite of efforts to protect them they are threatened by poaching and encroachment on their habitat.

(Modified from Talbot LM. A look at threatened species. A report on some animals of the Middle East and southern Asia which are threatened with extermination. London: Fauna Preservation Society for IUCN, 1960; Caughley G. Wildlife and recreation in the Trisuli watershed and other areas in Nepal. HMG/FAO/UNDP Trisuli Watershed Development Project, 1969; Dinerstein E, Jnawali SR. Greater one-horned rhinoceros populations in Nepal. In: Ryder OA, ed. Rhinoceros biology and conservation. Proceedings of an international conference. San Diego: Zoological Society, 1993:196–207; Khan MK, Nor BHM, Yusof E, Rahman MA. In-situ conservation of the Sumatran rhinoceros (*Dicerhinus sumatrensis*) a Malaysian experience. In: Ryder OA, ed. Rhinoceros biology and conservation. Proceedings of an international conference. San Diego: Zoological Society, 1993:238–247; Martin EB. The present-day trade routes and markets for rhinoceros products. In: Ryder OA, ed. Rhinoceros biology and conservation. Proceedings of an international conference. San Diego: Zoological Society, 1993:1–9; Ramono WS, Santiapillai C, Mackinnon K. Conservation and management of Javan rhino (*Rhinocerus sondaicus*) in Indonesia. In: Ryder OA, ed. Rhinoceros biology and conservation. Proceedings of an international conference. San Diego: Zoological Society, 1993:265–273; Santiapillai C, Giao PM, Dung VV. Conservation and management of Javan rhino (*Rhinocerus sondaicus annamiticus*) in Vietnam. In: Ryder OA, ed. Rhinoceros biology and conservation. Proceedings of an international conference. San Diego: Zoological Society, 1993:248–256; Sinha SP, Sawarkar VB. Management of the reintroduced great one horned rhinocerus (*Rhinocerus unicornus*) in Dudhwa National Park, Uttar Pradesh, India. In: Ryder OA, ed. Rhinoceros biology and conservation. Proceedings of an international conference. San Diego: Zoological Society, 1993:218–227.)

disturbance, hunting, and early dry-season burning of the grass have furthered the hare's decline. The best approach is a compromise worked out at the local level to ensure support among the people affected by strictures on use of the reserves. Harvesting of grass for thatch is carried out in Royal Chitwan National Park (Figure 4.21) under supervision but with difficulty as some 80,000 to 90,000 people are involved. The park staff are aware of the problems and have re-

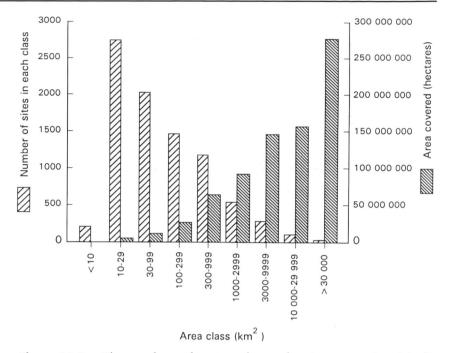

Figure 10.3. The numbers of protected areas by size categories. (Modified from Groombridge B. Global biodiversity: status of the Earth's living resources. London: Chapman and Hall, 1992.)

duced access to thatch collecting (Bell and Oliver 1992). The problems are not only for the hispid hare. Mishra et al (1987) recount the problems for tigers and people in Royal Chitwan National Park.

Reserves in practice vary widely in size (Figure 10.3), objectives, and time of existence. The larger the better is not automatically a valid criterion for reserves (Sections 10.2.2, 10.2.3) as ecological factors interplay with size. Kauri snails, *Parphanta busbyi busbyi*, are large endemic snails in New Zealand and they fared better in smaller than larger remnants of forest (Ogle 1987). Rats and pigs eat the snails, but the pigs do not occupy the smaller patches (Ballance 1986, Ogle 1987).

There are few very large reserves (exceeding 10,000 km²), and biogeographical theory and population viability models have been used to argue that most national parks in North America are too small to retain a full complement of species indefinitely (Schonewald-Cox 1983, Newmark 1985, Salwasser et al 1987, Grumbine 1990). Simberloff (1988) then argued that intensive manipulative management in smaller reserves was the way of the future. If that is indeed true, then it does not bode well for retaining large carnivores such as grizzly bears in all the parks, but equally the predictions suggest a few parks are large enough. Using persistence of large animals in the few long-established parks is an insufficient test for the success or otherwise of the reserves because the variance of minimum viable population estimates is large.

Intensive management of smaller reserves would bring its own set of problems. First, we would be struggling against the frontiers of our current ecological understanding. Second, management slides alarmingly quickly into management for its own sake unless it adheres to tightly written objectives with stopping rules (Box 12.2). Third, public controversy, especially as it slips over into the jurisdiction of the courts, cannot be lightly dismissed. Reserves, especially national parks, have already become emotionally charged battlegrounds when sight is lost of their objectives or objectivity. Yellowstone National Park is an example (Chase 1987), or we could turn to some of the African national parks (Owen-Smith 1983).

Large carnivores bring their own particular set of problems to reserves. Herdsmen in the Annapurna Conservation Area, Nepal (Figure 4.21), reported snow leopards, *Panthera uncia*, sallied forth from the reserve and took livestock to the value of about 25% the average per capita income (Oli et al 1994). The herdsmen consequently kill leopards, but a cash compensation scheme could protect both the interests of herdsmen and leopards. Innovative methods helped solve conflict when people needed to continue using a reserve that was established for tigers. The Sundarbans (6,000 km^2) forest is the only refuge for the Bengal tiger in Bangladesh (Figure 4.21) and was set up under Project Tiger. The tigers have persisted despite management of the forests for timber, firewood, and honey. Woodcutters and honey collectors who enter the forest ran the risk of being killed by one of the estimated 250 tigers in the area (Figure 10.4). Measures to reduce the human kill have included alternative water and food for the tigers (fresh [sweet] water rather than the brackish water and farm-bred wild pigs released in the buffer zone), removal of known maneaters, reduction of permits for collecting in high-risk areas (tiger dens), and aversive conditioning of the tigers by electrified dummies (Sanyal 1987). The measures are apparently successful as fewer people were killed, although the number of people confident enough to enter the reserve doubled between 1983 and 1985.

Reserves can, if we are not alert, lull us into a sense of false security. Assuming that a species is secure in a reserve network is risky unless its populations are monitored. Examples where a species declined or became extinct in a reserve are given in the preceding text. But there is yet another side to this: because a relatively widespread and common species is thriving in a reserve does not mean that the same holds for outside the reserve. The African elephant (Section 4.8) is an example of delays in detecting a decline because most research was in reserves where elephants thrived. Outside reserves, elephants declined wherever they competed with people trying to make a living from the same land.

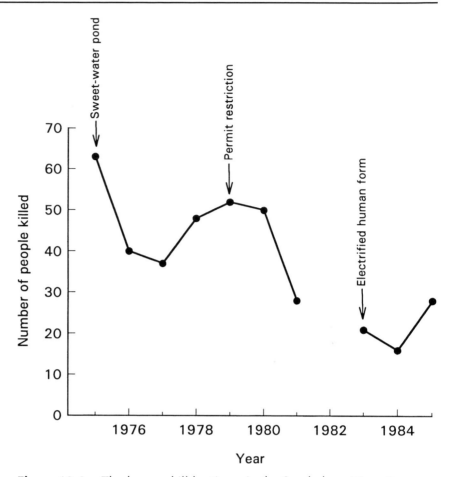

Figure 10.4. The human kill by tigers in the Sundarbans Tiger Reserve. (Modified from Sanyal P. Managing the man-eaters in the Sundarbans Tiger Reserve of India—a case study. In: Tilson RL, Seal US, eds. Tigers of the world. The biology, biopolitics, management, and conservation of an endangered species. Park Ridge NJ: Noyes, 1987:427–433.)

10.4   OFF-RESERVE CONSERVATION

Reserves are one end of a continuum of ways to manage areas for conservation. Only a small area of land can and will be held in reserves where land is specifically managed or held for conservation as the primary objectives, and even that will not proportionately represent all habitats (Groombridge 1992). Off-reserves are managed lands where conservation is not the overriding priority but control of specific activities protects some habitat. This can be achieved on privately and publicly owned land through legal means (Chapter 12). Off-reserves roughly fall into three categories: land linking individual reserves to form a reserve network, land to surround reserves to buffer them from surrounding land uses, and multiuse land where a secondary objective is conservation.

Biosphere reserves are an international program that endeavors to integrate these three categories to maintain conservation objectives compatible with people making their living in the same area. The United Nations Educational, Scientific and Cultural Organization (UNESCO) is behind the program, which essentially relies on zoning adjunct areas united under a single management plan and often with a complex management structure. Core areas are for conservation, buffer areas for management, and transition areas for sustainable use. Many biosphere reserves are only a few years old, and the program itself was only conceived in 1968 (Batisse 1993). An example is the 1.6-M ha Maya Biosphere Reserve in Guatemala (Santiso 1993). Half the area covers four national parks and three protected biotopes. The other half is a buffer area for those areas, and about 6,000 people make their living from the forestry and forestry products.

The safety net that an off-reserve system can provide is evident from the distribution of endangered mammals in Australia (Table 10.2). Eight mammals are found in only a single reserve, although two of these occur elsewhere, but on private land. Those species would benefit from off-reserve protection. Ten species are found in the off-reserves as well as reserves. Four species and their habitat are unprotected, being only on private land. Most of the species are semiarid or arid land fauna where extremes in environmental variation are unpredictable. In times of drought, the survival likely depends on refuges (Morton 1990).

Off-reserves usually serve more than one objective, and forest and grazing reserves are common examples. In South Australia the arid zone reserve system jumped from 3.7 M ha in 1982 to 19.2 M ha in 1992 (Cohen 1992). The reserves are multiuse, but it has yet to be demonstrated that grazing by domestic stock is not detrimental to wildlife. The conflict between grazing and conservation (Caughley 1986 and Curry and Hacker 1990 offer differing viewpoints) has its counterparts in Africa (Prins 1992) and North America (Fleischner 1994). The consequences of grazing in rangelands are complex (Walker 1993), usually underestimated, and often far-reaching, and management for rangelands for stock grazing and conservation of wildlife will be tricky. Meantime, it would be prudent to maintain ungrazed reserves for wildlife.

Progress has been made, at least in planning, to manage forestry reserves to combine forestry and wildlife objectives. Recovery plans for the red-cockaded woodpecker and the northern spotted owl call for management on reserve and off-reserve lands. Doubts about whether enough is known about the habitat requirements of either species add urgency to the need to take an experimental approach (Irwin and Wigley 1993). Conflicts already abound that management in national forests to maximize lumber yield reduces habitat for the red-cockaded

Table 10.2. The distribution of endangered Australian mammals on reserves, off-reserves, and private lands

	Reserve		Off-reserve		
	National park	Nature reserve	State forest	Other	Private lands
Western barred bandicoot *Perameles bougainville*	—	●	—	—	—
Northern hairy-nosed wombat *Lasiorhinus krefftii*	●	—	—	—	—
Banded hare-wallaby *Lagostrophus fasciatus*	—	●	—	—	—
Bridled nailtail wallaby *Onchogalea fraenata*	—	●	—	—	—
Greater stick-nest rat *Leporillus conditor*	●	—	—	—	—
Shark Bay mouse *Pseudomys praeconis*	—	●	—	—	—
Kowari *Dasyuroides byrnei*	●	—	—	—	●
Rufous hare-wallaby *Lagorchestes hirsutus*	—	●	—	—	●
Dibbler *Parantechinus apicalis*	●	●	—	—	—
Burrowing betong *Bettongia lesueur*	●	—	—	—	—
Golden-backed tree-rat *Mesembriomys macrurus*	●	●	—	—	—
Numbat *Myrmecobius fasciatus*	—	●	●	—	—
Western quoll *D. geoffroii*	●	●	●	●	—
Red-tailed phasogale *Phascogale calura*	●	●	●	—	—
Sandhill dunnart *Sminthopis psammophila*	—	●	—	●	●
Golden bandicoot *Isoodon auratus*	—	●	—	●	—
Leadbeater's possum *Gymnobelideus leadbeateri*	●	—	●	—	—
Mountain pygmy-possum *Burramys parvus*	●	—	—	●	—
Brush-tailed bettong *Be. penicillata*	●	●	●	●	—
Northern bettong *Be. tropica*	●	—	●	●	●

Table 10.2.   (Continued)

	Reserve		Off-reserve		
	National park	**Nature reserve**	**State forest**	**Other**	**Private lands**
Long-footed potoroo *Potorous longipes*	●	—	●	—	—
Julia Creek dunnart *S. douglasi*	—	—	—	—	●
Dusky hopping-mouse *Notomys fuscus*	—	—	—	—	●
Heath rat *Ps. shortridgei*	—	—	—	—	●
Central rock-rat *Zyzomys pedunculatus*	—	—	—	—	●
Desert bandicoot *P. eremiana*	—	—	—	—	?

Modified from Kennedy M. Australasian marsupials and monotremes. An action plan for their conservation. Gland, Switzerland: IUCN, 1992.

woodpecker—63% of all known woodpecker colonies occur in national forests (Jackson 1986). The ecological research on the habitat requirements for Leadbeater's possum (Box 9.3) is an example of information that could underpin the successful conservation on both reserves and off-reserves.

Management of multiuse reserves is complex and essentially untested. Hansen et al (1993) outline possible steps and their modeled consequences to manage a watershed in Oregon for lumber production and bird habitats. They emphasized the need to be armed with clear objectives and to break down the necessary management actions into components that relate to the suite of species and their habitat requirements.

The concept of an off-reserve system to bolster and buffer existing reserves was an attractive idea promoted in the 1980s. But it has struggled through unrealistic expectations and administrative and legislative problems compounded by difficulties of determining what constitutes sustainable use (Wells and Brandon 1993). An example is the Kosi Tappu Wildlife Reserve which holds the last Nepalese population of wild water buffalo (Figure 4.21). It was established as a sanctuary for the buffalo in the 1960s. A survey in 1968 identified the conflict between people making a living in the area and the park (Caughley 1969), but then this remained unresolved. Local people found the reserve restricted their access to harvest thatch, fuel wood, and fishing and the wild buffalo raided surrounding croplands (Heinen 1993). Buffer areas around established reserves now extend the habitat of the reserve while local people could use the buffer zones in ways compatible with the reserve's objectives. Time will tell whether the buffers will work.

Off-reserve sites run into the same problems as reserves. Objectives have to be specified, and protection from a specified threat is rarely enough as habitat requirements have to be met. In Britain, in response to the steep decline in the natterjack toad, *Bufo calamita*, most population sites were declared as reserves. Between 1985 and 1989, other sites were given legislative protection as off-reserves and were successfully defended against planned agricultural or tourist developments. But habitat degradation from shrub encroachment was similar between protected and unprotected sites, and shrub encroachment has coincided with the decline of several populations (Banks et al 1994).

An off-reserve system specifically to provide a species—the black rhino—protection from a single threat—poaching—is working in Kenya (Milliken et al 1993). The Kenyan sanctuaries which include private lands were in 1992 defined as fenced areas (55–390 km²) where rhino populations would be intensively managed. The program started in 1984 after poaching had reduced the rhinos to the level where it was no longer profitable to poach. Since then numbers have increased to 521 black rhinos of which a third are held on private land. The costs are huge: the outside agencies' support alone was $5 (U.S.) million between 1984 and 1991. A new outbreak of poaching is an ever-present threat as the rhinos are concentrated in 17 small predictable locations.

Long-term dependence on private land without a covenant or easements to ensure protection is risky. The southern white rhino's recovery in the Umfolozi Game Reserve and the Hluhluwe Game Reserve led to the translocation of 2,000 rhinos to public and private land. The 1,291 rhinos sold to 149 privately owned ranches from 1961 to 1987 had decreased to 931 by 1987. Some ranches were unsuitable range, and breeding success in small groups was low. The dominant reason was that the purchase price was lower than the income from "trophy" hunting and replacement rhinos could be purchased. Rhinos moved to public land fared better and increased (Anderson 1993). Our point is that dependence on private land to conserve a species may be risky when the impetus for the private landowner is commercial.

A recent development in the last two decades is legal settlement of the land claims of indigenous peoples. Transfer of large areas of land to the titled ownership or management of indigenous people such as the Inuit of Canada's Northwest Territories is occurring. Those lands can be expected to function as an off-reserve system for conservation within the priorities of the empowered landowners. The problems come with difficulties of balancing development to improve living standards and sustainable utilization of resources (Sections 11.4, 11.5).

Off-reserves can function as multiuse extensions of existing reserve systems or buffers around existing reserves or have specific roles on private or leased lands. Regardless of the objective, it has to be clearly defined and the ecology has to be well enough known to ensure that the off-reserve can provide the necessary resources for the species in trouble. In this way reserves and off-reserves require the same ecological approach.

10.5   SUMMARY

The design and management of reserves flows from whether the reserve aims to conserve states (a species) or ecological processes. Yet this logical first step is often overlooked. The theory of reserve design is largely derived from island biogeography despite indifferent success in testing its premises. Reserve size and number—whether one large or several small reserves will foster a species' persistence—shape and connecting corridors are topics that have preoccupied discussions. Metapopulation dynamics with its emphasis on habitat quality may be a more fruitful basis than the island biogeography's use of area as the dependent variable.

New reserves or modifications of existing reserves will in practice be constrained by social and economic factors as much as any theoretical requirements. Knowing the ecology of a species may be a more secure way to determine the necessary size and number of reserves as species have gone extinct in reserves which did not provide their habitat requirements. Vigilance against predators, be they human or otherwise, is often necessary especially when reserves are isolated enclaves in settled areas. Conflict between a reserve and needs of neighboring people is common, and one of the solutions is multiuse buffer zones.

A myriad of schemes to extend reserves and their functions are lumped as off-reserve systems. They may be buffer zones around existing reserves or multiuse areas with conservation as only one of their objectives. They run into many of the problems of reserves as well as complications depending on their ownership and security of their land tenure.

CHAPTER *11*

Economics
and trade

Trade in wildlife and its products understandably can make conservationists nervous. The collapse of the North American seal, sea lion, and fur seal trade is a single and adequate illustration of what can happen through overharvesting for monetary profit (Figure 11.1). Modern trade in wildlife is vast and its international dimension was annually valued at $5 (U.S.) billion in the 1980s (Fitzgerald 1989). The main problem for half the world's threatened animals (Groombridge 1992) is, however, the less-well-documented national trade in wildlife products, which includes overharvesting for locally consumed meat.

While a negative argument toward using wildlife for profit is well grounded in declines attributed to trade, counterarguments exist. Their point is that if wildlife is harvested sustainably, any profits generated will strengthen the case for conservation. Some take the argument further, claiming that if wildlife does not pay its way, particularly in poor countries, it will not be conserved. There is logic in both cases, but neither is complete. If profits can be made through conserving species, they may contribute to its conservation. But as we shall show, the logic of economics will, with many species, favor unsustainable, as compared to sustainable, wildlife use.

In this chapter, we review concepts of sustainable use and economics as they impinge conserving wildlife. We also briefly touch on ownership, as it too influences conservation.

11.1   THE DYNAMICS OF HARVESTING

Animal populations offer two routes to sustainable use: one is when the animal's death is not required—for example, down collected from nests of eider ducks, *Somateria*, or viewing wildlife—and the other is when the animal has to die. In the first, highest sustainable productivity comes from populations at their largest size. In the second it derives from populations whose growth is highest when they are below maximum size, and this section is concerned with those populations.

We illustrate the principles of sustained-yield harvesting by way of the logistic model. The growth of few species will conform to it precisely, but the principles it illustrates are general enough (Caughley 1977). The equation takes the form

$$\frac{dN}{dt} = r_m N \left(1 - \frac{N}{K} \right)$$

where dN/dt is the instantaneous increment to N, N is population size, r_m is the intrinsic rate of increase, and K is maximum population size. If we now harvest the population at an instantaneous rate H sufficient to hold N constant, the instantaneous offtake rate must equal the instan-

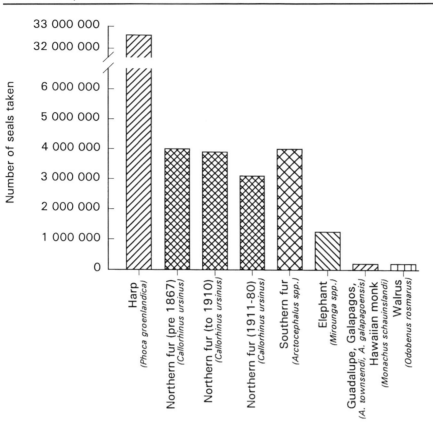

Figure 11.1.   Levels of the commercial harvest (to the nearest 100,000) taken by North American sealers between 1790 and 1980. (Modified from Busch BC. The war against the seals. A history of the North American seal fishery. Montreal: McGill and Queen's University Press, 1985.)

taneous growth increment. Hence, the instantaneous sustainable offtake *HN* from a given population size *N* is the same as *dN/dt*, the instantaneous growth increment, for that population size:

$$HN = r_m N \left(1 - \frac{N}{K} \right)$$

or as

$$HN = r_m N - \frac{r_m}{K} N^2$$

The second equation has the form $y = bx - cx^2$, which is the algebraic definition of an upwardly convex parabola passing through the origin (Figure 11.2). Because the sustained yield *SY* is the same as the harvesting increment *HN*, that equation informs us that *SY* is zero when the

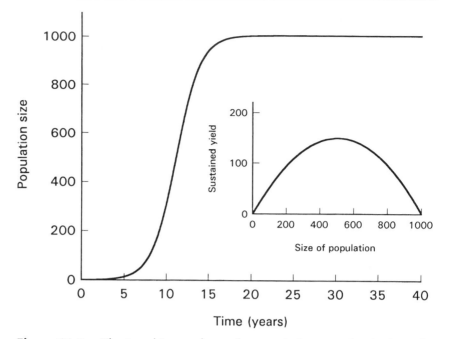

Figure 11.2.   The trend in numbers of a population growing logistically and the parabolic relationship between sustained yield and population size. (Modified from Richter WE. Handbook of computations for biological statistics of fish population. Fish Res Board Can Bull 1958,119:1–300.)

population size is zero (obviously if there is no population, there can be no yield), but that it is also zero when the population is at its maximum unharvested size K (because the realized growth rate at $N = K$ is zero). Thus any offtake from a population size of K will decrease the size of the population. Between $N = 0$ and $N = K$ the SY first rises and then falls (Figure 11.2). The maximum sustained yield MSY is taken from a population size of $N = \frac{1}{2}K$ at the instantaneous rate $H = \frac{1}{2}r_m$.

If an unharvested population of size K is harvested at its maximum sustained yield, it will progressively fall in numbers, first precipitately and then more gradually, until stabilizing at $N = \frac{1}{2}K$. Figure 11.3 shows the trend and what happens if the chosen yield is either above or below maximum sustained yield. It makes two points: first, any sustained offtake greater than MSY will drive the population to extinction. Second, the trend of an overharvested population heading for extinction is initially indistinguishable from the trend of a population declining to that level for which the offtake is a sustained yield.

That example, in which offtake (whether sustainable or not) is set at a fixed number of animals per year, leads to a sustained yield equal to or less than the MSY taken from a population size equal to or greater than furnishing the MSY. There is no way the population can settle on a size below that providing the MSY. If it is pushed into that zone by an

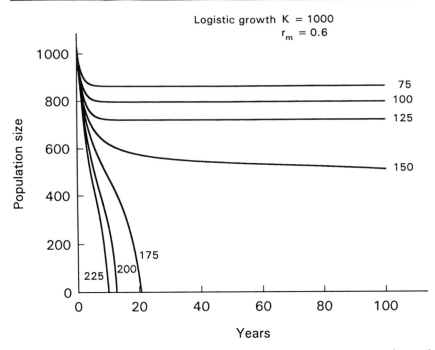

Figure 11.3. Trend of population size when a constant number of animals are removed each year. The maximum sustained yield is 150 per year. (Modified from Caughley G. Analysis of vertebrate populations. Chichester: Wiley, 1977.)

annual offtake that exceeds the *MSY* it will continue to decline unless the offtake is adjusted downward sharply.

However, authorized offtake is often determined, not as an absolute number but as an allowable effort. It might be set, for example, as the number of fish caught by 100, and no more than 100, trawlers working at maximum capacity. This system of regulation has an automatic negative feedback loop built into it. When the population is large (as might happen when a particularly strong year-class enters the fishery) the capped effort will take a large yield. When the population declines (as might happen with a year-class failure or a shift in the boundary between water masses) the constant effort will harvest a lower yield. As a close approximation a constant effort takes the same percentage of the population whether it is large or small. Thus the yield from constant-effort harvesting tends to track population size. However, in contrast to constant-effort harvesting, constant-yield harvesting can, and often does, push a population below the size yielding the *MSY* and can stabilize it at that reduced size which, by the logistic model, is $\frac{1}{2}K$. Figure 11.4 shows the trajectories of populations harvested in this way where harvesting effort is held constant and each harvesting regimen takes a different constant proportion of the population each year.

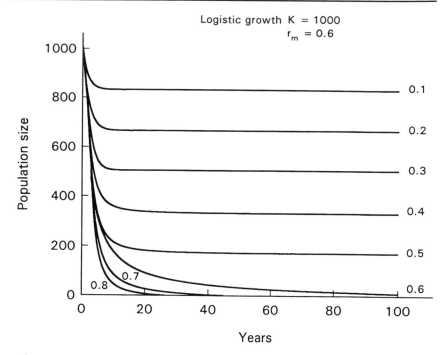

Figure 11.4. Trend of population size when a constant proportion of the population is removed each year. The maximum sustained yield is removed at a rate of 0.3. (Modified from Caughley G. Analysis of vertebrate populations. Chichester: Wiley, 1977.)

By this system of regulation the population size can easily end up on the wrong side of that providing the *MSY*. In Figure 11.2, an effort providing a sustained yield of 96 per year would stabilize the population at a size of either 800 or 200. Both are taking sustained yields, but the first takes a proportion 0.12 of the population (96/800 = 0.12), whereas the second takes a proportion 0.48 of the population (96/200 = 0.48). The second harvesting system, while taking the same sustained yield as the first, is taking it with four times the harvesting effort. That predicament, whereby the harvester is working more and enjoying it less, describes the present state of the majority of the world's fisheries.

11.2 ECONOMICS AND HARVESTING

Economics in itself has nothing to do with conservation other than studying its costs and benefits and how these are allocated. There is no doubt that some instruments of resource economics can be applied to conserving a species. Taxation policy, pricing policy, subsidies, and ownership of natural resources and harvesting rights can all be set to encourage protection of a species directly or to conserve its habitat. Resource economics, however, has two tendencies that work against conservation. The first is specific to population harvesting, the second is general.

The first tendency is a failure to regard sustained yield as a mandatory precondition to be met by any economic analysis of population harvesting questions. Resource economics literature has little on sustained-yield harvesting, perhaps for the reason that in many circumstances this option is less rewarding than harvesting the whole stock immediately and reinvesting the realized sum in an enterprise yielding a higher rate of return than that of a naturally growing stock. There are two quite separate problems posed by harvesting. One is calculating the appropriate offtake rate needed to achieve a sustained yield and so to protect the stock. The other is who should pay the costs and who should reap the benefits of the harvesting operation. One can find papers and even whole books devoted to the question of costs and benefits of harvesting that lack any mention of calculating the appropriate offtake. A good example is the discussion of the skipjack tuna, *Euthynnus pelamis*, fishery in the Pacific (Grynberg 1993).

The second, and more general, tendency is treating conservation problems as not ecological problems at all but as problems in applied economics. The rationale goes that market forces are so powerful the market will provide the desired outcome so long as the economic settings with which the market operates are right. The authors of the publication *Blueprint for a Green Economy* (Pearce et al 1989), also known as "The Pearce Report," argue that conservation is largely an economic problem and that its ills will be remedied by the correct pricing policy. As with others (e.g., Bowers 1990, Cocks 1992), we do not accept such reductionism: the issues are a great deal more complex.

It might seem intuitively obvious that the way to make the most money out of a renewable resource is to harvest the population down to a level somewhat higher than that yielding the *MSY* and then to harvest year after year at the *SY* (defined in terms of effort or numbers) appropriate to that population size. We would not try to take the estimated *MSY* itself for two reasons. First, the estimate of *MSY* is always an approximation and so a substantial margin of error must be built into the calculation of an appropriate *SY*. Second, the population size will fluctuate under the effect of environmental stochasticity, entirely independently of the effect of the harvesting on the population's size, to disrupt any neat relationship between the sustained yield and the population size from which it is taken. For both these reasons the *MSY* as such is a dangerous objective. But while keeping those caveats in mind we would maximize offtake to an upper limit set by a healthy margin of error. That roughly describes the way sport hunting is regulated in North America, parts of Europe, Australia, New Zealand, and southern Africa.

Any economic analysis of sport hunting is complicated by its nonmarket value. This follows because in the eye of the sport hunter its value exceeds the immediate products be they hide, meat, or antlers. In this respect the "products" from sport hunting fall within the same esthetic realm (although we realize not everyone would agree) as those from watching wildlife. Until *why* people pay to indulge themselves in those activities is explained satisfactorily in behavioral terms, their values will remain intangible. This has not prevented some economists from trying to do so, and their efforts are spreading to conservation biology.

The concept of "value" has broadened in the last two decades away from a narrow definition based on the marketplace (Stoll and Johnson 1984, Cocks 1992). As the definitions of nonmarket values (intangible human values) developed, so did the techniques for estimating them. The best known is contingent valuation, which seeks to convert nonmarket values into their equivalent dollar values.

Stoll and Johnson (1984) attempted to value the embattled flock of whooping cranes, *Grus americanus*. They found from questioning visitors to the Aransas National Wildlife Refuge in Texas (where the cranes winter, Figure 4.4) that the average visitor would pay an extra dollar for a permit to enter the refuge if the cranes were present. The public was then approached through a mail questionnaire which asked those who expected to visit the refuge at some time what they were prepared to pay to enter it, and those who did not expect to ever enter what they, nonetheless, felt would be a reasonable admission fee. With these data the authors then extrapolated the results from their sampling to the entire population of the United States and came up with $1.6 billion for the existence of whooping cranes. That approach—called contingent valuation—is vulnerable to problems that beset polling generally. It is one thing for the polled to speculate on values and quite another to get them to "put their money where their mouth is." While appreciating that economists need to reduce all values to monetary denominators for comparisons, we feel that such approaches show their limitations more than they answer questions. To a degree, the value of whooping cranes depends on their rarity—the same quality that led collectors to exterminate the great auk (Box 3.1). Decrease rarity and the whooping crane will become less valuable which, in turn and inter alia, will decrease the value of the Aransas National Wildlife Refuge. Sophist though this may seem, it illustrates incompatibility between what makes a species monetarily valuable, which is what the economist seeks, and its abundance rather than rarity (therefore low intangible value), which is what the biologist wants. The difference between objectives seems to have been overlooked by Pearce et al (1989), who argued that conservation is largely an economic problem whose ills can be remedied by correct pricing policies. They did not mention that nothing will drive up prices better than rarity.

The foregoing indicates our attitude toward the many unresolved technical problems inherent in estimating and comparing sets of economic values, particularly those which are straightforward market measurements and the intangibles, which brings us back to sport hunting. Returns from this activity and thus, profitability, are not absolutely tied to the yield as intangibles have to be factored in.

Initially it seems so intuitively obvious that commercial harvesting could follow the same approach as sport hunting of taking a sustained yield that it is strange that this is not done widely. The one field where it is attempted, although rather unsuccessfully, is fisheries. Sustained-yield harvesting does not hold for commercial hunting because of commerce's basic objective of maximizing monetary returns.

The principles of free-market renewable-resource economics, greatly simplified here, were derived in rigorous detail by Clark (1976). May (1976) applied them to the vexed question of why the whaling industry never went for a sustained yield, showing that although the decision was biologically inappropriate it was nonetheless economically rational. Thus the recent international quarrel (Conrad and Bjørndal 1993) as to whether in theory minke whales, *Balaenoptera acutorostrata*, can be harvested on a sustainable basis debated the wrong question. The answer is a clear yes. The right question, given the history of whaling over the last 40 years, is whether they would, in practice, be harvested for sustained yield.

The lifetime profit from any enterprise is the money earned minus the costs incurred in earning it. That is not the simple calculation it might appear because a dollar earned this year has a different value from a dollar earned in 10 years, quite apart from inflation. The reason lies in the present value of future earnings. A dollar earned now is worth appreciably more than one earned 10 years down the track because it can be invested for 10 years. The interest it earns over that period is a component of its real value compared with the dollar to be earned in 10 years. Money in the hand is worth more than a promissory note for the same amount. The present value of a future dollar, when we ignore complications such as the cost of repairing and replacing equipment, servicing loans, and so on, is approximately the dollar value corrected for inflation minus the interest that it would generate if we had access to it now. In many investments capital tends to grow (real growth plus inflation) at about 10% per year. Thus the present value of a dollar earned 10 years hence is approximately $1/1.1^{10} = \$0.39$. The present value of a dollar deferred for 20 years is $1/1.1^{20} = \$0.15$, almost nothing.

This is the one reason why sustained-yield harvesting is not necessarily the best economic strategy for utilizing a renewable resource. Because money earned now is worth more than money earned later, it may make good business sense to take a higher harvest now rather than a lower harvest later. What matters in this calculation is how profitably the returns from the higher harvest now can be reinvested elsewhere.

The extent to which sustained-yield harvesting may be economically inefficient depends on the alternative uses to which capital can be put. Think of whales as waterproof bank notes. The standing crop of whales represents a principal sum, and the sustained yield that can be harvested from it is the interest generated by that principal. The annual interest generated by a population of whales is less than 5% per year. We can now contrast two strategies of utilization: either harvest at a little below 5% per year indefinitely, or gather in the principal as rapidly as possible and reinvest the cash in an enterprise yielding a higher rate of interest. The first is the correct conservation decision. The second generates the most money over any time period, short or long. Thus the economic case for any species with an intrinsic rate of increase of less than about $r_m = 0.2$ would reduce it rapidly to economic extinction and invest the returns elsewhere rather than harvest it for sustained yield. The case strengthens with the value of the product. This high-risk category includes elephants, rhinos, whales, some birds such as parrots, some fish, and most timber trees. A species with an intrinsic rate of increase of $r_m > 0.2$ would more likely be harvested sustainably, but not necessarily at close to *MSY* because the calculation of the present value of future earnings dictates that more of the offtake should be taken earlier rather than later in the operation. The population may be reduced to below the level generating the *MSY* before it is harvested by a constant offtake per year. Thus a temporary higher early offtake is traded off against a permanently lowered future offtake, and this will maximize total earnings when transformed to their present value. The lower the r_m of the species the lower will the size of the population be reduced below the level generating *MSY* and the greater will be the disparity between earlier and later annual offtakes.

The harvesting above the sustainable yield may run down the natural capital which the resource represents, but it will not necessarily hamper the accumulation of gross capital. Estimates of national income and wealth do not account for the depreciation of a country's natural resources or recognize who incurs the cost of that depreciation (Cocks 1992). This question of resource property rights—the distribution of rights to control effort, to profit from it, and the risks associated with using it—is fundamental to conserving and harvesting wildlife.

While the question of who owns wildlife may not have exercised biologists much until recently (they are themselves a relatively new breed), it has exercised lawmakers for millennia. The core of the problem is that what is wild is by definition unownable, as to be owned the owner must be able to control an animal and be responsible for it. The inapplicability of this to wild animals was recognized by most societies from the earliest time. That recognition led to the view that it was the landowner, individual, or community who was the de facto owner of the wildlife on the land.

The type of property (private, common, state) and the type of access to the resource have to be separately considered when predicting the impact of economic incentives. Private ownership is having exclusive access to resources. Aguilera-Klink (1994) corrected Hardin's (1968) account of the tragedy of the commons: common property is not open access but access is restricted to those who are part of a group of co-owners of the resource. Thus, as such, they may have an interest in the conservation of the resource to which they have property rights. Completely open access to resources impedes conservation as the harvester has no incentive to invest his efforts if others will benefit.

A change from public to private ownership is often promoted as all that is needed to conserve; Rasker et al (1992) suggested that was the solution for African elephants, but we disagree (e.g., Caughley 1994). Economic incentives (e.g., McNeely 1993) are a common argument for game ranching and other types of private ownership (e.g., Grossman et al 1992). Chisholm and Moran (1993) also argue for market-based conservation and push for private ownership. They treat ownership as a simple phenomenon which, at least where it relates to wild animals, we believe is wrong.

Even if a landowner looks after the wildlife and ignores the rational economic reason for overharvesting there are still three limitations. First, private ownership is unlikely to promote the conservation of species that yield no tangible or economic rewards. Second, the ranges used by migratory species or other species needing large areas may exceed the units of ownership be they private or communal. It seems no coincidence that it is among the oceanic species where no possibility of ownership can be entertained that the greatest overharvesting has taken and is taking place.

The third limitation is that, while the owners reap the benefits of ownership, they are also by the same token responsible for any untoward consequences. When one's kangaroo leaps the boundary fence and nibbles the neighbor's prize-winning blooms, the issue of compensation can be straightforward. But suppose overfishing causes fish stocks to collapse, triggering a seabird's decline and that in turn removes a source of eggs normally harvested by people who live in a different jurisdiction from the overfishers. Then how can responsibility be enforced?

The great variation in forms of land tenure and pragmatic aspects of private ownership, where wildlife is concerned, means it is not a blanket panacea for conservation. Indeed there are many situations where it is irrelevant and in which the right structure is mixed and nested. The options then are to either rebuild a "community" capable of self-management or to allocate private harvest rights (such as leases) that are constrained (nested) by state- or community-determined regulations. Harvesting concessions in the form of long-term leases to harvest particular areas are suggested for the take of parrots in Indonesia (Lambert 1993). However, no resource ownership and rights are simple. Some might argue that exclusive concessions granted as a monopoly to an individual or company would confer the economic characteristics of private ownership. In our view that would only acquire such characteristics if it was granted in perpetuity, which would be unacceptable to most societies. A time limit on exclusive use concessions would alone tilt the user toward the economic rationale for overharvesting.

Another antidote is to place the harvesting of a resource under the control of a government department. The rationale is that the people running the department do not themselves profit from the harvesting and so will regulate the harvest in the interest of conserving the stock rather than in the interests of maximizing the immediate profit to the harvesters. It seldom works that way. The department and the industry, particularly if the department's focus is on production rather than on conservation, usually have a strong symbiotic relationship. The issue is not simple: governments are uneven in their ability to enforce their decisions, and even if their motives and intentions are conservation minded, they may be overruled by other lobbies.

11.3 EXAMPLES

The annual trade in reptile leather peaked in the 1950s and early 1960s at 5–10 million crocodile skins. As one species became economically extinct, the trade shifted to another (Luxmoore 1992). The annual trade of live wild birds has declined from a peak of 7.5 million birds in the early 1970s to between 2 and 5 million in the 1980s (Thomsen et al 1992). Taiwan annually sells 15–500 million butterflies with a value of $2–30 (U.S.) million (New 1991). The figures are impressive, but little is really known about the populations from which these harvests came, whether they are sustainable, or details of the harvests themselves. Nonetheless there are examples of wildlife trade about which considerable detail is available even if it is not always sufficient to indisputably demonstrate the effects of trade. We have selected three oceanic examples whose conservation issues are clearly economic in origin. We have included the rhinos because their example is, we suspect, common in that economics is assumed to be the problem, but other factors are contributing if not more important.

11.3.1 Example: the humpback whale

The harvesting history of the population of humpback whales that migrates up the west coast of Australia (the "Group IV" population) has been described by Chittleborough (1965). The population was harvested in Antarctic waters at an insignificant rate before 1934. Hunting intensified during the southern summer of 1934–1935 to that of 1938–

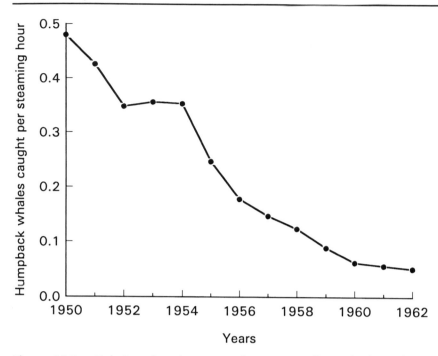

Figure 11.5.   Relative abundance (catch-per-unit effort) of a humpback whale population (Group IV) as measured each winter on the west coast of Australia. (Modified from Chittleborough RG. Dynamics of two populations of the humpback whale, *Megaptera novaeangliae* (Borowski). Austral J Marine Fresh Res 1965;16:33–128.)

1939 when 7,244 were taken by the Australian shore stations and 5,429 by pelagic whalers. The population collapsed but was immediately given de facto protection by World War II and apparently increased markedly in the 10 years before whaling recommenced in 1949. Thereafter quotas were set by the International Whaling Commission (IWC), whose charter was to preserve the stocks of all harvested whales and to guard the interests of the whaling industry. By 1960 the population was severely depleted, and the IWC proposed to close this population's Antarctic feeding ground to pelagic whaling for a period of 3 years. Objections lodged by the governments of Japan, Norway, U.S.S.R., and the United Kingdom effectively scuttled that plan.

The population collapsed to economic extinction in 1963. Figure 11.5 shows the decline in catch-per-unit effort recorded for the Australian shore stations and indicates that the population had been overharvested from at least 1955. Table 11.1 shows the trend in quota and offtake for the Australian shore stations. The Australian Department of Primary Industry negotiated their quotas with the IWC. Note that the catch after 1957 only reached the quota in 1961. As a consequence of that annual quota being achieved, the quota for the next year (1962) was raised. These figures did not take into account possible increases in

Table 11.1. Catch and quota for shore-based humpback whaling off the coast of Western Australia; the catch in Antarctic waters by other whaling nations harvesting the same population ("Group IV" of the whaling literature)

Year	Australian quota	Australian catch (shore-based)	Antarctic catch (pelagic)	Total recorded catch
1949	600	190	0	190
1950	1,200	388	779	1,167
1951	1,250	1,224	1,112	2,336
1952	1,250	1,187	1,127	2,314
1953	1,303	1,303	193	1,496
1954	1,320	1,320	258	1,578
1955	1,126	1,126	28	1,174
1956	1,120	1,119	832	1,951
1957	1,120	1,120	0	1,120
1958	1,120	967	0	967
1959	1,175	700	1,413	2,113
1960	870	545	66	611
1961	580	580	4	584
1962	640	543	56	599
Totals	14,674	12,312	5,868	18,180

Modified from Chittleborough RG. Dynamics of two populations of the humpback whale, *Megaptera novaeangliae* (Borowski). Austral. J. Marine Fresh. Res. 1965;16:33–128.

the industry's ability to take whales (and commerce constantly refines such abilities) and suggest the quotas were being set according to estimates of what the shore stations could take when working at full capacity rather than according to the state of the stock.

Recent revelations suggest another factor was at work and that was deliberate underreporting. Soviet factory ships only reported 2,710 humpbacks compared to the 48,477 actually processed from the Southern Hemisphere between 1948 and 1973 (Baker and Palumbi 1994).

11.3.2 Example: the bluefin tuna

The bluefin tuna is a powerful fast predatory fish growing to 700 kg. There are two species of bluefin: the Atlantic, *Thunnus thynnus*, and southern Pacific bluefin, *T. maccoyi*. Bluefin tuna fishing is a tiny fraction of the world's fisheries—it only accounts for <1% by volume, but its value is high and increasing as the stocks fall to where extinction is a possibility. The bluefin tuna fisheries have achieved prominence as international debate focuses on that possibility. We have followed Safina's (1993) review of the recent history and present state of this stock and Gaski's (1993) compilation of the international trade in tuna.

Japan is the leading market for price and quality and consumes about 40% of the world's catch of bluefin (Gaski 1993). A single fish can earn a fisher up to $30,000 (U.S.) and it retails in Japanese sushi restaurants for $770 (U.S.) per kg. The trend in market value has been upward while imports have shown a downward trend as catches have fallen off (Figure 11.6). The Japanese market was only "discovered" in 1970. Until then, the Japanese fishing fleet supplied its own national market, but such is the momentum of the valuable market that now 59 countries export bluefin tuna to Japan: 78% comes from Australia, United States, Taiwan, Spain, and Canada.

The by-catch of dolphins brought publicity to the tuna fishery. That debate was about the tropical yellowfin tuna, *T. albacares*, of the eastern Pacific. Neither the Australian purse seine fishery nor the long-line bluefin fishery incidentally catch dolphins. The incidental catch for long-lining is, however, albatrosses, *Diomedea* spp., and other seabirds caught while trying to scavenge baits. Murray et al (1993) report evidence for the role of long-lining in the decline of some seabirds in the southern oceans and the effect of mitigative measures designed to reduce the by-catch of the birds.

The body managing the offtake of the Atlantic tuna is the International Commission for the Conservation of Atlantic Tunas (ICCAT),

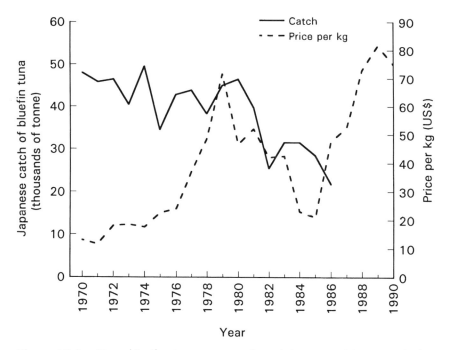

Figure 11.6. Trend in the Japanese catch and the annual import wholesale price of fresh bluefin tuna. (Modified from Gaski AL. Species in danger. Bluefin tuna: an examination of the international trade with emphasis on the Japanese market. Cambridge: TRAFFIC International, 1993.)

which consists of representatives of 20 Atlantic-rim nations, Korea and Japan as major importers, and is an important harvester in its own right. The convention establishing the commission was signed in 1966 after the introduction of purse seine or pelagic long-line fisheries in the 1950s and 1960s had been followed by the decline of several Atlantic and Mediterranean stocks. ICCAT's mandate is to regulate the harvest to or below the *MSY* level, but it oversaw the reduction of the stock below the level generating the *MSY* and did not adopt a recovery strategy until the 1990s. Its initial steps to stem the decline were the introduction of a minimum size limit of 6.5 kg in 1974 and, in 1982, a catch limit of 1,160 tonnes for scientific monitoring, which was raised to 2,660 tonnes in 1983 (ICCAT 1994). The decline of the breeding stocks continued. Also in 1982 a decision was made that, despite differing opinions as to the validity of the existence of different stocks, the tuna were divided into eastern and western, with regulations to apply to the western stock.

Figure 11.7 shows the trend of the breeding stock (adults >7 years of age) estimated by ICCAT's scientific committee. The *MSY* would be taken from a total fishable biomass equivalent to 273,000–546,000 individuals, but the stock had been pushed through that level by about

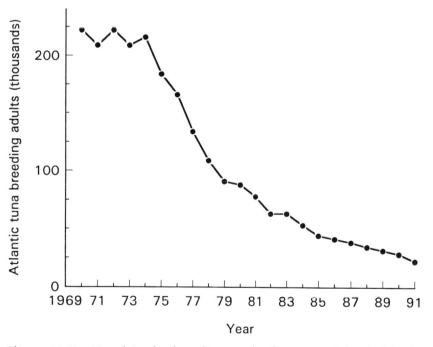

Figure 11.7. Trend in the breeding stock of western Atlantic bluefin tuna as estimated by the International Commission for Conservation of Atlantic Tunas (ICCAT). (Modified from Safina C. Bluefin tuna in the west Atlantic: negligent management and the making of an endangered species. Conserv Biol 1993;7:229–234.)

1977. ICCAT uses three models to estimate *MSY* based on the 1950 biomass, and those median estimates were 3,942–6,755 tonnes (ICCAT 1994). In 1990 the Commission's scientific committee noted that the catch quota would reduce the breeding stock further, and in 1991 it estimated that the breeding stock had declined by 24% over the previous year. By 1993 the biomass was estimated to be only 6–12% of that which could yield the *MSY*. At the ICCAT 1991 meeting Japan suggested a ban on trading in non-ICCAT tuna (a proportion of the catch, perhaps more than half, is taken by non-ICCAT countries) but it was opposed by the European Economic Community. The United States suggested a 50% reduction in annual quota, but eventually the agreed reduction was set at 17.5% over 4 years.

In 1991 Sweden suggested that the western Atlantic bluefin tuna be placed on the CITES Appendix I, which would have restricted imports into Japan. Normally CITES does not list species if they are the subject of a convention that predates CITES. However, CITES can intervene if it appears the species is in trouble (from international trade). The proposal was withdrawn at the 1992 Conference of the Parties to CITES in Kyoto, Japan, when ICCAT nations made commitments to take steps to conserve tuna. In late 1993 ICCAT reached agreement to reduce the 1994 quota 15% from the 1993 quota and, in 1995, to take 50% of the 1993 quota, a reduction to 1,200 tonnes (Gaski 1993). The figure of 1,200 tonnes comes from ICCAT's estimate of a catch ". . . as providing a 50% chance that there would be no further decline in the spawning stock" (ICCAT 1994). Despite the apparent authority of the 1,200 tonnes it is important not to lose sight that it is a model prediction and the model is generated from abundance indices and catch. ICCAT's aim is by 2008 to have increased the breeding stock biomass by 50%, although that is still not the level from which an *MSY* can be taken. In 1994 the United States challenged ICCAT: they argued that the decline was not real and that part of the confusion was an estimated net exchange of 1–2% between the eastern and western stock (Holden 1994)—all of which goes to show the wrangles that follow when national interests clash and the management data are less than adequate.

The second species of bluefin is the southern bluefin tuna—the Southern Hemisphere equivalent of the Atlantic bluefin tuna. The circumpolar stock spawns in the tropical waters between Australia and Indonesia. Most of the catch is taken by Japan and Australia, with significant catches by Indonesia and Taiwan and minor catches by Korea and New Zealand in waters south of latitude 35°S. This account is based on those of Caton et al (1993) and Nishida (1993).

The fishery expanded in the 1950s and 1960s, and Japan's catch topped 1,000 tonnes for the first time in 1953, and Australia's in 1957. Japan's maximum catch was taken in 1961 (77,000 t) and Australia's in 1982 (21,000 t). The Japanese catch from the breeding stock (>7 years of age) is taken across parts of the southern Atlantic Ocean, Indian Ocean, and western Pacific Ocean. The Australian catch historically was juv-

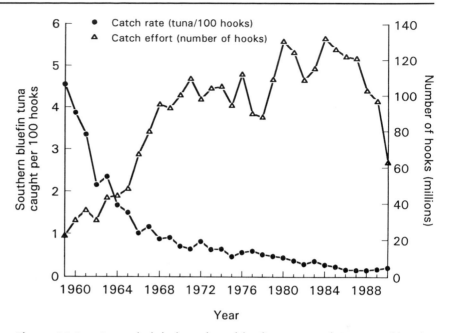

Figure 11.8. Annual global southern bluefin tuna catch (rate) and hooks (effort) set by the Japanese longline fishery. Comparable data from other nations are not available. (Modified from Nishida T. Status report on Japanese southern bluefin tuna (*Thunnus maccoyii*) fisheries. Hobart: CSIRO Marine Laboratories, 1993.)

eniles taken by surface fisheries off the southern coast of Australia. The surface fishery catch off the southeastern coast declined after 1983 as surface schools of young juveniles ceased to occur there.

The decline in abundance of the global breeding stock of the southern bluefin tuna is shown in Figure 11.8. The catch-per-unit effort, measured as fish per 100 hooks, declined progressively from 1960, whereas the fishing effort, measured as hooks set, increased progressively.

Table 11.2 gives bluefin tuna quota and catches for the Australian and Japanese fisheries. The quotas were arrived at by negotiation between the two governments. The Australian catch appears to have been constrained by quota from 1985 onward (the catch is close to the quota), but initially quotas for Japan have had no constraining influence and the industry working at full capacity could not reach the annual quotas.

The figures illustrate that the species has been overharvested severely and that future yields will go on falling to that point at which it is not worth catching them (economic extinction) or they will become extinct. Beverton (1990) explains this: the escalation of catchability q is the ratio of fishing mortality F to fishing effort f and it increases as population size decreases. Technical enhancements, such as satellite

Table 11.2. Catch and quota (tonnes) for the Australian and Japanese southern bluefin tuna fisheries

Year	Australian		Japanese	
	Quota	**Catch**	**Quota**	**Catch**
1984	21,000	15,843	None set	23,323
1985	14,500	13,486	None set	20,393
1986	14,500	13,237	None set	15,522
1987	11,500	11,308	23,150	13,955
1988	11,500	10,976	19,500	11,422
1989	6,250	5,984	8,800	9,150
1990	5,265	4,848	6,065	7,056
1991	5,265[a]	4,315[a]	6,065	6,774*
1992	5,265[b]	4,894[b]	6,065	6,973*
1993	5,365[c,d]	5,212[c]	6,065	n.a.

[a] 1,000 t voluntarily withheld, except for 300 t allocated to a program monitoring Japanese longline catch, effort, and size composition in real time (RTMP). The Australian catch includes the 300 t Japanese RTMP catch.
[b] Includes 800 t Japanese RTMP allocation and catch.
[c] Includes 650 t Japanese RTMP allocation and catch.
[d] 100 t of 1994 Australian quota was brought forward in case of quota "overshoot." Even though initial quota was not exceeded, the 1994 quota was reduced to 5,165 t.
* = estimate; n.a. = not available.
Modified from Caton A, Williams MJ, Ramirez C. The Australian 1989–90 to 1992–93 southern bluefin tuna seasons. Hobart: CSIRO Marine Laboratories, 1993; Nishida T. Status report on Japanese southern bluefin tuna (*Thunnus maccoyii*) fisheries. Hobart: CSIRO Marine Laboratories, 1993.

imagery to monitor ocean surface temperatures to locate upwellings where the tuna schools concentrate, boost fishing effort. That type of short-term biological irresponsibility (note that it is not necessarily economic irresponsibility) would be less of a surprise originating from an industry operating on the edge of commercial constraints. This is different. It is governments working in cooperation with their fishing industries to maximize the yield of dollars and yen over a relatively long period while minimizing the yield of biological products over the same period. That would not be much of a surprise in the 1930s. But it is still going on in the 1990s and for the same reasons. The total allowable catch was unchanged between the 1992–1993 and the 1993–1994 seasons despite the concerns raised by the scientists that the decline was continuing (Gaski 1993). Australia, New Zealand, and Japan signed a convention in May 1993 to conserve the southern bluefin, and it will have to reconcile the views of the fishery industry and the scientists. Industry holds that the stocks are recovering. The scientists report that the parental biomass of the tuna in 1992 was 10–20% of that in 1965 (TRAFFIC 1994).

11.3.3 Example: African and Asian rhinos

The decline in rhinos requires more than an understanding of economics, and thus the story of their decline parallels the elephant (Section 4.8). Rhino numbers have plummeted and their ranges contracted over the last few centuries (Figure 11.9). Conversion of their forest and grassland habitats for agriculture and forestry as the numbers of people grew took its toll. The longevity of the decline is often seemingly overlooked. This century, rhinos have been mostly reduced to scattered populations in reserves (Figure 11.9), and by the 1990s no more than 11,000 individuals of all five species are left. The trends of increasing populations (Figure 11.10) have to be seen as small flurries of hope against a woeful history of decline encapsulated in the maps of the contracted ranges.

The high value of rhino horn has maintained trade despite regulations to reduce the horn supply to the market by protecting the rhinos from illegal harvesting. Legislation to reduce demand has had limited success, and legislation to restrict trade in horn essentially drove it underground. The structure of the market controls the supply and demand, and that is where efforts will have to be directed to halt the trade in rhino products, of which horn is the most used.

The black rhino has declined rapidly this century. By 1900, it had disappeared from most of western and southern Africa, and its range has continued to contract and fragment. Even so, in the 1960s there may have been 100,000 left, but then a rapid slide began (Figure 11.10). Three decades later, 81% of the rhinos were gone, leaving a scant 2,500 (1992). If the rate of decline remains similar, numbers are halving every 5–6 years. As numbers tumble, the effect of illegal harvesting increases.

The southern white rhino sank to its nadir in numbers by 1910, but since it has been protected in sanctuaries numbers have built up from 20 to 30 rhinos to 6,000 (Section 10.3). The status of individual populations of African rhinos is summarized in the IUCN/Species Survival Commission Action Plan (Cumming et al 1990). The three species of Asian rhinos have shrunk to remnants of their former distribution under the loss of their habitat and harvesting. Information on Asian rhinos is more scattered, and thus we have compiled a summary (Table 11.3 and Box 10.5).

The trade in rhino parts has continued despite herculean efforts of international and national legislators to regulate the import and possession of rhino horn. For their part, the exporting countries have strived to prevent illegal harvesting through antipoaching patrols, fenced sanctuaries, and, in Namibia, Swaziland, and Zimbabwe, capturing and dehorning black rhinos. Success has been mixed. South Africa reversed its downward trend in southern white rhinos and for

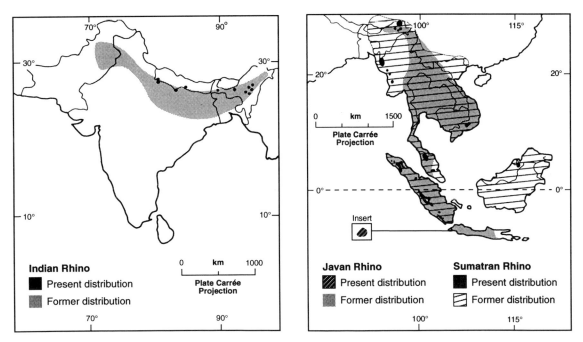

Figure 11.9. Contraction in the distribution of five species of rhino. (Modified from Fitzgerald 1989; Santiapillai et al 1993; Satiapillai and MacKinnon 1993; and Romono et al 1993.)

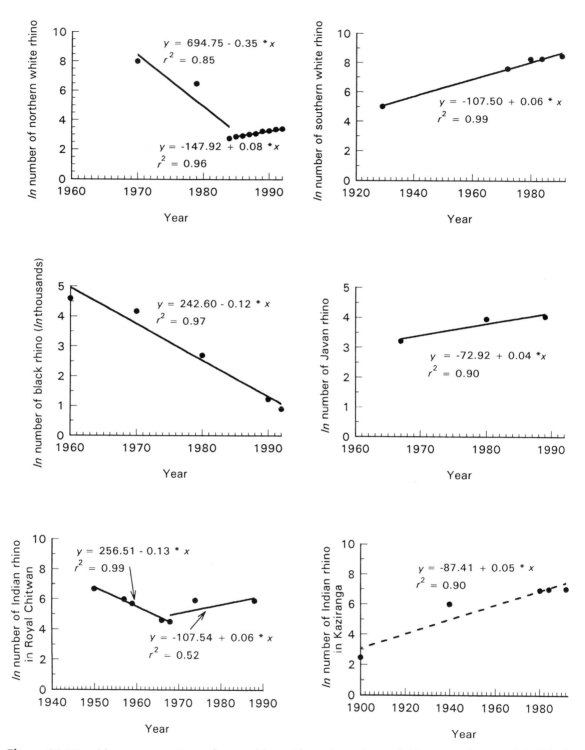

Figure 11.10. Linear regressions of natural logarithm of numbers of rhino populations. (Modified from Anderson 1993; Bhattacharya 1993; Caughley 1969; Gakahu 1993; Fitzgerald 1989; Martin 1993; Milliken et al 1993; Ramono et al 1993; Santiapillai et al 1993; Smith et al 1993; Vigne and Martin 1991.)

Table 11.3. Locations and numbers of Asian rhinos

Country	Location	Rhino status			Area	
		Estimate	Year	Poaching	Status	km²
Great one-horned rhino						
Bhutan/ India	Manas	60	1989	1985–89, 18 killed	WLS	391
India	Dudhwa[a]	9	1989	Unknown	NP	490
India	Kaziranga	1,080	1989	1983–89, 235 killed	NP	430
India	Laokhowa	5	1989	All killed	WLS	70
India	Orang	65	1989	1985–89, 21 killed	WLS	73
India	Pobitara	40	1989	1985–89, 8 killed	WLS	39
Nepal	Royal Bardia[b]	38	1991	Unknown	WLS	968
Nepal	Royal Chitwan	375	1989	1973–90, 35 killed	NP	932
Sumatran rhino						
Malaysia	Taman Negara	26–40	1987	Some poaching at 6 sites	NP	434
	Endau Rompin	10–25	1987		State Park	
	Others	48–68	1987		12 unprotected sites	
Sumatra	Gunung Leuser	132–200	1986	10–20 killed/yr 1980s	NP	7,927
	Kerinci-Seblat	250–500	1986		NP	14,847
	Barisan Selatan	25–60	1986		NP	365
	Others (4 sites)	20–35	1986			
Javan rhino						
Java	Ujung Kulon	57	1990		NP	784
Vietnam	Lam Dong	8–12	1991		Province	350

[a] Translocated from Assam (5) and Nepal (4) 1984–1985.
[b] Translocated from Royal Chitwan: 13 in 1986 and 25 in 1991.
NP = National park; WLS = Wildlife sanctuary/reserve.
Modified from Dinerstein E, Jnawali SR. Greater one-horned rhinoceros populations in Nepal, 1993:196–207; Khan MK, Nor BHM, Yusof E. In-situ conservation of the Sumatran rhinoceros (*Dicerhinus sumatrensis*) a Malaysian experience, 1993:238–247; Martin EB, The present-day trade routes and markets for rhinoceros products, 1993:1–9; Ramono WS, Santiapillai C, MacKinnon K. Conservation and management of Javan rhino (*Rhinocerus sondaicus*) in Indonesia, 1993:265–273; Santiapillai C, Giao PM, Dung VV. Conservation and management of Javan rhino (*Rhinocerus sondaicus annamiticus*) in Vietnam, 1993:248–256; Sinha SP, Sawarkan VB. Management of the reintroduced great one horned rhinocerus (*Rhinocerus unicornus*) in Dudhwa Natural Park, Uttar Pradesh, India, 1993:218–227. All of these articles are taken from Ryder OA, ed. Rhinoceros biology and conservation. Proceedings of an international conference. San Diego: Zoological Society, 1993.

two major populations of black rhino. Zimbabwe is losing rhinos to poaching even after they have been dehorned (Milliken et al 1993). Kenya has depended on small heavily guarded reserves to rebuild its black rhino populations after almost losing them to poaching. Even where the sanctuaries have been effective, poaching still occurs, which raises the specter that the protection could be quickly undermined. Elsewhere, since 1981, six African countries have lost their black rhino populations. Nine countries have had a decline of 50% or more of their rhinos.

In the Royal Chitwan National Park, Nepal, numbers of the Indian rhino sank from 800 in 1950 to about 100 in the late 1960s. Then protection became tighter under the watch of several hundred guards and the downward trend reversed, and by 1990 some 400 rhinos inhabited the park (Box 10.5). Indonesia and Malaysia have both relied on anti-poaching efforts in national parks with success.

In the following we summarize the changes in the trade of rhino parts following international and national legislation. We describe the market and how it is structured. TRAFFIC (Section 12.1.2) reviewed trade in rhino horn and we have drawn on those reports in this brief account (Leader-Williams 1992, Nowell et al 1992, and Milliken et al 1993). Another valuable source is Martin (1993).

The market is centuries old. Chinese, Burmese, Thais, Nepalese, and, more recently, Japanese and Koreans use rhino parts in traditional medicines. Powdered horn is a lifesaving medicine for persistent fevers, strokes, and convulsions. Another market for rhino horns was the manufacture of handles for traditional daggers in Yemen (Figure 4.9). In times bygone Chinese carved highly valued knife handles, scabbards, and cups. The cups were part of religious ceremonies and were used by rulers to detect poisons, which is likely effective for strongly alkaloid poisons. Those antique carvings are now more valuable ground to a powder and marketed for their medicinal properties.

The volumes of horn that were exported from east Africa in the 1800s were high. The annual average was 11,000 kg/year between 1849 and 1895, which may have totaled 170,000 rhinos. The most complete records of exports for this century are those of horn exported from the east African countries of Kenya, Uganda, and Tanzania (Figure 11.11a). In the 1970s, 8,000 kg of horn (2,800 rhinos) were legally annually traded from Africa, which dropped to 3,000 kg/year by the early 1980s after legal trade between CITES nations stopped in 1977. Less is known about the earlier Asian trade, but volumes traded are lower than those from Africa and products are used domestically as well as exported. In the early 1900s the average annual export from Sumatra and Borneo was 90 kg (350 Sumatran rhinos). Later figures are complicated by the role of countries, such as India, in reexporting horn from African countries.

CITES perceived the decline in rhino numbers as serious enough in

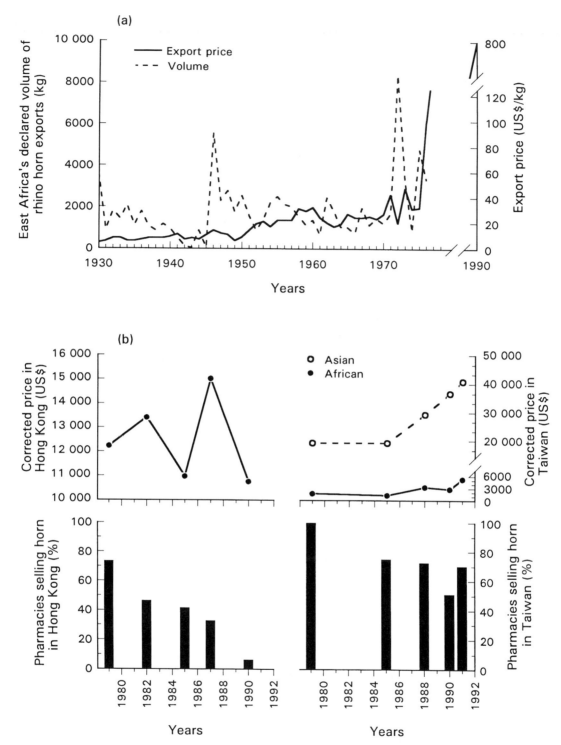

Figure 11.11. (a) Trends in the declared volume of exports of black rhino horn from East Africa and the export price. (b) Trend of changes in the wholesale market and value of rhino horn as measured by the proportion of pharmacies selling horn in Hong Kong and Taiwan. (Modified from Leader-Williams N. The world trade in rhino horn: a review. Cambridge: TRAFFIC International, 1992.)

1977 to prohibit international trade in rhino parts by listing the five rhino species in Appendix I. Records of the volumes traded legally continued after 1977 as not all importers were signatories to CITES and it took countries time to amend their national legislation to support CITES. Subsequent volumes are unknown after 1987; only North Korea lacked national legislation to regulate rhino products. Trade disappeared underground. Evidence for its continuation includes arrests of poachers, carcass counts, declines of rhinos, seizures, and covert market surveys.

The known figures for poaching are a minimum, but we report them to convey the scale. In Zimbabwe between 1984 and 1990, 782 rhinos were killed. Between 1985 and 1989, 243 rhinos were poached in Assam. Covert market surveys document any changes in the proportion of outlets selling the rhino products, but the relationship between that and the volume of demand is murky.

The motive for poaching can be seen in changes in the value of rhino horn. The value of the horn increased steadily until it jumped sharply in the late 1970s after CITES started to regulate international trade (Figure 11.11a). Asian horn has a premium price in the market, where it is three times more valuable than African horn. Horn salvaged from Indian rhinos in Assam was sold legally between 1965 and 1979, and the wholesale price tripled in 1979. The prices stabilized between 1979 and 1987, although the amount of horn traded decreased. Then in 1987 the prices started to increase again. In 1986 an Asian poacher received $2,600 (U.S.) per kg, and $6,250 (U.S.) per kg only 3 years later. By the time the Indian horn reaches Taiwan, the wholesale price in 1991 was $62,400 (U.S.) per kg compared to $8,000 (U.S.) per kg for African horn.

Tackling the trade in rhinos at the supply end through repression of poaching and sanctuaries has had only limited success and is threatened by dependence on outside aid and political and economic instability (Martin 1992). Dehorning rhinos has also had uncertain success, although it may just be too soon to evaluate it (Section 9.4.1). Regulation of trade and changing marketing of rhino products were considered a more effective approach, but success has been partial. The amount of horn reaching the market fell in the 1980s compared to the 1970s, which is assumed to reflect reduced demand as the prices had not increased in some places between 1980 and 1986. By the mid-1980s, consumption had only unambiguously declined in Japan and North Yemen with acceptance of substitutes for the rhino horn. In Hong Kong the proportion of pharmacies stocking horn decreased and the price tended not to increase (Figure 11.11b), which suggests that national legislation may have been effective (Section 12.2.1). India's consumption decreased when horn was exported to Taiwan where demand had raised the price. Surveys of price and quantities stocked suggest an increase in demand in the early 1990s in Taiwan (Figure 11.11b). Taiwan's legislation to prohibit the use and display of rhino horn passed in 1992 is modeled on similar national legislation as Hong Kong's (Section 12.2.1). Subsequent sampling of traditional pharmacies in 1993 revealed that, although the price had fallen, rhino horn was still available (Loh and Loh 1994). Under prompting from CITES (although Taiwan is not a signatory to the convention) and the United States, Taiwan started to take more stringent steps to stop the trade in rhino horn.

A decline in the traditional medicine shops stocking rhino products in 1993 in South Korea was just as likely to reflect the trade going underground as it was a decline in demand. Given that China and Taiwan both hold stockpiles probably totaling 18,500 kg, supply will meet and maintain demand for 10–25 years. The retail price is artificially high now because middlemen have speculatively invested in stockpiles of horn.

The problem then remains that a demand exists for a valued product by millions of people. Its use was established centuries ago and is part of several cultures and will not be easily changed by pressure from outside those countries. Rhino products are valuable, easy to smuggle (being low volume), and, when horn is ground into powder, difficult to identify. The powder is mixed with other ingredients, which compounds the difficulty of identification and is dispensed by a few companies selling packaged medicine or by individual practitioners. Smuggling routes are well entrenched as, even before legislation to ban trade, an illegal trade already existed probably to avoid import duties.

The network of middlemen who export from the producing countries to the importing countries has been the least-scrutinized link in the rhino trade. An idea of the intrinsic value of the horn that is "captured" during each transaction can be gauged from a comparison of prices of the horns from African rhinos. The consumer pays $3,250–10,000 (U.S.) per kg to the retailer, who pays the wholesaler $1,600–3,000 (U.S.) per kg. The wholesaler will pay an importer/smuggler/exporter $800 (U.S.) per kg. The currency then shifts to foreign currency, and the price is per horn (about 2 kg) as another middleman is paid $250–500 (U.S.) per horn, double what they paid the poacher in local currency. The expenses at each stage are unknown, but those prices suggest that the smugglers make the highest profit and they can control both ends of the market by regulating supply and price.

Leader-Williams (1992) points out that legalizing the horn trade should in theory depress the price through the sale of confiscated horn and the stockpiles. The control of the middle market in dictating prices will be uncoupled. Once the value drops, the incentive to poach will be dampened. We do not do justice here to all the complexities of the consequences of legalizing the trade in rhino products. What we have tried to do is to reveal the complexities and suggest that simply imposing legislation to stop hunting or trade is rarely enough. An understanding of the economics is needed.

Our final point is an ecological one: the role of habitat decline almost certainly exceeded the effect of harvesting until the populations were "trapped" as small populations. The mere fact of being in a reserve adds an element of access and predicability that helps illegal harvesting as well as protection. The role of habitat changes in the decline of the African rhino has been less described but will remain when the illegal harvesting problem is resolved.

11.4   SUBSISTENCE AND COMMERCIAL HARVESTING

Rationales for harvesting wildlife appear to vary from a hunter needing food (usually called subsistence harvesting) to needing to sell a wildlife product to raise cash (often called commercial harvesting). Subsistence harvesting, especially if traditional techniques are used, is seen as more respectable than a corporation's whaling fleet—the epitome of commercial harvesting. But in reality commercial and subsistence harvesting have the same basic motivation and differ only in scale (although we acknowledge that subsistence harvesting may have an intangible value to indigenous cultures). Nonetheless for the sake of convenience, we use the term "subsistence" for the harvesting by an individual when the direct products of the hunting are consumed or utilized by the hunters and their dependents. Commercial harvesting means operations where the products are sold before their benefits accrue to the hunter. Where the conservation biologist is concerned, the central issue is that the harvest is sustainable regardless of who takes it or how it is taken.

Despite the differences in scale between an individual hunting for meat to feed his household and a corporation owning a whaling fleet, a similar economic rationale drives the harvesting, namely to maximize the return. There is sometimes a difference between the extremes of the continuum between subsistence and commercial harvesting, namely the rate of harvesting. Subsistence harvesting is more likely to develop gradually, whereas a corporation's access to technology and influence with legislators can escalate a rate of harvesting. It is this dichotomy in the pace of impact and influence on regularity processes that the concept of subsistence harvesting being "good" and commercial harvesting "bad" has genuine cause. Yet subsistence harvesting can develop rapidly, particularly when the product is valuable, and for an individual the economic incentive to harvest can be high. The musk from only two male musk deer is the annual income for a family in Nepal (Green 1987), and the value of the skin of a single tegu lizard, *Tupinambis* spp., is equivalent to a day's wages in Argentina (Fitzgerald et al 1991).

11.5 WILDLIFE UTILIZATION
AS AN AID TO CONSERVATION

An idea presently with wide and growing popularity is that of using the revenue generated from the utilization, usually by harvesting, of an endangered species to fund its conservation. This has merit but carries dangers. It can constitute a positive feedback loop between the severity of the conservation problem and the amount of utilization to generate the revenue needed to fix it. The loop starts with a species in decline and various measures are urgently needed to halt it. A proportion of the population is then removed and sold. That removal steepens the decline and the conservation problem is thereby exacerbated. And so a further proportion of the population is removed and the cycle repeats itself. Such dynamic interactions are powerful, of more than theoretical interest, and not to be trifled with. The essential point is that it is *the rate of the decline which is the critical factor and not the initial size of the population.*

Yet this notion of utilizing wildlife to pay for conservation is becoming firmly entrenched in conservation biology. It has the potential to be beneficial if the utilization creates public interest in future wildlife populations. In 1992, CITES (Section 12.1.1) passed the resolution "that commercial trade may be beneficial to the conservation of the species and ecosystems and/or the development of local people when carried out at levels that are not detrimental to the survival of the species in question." The intent is that trade will be beneficial if it is monitored and sustainable; the returns are used to support the management of its habitat and there are no deleterious effects on look-alike species. There are very few species for which this information exists. Under schemes where the revenue is returned to the government, it is difficult to restructure government administration to ensure that income is returned not to general revenue but to a particular department and program. This carries its own risks.

The aim to utilize wildlife to support conservation is almost always hedged with the proviso that the use is sustainable or that it meets the criteria of scientific management. The criteria for sustainable use or scientific management are not specified (Section 12.1.1). The danger is that those criteria will be developed during a harvesting program that will have its own economic momentum. Utilizing wildlife to support conservation may be the beginning of a slippery slope to biological extinction unless management criteria with allowance for uncertainties in estimation of parameters are first established. Management criteria are not the only safety net: the structure of the market and access have to be under control.

The utilization of wildlife to generate funds for conservation is an offshoot of the "use it or lose it" philosophy. The economic approach to this is covered in Swanson and Barbier (1992). In theory, the utilization of a resource will ensure the conservation of its habitat, and thus other ecological components will be conserved willy-nilly even if we do not appreciate their existence or value. The real world is not that simple, and conservation of either the resource or its habitat is only likely under certain conditions as described in the next section.

Given the economic and political problems in securing sustainable yields while harvesting wildlife, the wish to ban trade is understandable. Yet it, too, has attendant problems. On land a resource that cannot be used is not a resource, and in most cases it will be replaced by domestic animals or crops. Reserves are not the solution as they are no more than 5% of the earth's surface, and that brings us back to the logic of using wild resources.

The possibility that sustainable harvesting of wildlife may not be economically competitive with alternative land uses sets an imperative. It is that such use should not be presented as the long-term raison d'être for conservation as, if it does rest on economic competitiveness, the case for conservation vanishes as soon as a better economic option appears. At best such harvests should not be presented as more than aids to conservation. The fundamental reason for conserving lies in the intangible domain of our minds—relying on materialistic justifications weakens the case.

11.6   SOCIAL TRAPS, INCENTIVES, AND ECONOMICS

The longer-term treatment of species in trouble is to deal with the chronic shortsightedness of how people use resources. The tendency of people to make their decisions based on local and short-term conditions is termed a "social trap" and applies to conservation (Costanza et al 1993). The short-term economic tendency to harvest beyond the sustained yield is one example. We all fall into them at one time or another.

People avoid social traps in the use of resources often through their cultural strictures and taboos (e.g., Parker 1983, Gadgil et al 1993). Pressures of overpopulation or immigration can change attitudes toward resources. Government control through regulation and penalties is usually ineffective, as illustrated by the prevalence of illegal harvesting. Education works (Box 9.6) but is limited by the ratio of "teachers" (conservationists) to audience (users). The other approach to social traps is to use either incentives or disincentives to modify short-term goals.

Incentives include involvement of local people in the regulation of the decisions over the use of resources. Local people are often affected by imposition of reserves or harvest regulations. Although local involvement has frequently fallen short in practice (Wells and Brandon 1993), it still has to be pursued for the long-term hope it offers to conservation.

An example of using economic incentives at the local level comes from efforts to halt the decline in the one of the world's largest and rarest eagles (Salvador 1994). Numbers of the Philippine eagle, *Pithecophaga jefferi*, declined to 100–200 birds despite sanctuaries, surveys, and education. The pressing needs of people to survive were why they were clearing forests—the cause of the eagles' decline. Modification of farming practices increased income and created an incentive to start planting trees instead of felling them. It is too soon to measure the success for the eagle, but at least the first one has been sighted in the area for the first time in 4 years.

Incentives and disincentives could play an important role in regulating trade in wildlife. First, the relationships between costs and intrinsic value have to be teased apart. Every time the product changes hands the economic rationale differs for each level. In the live-bird trade, costs are high compared to received benefits for the trappers compared to the middlemen, which leads logically toward a higher harvest rate with greater loss of birds before export (Box 11.1). The same logic explains the disincentive for local people to conserve habitat of the birds. The ratio of costs to benefits is asymmetrical between the trappers and middlemen. Thus to create an incentive for the trappers the role of middleman has to be changed. The same arguments apply to the trade in rhino horns. More direct linkage between the harvesters and their market through cooperatives or their own marketing agencies are possible models that would bypass the middlemen. These arguments are the rationale for a CITES project in Paraguay. The hunters of tegu are to form a cooperative to sell the lizard skins direct to the tanneries (Fitzgerald 1994).

Our example of parrots and rhino horns is the two sides of the same coin. For the parrots the intent would be to increase the price to the hunter as an incentive to treat the parrots well and to conserve their habitat. With the rhino horn, first trade would have to be legalized and then the middleman bypassed. This would allow the price to be dropped at the retail end of the market, which would then reduce the price paid to the hunters. Once the price drops below a threshold the incentive to poach would stop (Leader-Williams and Milner-Gulland 1993). Removing the incentive to poach by dropping the price would run counter to various proposals by South Africa and Zimbabwe, which want to maximize revenues from the sale of live rhinos and trophy hunting (immobilizing/dehorning) to raise funds for rhino conservation (but see Section 11.4). In reality, any solution will be more complex but nonetheless has to be attempted.

Box 11.1. An economic model to explain how sale price, costs, and intrinsic values relate to sustained harvesting and humane treatment in the trade in live birds

Problem

The trade in live birds is a lucrative business but has become controversial. The issues are whether the harvests are sustainable and the high rate of losses of birds before sale. Estimates of the annual trade of birds were 2–5 million in the 1980s. The ecological effects are unknown but the trade threatens some. The control of the trade by exporters and importers ("middlemen") effectively acts as a disincentive for the harvesters to humanely care for the birds and to safeguard their populations and habitats.

Market structure and characteristics

The market structure for the trade in live birds starts with the trappers who sell the birds to the middlemen. The middlemen typically have agents who may go to villages or local markets and collect the birds to a port of export. Exporters are few and importers even fewer and more organized. Their knowledge of both the market and the producers gives them control.

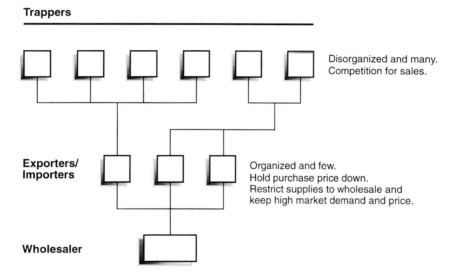

Structure of prices, costs, and values and resulting incentives and disincentives

Intrinsic value is the net benefit after costs (labor, cages, and so on) have been subtracted from the sale price. Economists refer to the intrinsic value as the rent to be captured at each transaction. The higher the rent captured the greater is the incentive to care for the resource. Harvesters capture some 33% compared to almost 700% for the middlemen. The greatest proportional loss of birds is while their value is still low. Education and facilities may count, but the overriding factor is their low value in a competitive market. The same reasoning explains the lack of incentive to either harvest sustainably or protect habitat.

(Continued)

Box 11.1. (Continued)

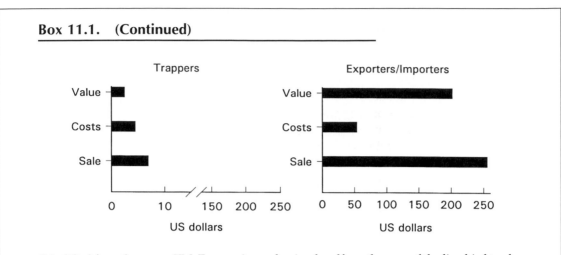

(Modified from Swanson TM. Economics and animal welfare: the case of the live bird trade. In: Thomsen JB, Edwards SR, Mulliken TA, eds. Perceptions, conservation and management of wild birds in trade. Cambridge: TRAFFIC International, 1992:43–57; Mulliken TA, Broad SR, Thomsen JB. The wild bird trade—an overview. In: Thomsen JB, Edwards SR, Mulliken TA, eds. Perceptions, conservation and management of wild birds in trade. Cambridge: TRAFFIC International, 1992:1–42.)

11.7 SUMMARY

Many, if not most, extinctions have a critical, if not causative, economic rationale. To explain that, we have illustrated harvesting dynamics and techniques to estimate maximum sustainable yield and the advisability of opting for lower sustainable yields. Estimating the sustained yield as a precondition is usually ignored in economic analyses of harvesting. Instead economics emphasizes the relationship between current value discounted against future value, which works as an incentive to overharvest. This is not an aberration of the past but is still going strong today as our account of whales, tuna, and rhinos illustrates. The rules for sustained harvesting and the economic impetus to overharvest are similar regardless of the type of harvesting, be it commercial or subsistence.

Many wildlife uses (sport hunting, viewing wildlife) span both tangible (cash value of meat, etc.) and intangible (human perceptions) values. Resource economics has techniques to estimate nonmarket values, so, for example, the value of an area for an endangered species can be compared to its value for forestry. Best known of the techniques is contingent valuation, which uses a bidding approach—asking people how much they would be prepared to pay to see an area held as a reserve. But such approaches have shortfalls since they depend on unverifiable biases.

The question of ownership—who pays the costs and gains the benefits of harvesting—is a key feature. There are those who recommend that the resource should be under private ownership and thus, with total control over the harvesting rate, the owners will act to conserve their property rather than allow it to be depleted. This does work but it has limits. Some aspects of conservation are so extensive as to be insoluble at anything less than the ultimate communal endeavor of international cooperation. Cooperative and time-limited concessions may offer workable options halfway between the poles of state and private control. The most appropriate tactics will depend on prevailing local circumstances, and strategies for sustainably harvesting wildlife have their greatest chance of success when tailored to the political structure of the societies involved.

The most vexing questions for conservation arise from the near inevitability with which individuals and societies take short-term advantages rather than long-term ones when deciding use of rsources. Incentives help in avoiding those social traps. For example, shifts in the market structure to increase the value to the hunter and reduce the intermediate steps in a marketing structure could decrease deaths among birds taken for the live-bird trade or decrease the illegal harvest of rhinos. Other economic incentives for conservation, such as private ownership or using endangered species to generate funds for conservation, are strewn with pitfalls. Given the multitudinous political and social circumstances in which conservation has to take place, we believe that no one set of rules is generally applicable.

CHAPTER 12

International and national legislation

Protection of wildlife and their habitat has over the centuries prompted measures in statute form and common law. The last 20 years has, however, seen an increase in legislation specifically to conserve species and their habitats. International conventions to protect wildlife have been in effect for decades but with mixed success. National and international efforts to regulate trade in wildlife have achieved a high profile, but legislative protection as often as not is incompatible with economics, which is why legal means to reduce or prohibit harvesting only have mixed success. The International Union for the Conservation of Nature–The World Conservation Union (IUCN) does not have legislative powers, but its pivotal role in organizing and supporting conservation throughout the world mandates its inclusion in this chapter. Finally we briefly describe the specific management (recovery) plans for endangered wildlife and how risk analysis can be used to help develop management options.

12.1 INTERNATIONAL TREATIES AND CONVENTIONS

There are three levels of international treaties and conventions. First are those that are worldwide in their scope and are ratified by many nations. The three most important multilateral treaties that relate to conservation are The Convention on International Trade in Endangered Species of Wild Fauna and Flora (CITES–1973), The Convention on Wetlands of International Importance (Ramsar–1971), and The Convention Concerning the Protection of the World Cultural and National Heritage (World Heritage–1972). It is too soon to gauge implications of two recent conventions for conservation as they are barely in place. The Biodiversity Convention came into force in December 1993, and the Convention on the Law of the Sea, which will play a strong role in coastal and marine conservation, was ratified in October 1994.

Second are the regional treaties specific to geographic areas such as the Antarctic. Third are the multilateral treaties which deal with groups of species such as the International Convention for the Regulation of Whaling–1946. Some have a long lineage, such as the 1918 North American Convention on Migratory Birds. Other multilateral treaties deal with single species such as the polar bear, *Ursus maritimus* (Agreement on the Conservation of Polar Bears–1973). This Agreement was signed by Canada, Denmark, Norway, United States of America, and the then Soviet Union. The trigger for its negotiation, which began in 1965, was an increase in bears killed for their hides in the 1950s and 1960s. The Agreement was not simply protective of bears; it laid down conditions for bear management, and, significantly, it expressly protected the bear's ecosystem (Stirling 1986).

Those international conventions can certainly chalk up some

successes in conserving species. The international polar bear agreement has been effectively used to manage the bears across international borders (Stirling 1986). The International Whaling Commission has seen increases in some whales, but caution is needed to attribute success to the convention itself rather than the economics of commercial harvesting. The convention was originally drafted with unreconcilable objectives (protection of trade and whales), and the shenanigans that ensued as the trading and protectionist nations wrangled are reviewed in Cherfas (1988) and Andresen (1993). The recent figures on deliberate underreporting of Soviet catches raise more doubts about the effectiveness of the Commission.

The only international treaty that deals only with the protection of habitat is the Ramsar Convention. Refreshingly, Ramsar is not an acronym but it is the city in Iran where the Convention on Wetlands of International Importance was negotiated in 1971. By 1994, 81 parties had ratified the Ramsar Convention (S. Stuart, personal communication, 1994). The convention does not legislate protection for the wetlands; responsibility is left at the national level. The obligations of member nations are that they list at least one wetland. Then the requirement is that the listed wetlands are managed such that human activities do not destroy their ecological characteristics. If a wetland is damaged, either the country nominates an alternative wetland or the offending country faces moral disapproval of the other parties to the conference.

The World Heritage Convention encourages an international interest in national resources on behalf of the international community. Sites are eligible for funding through UNESCO as an incentive to nations to safeguard sites for the international community's sake. This convention had 138 parties in 1994 (S. Stuart, personal communication, 1994).

12.1.1 The Convention on International Trade in Endangered Species of Wild Fauna and Flora

CITES is perhaps the best known and most wide-reaching of the international conventions. By the end of 1994, 126 nations had ratified the convention, which regulates trade of species (and items manufactured from their parts) perceived to be in danger of extinction partially or wholly through international trade. Trade, in this case, refers to any export/import between nations regardless of whether it is commercial or not. CITES regulates this trade through listing species on three "appendices" which have restrictions attached to them. CITES nations meet at biannual conferences to review policies and to consider additions and deletions from its Appendices I, II, and III.

Appendix I species are those species threatened with extinction as well as being actually or potentially affected by trade. Export and import of Appendix I species are not allowed by signatory nations except under specialized conditions of noncommercial use, such as scientific. In this case, both the importing and exporting nations must issue permits once they have determined that the trade is not commercial and not detrimental to the survival of the species.

Species listed on Appendix II could become threatened with extinction unless trade is regulated. Appendix II species can be traded only if the exporting nation issues a CITES export permit indicating that its CITES scientific authority considers the trade to be nondetrimental to the species. In practice, countries vary in their capabilities and incentives to provide these "nondetriment findings" on the basis of sound, scientific knowledge. Appendix II also covers a suite of species covered under "look-alike" and captive-breeding provisions. Look-alike species are regulated because of difficulties in distinguishing them from other regulated species. An example is the North American black bear listed on Appendix II in 1992. Black bears are not at risk in North America, but their gallbladders are indistinguishable from those of endangered Asian bears whose galls are sought for making traditional medicines. Galls from endangered bears could be labeled as coming from healthy North American populations, thereby putting Asian bears at increased risk. Other look-alikes are all parrots, cacti, orchids, cats, and primates, except those listed on Appendix I. Appendix I species which have been captive bred under strict supervision and rules can in some cases be exported as Appendix II.

Member nations can unilaterally declare their national population of a species as Appendix III if they feel that international cooperation is needed to assist them with their conservation efforts. There are no specific criteria for a listing as Appendix III, and the category is of questionable utility.

The listing of species on Appendices I and II is based on rather loose criteria developed at Berne in 1976 and known as the "Berne Criteria." Efforts are under way to place these listing criteria on an objective basis (Nelson and Sullivan 1993). However, development of these such criteria has proven to be technically difficult and politically controversial.

Nations may take out a "reservation" on the listing of a species if they choose to opt out of CITES controls. By so doing, they may still trade in that species with non-CITES states or with other CITES states also taking out a reservation. The provision for reservations was instituted to encourage nations to join CITES and remain as members despite listings not to their liking. Reservations do have the capacity to hamper efforts to ameliorate trade in endangered species as it seems to have done in the case of the musk deer. About 25 g of musk is extracted from the preputial gland of musk deer harvested in the Himalayas or farmed in China (Green 1987, Fitzgerald 1989). Japan took out a reservation on musk deer allowing musk to be imported. In 1987 Japan announced that it planned to drop the reservation in 1989, thereby closing up the market. The result was a doubling in harvest levels in 1987 to 900 kg of musk (worth $58 [U.S.] million) in an attempt to stockpile as much musk as possible before import restrictions went into effect.

CITES can only be effective in regulating international trade if all countries become members and if member nations have supportive national legislation, the means, ability and commitment to do scientific population assessments, and effective enforcement. Simply listing a species on a CITES appendix is itself not an effective conservation action.

The trade in neotropical parrots exemplifies this problem. The trade in these parrots is substantial. Between 1982 and 1986, 1.4 million individuals of Appendix II species were legally exported and were worth $1.6 (U.S.) billion (Thomsen and Mulliken 1992). Since the 1970s, most South and Central American countries have regulated commercial trade in parrots through prohibition, quota, or permit. However, in the absence of population data, quotas are often set too high. In the case of blue-footed Amazons, *Amazona aestiva*, the export of 45,000 between 1986 and 1988, in concert with habitat destruction, has caused the species to disappear from the southern part of its range (Thomsen and Mulliken 1991). Parrots are often channeled through nonmember states. The Appendix I hyacinth macaw, *Anodorhynchus hyacinthinus*, were once commonly smuggled from Brazil, where trapping and export is illegal, to Bolivia where export was legal. In 1984 Bolivia implemented legislation prohibiting export of wildlife. The result was an increase in the export of hyacinth macaws from Argentina (Thomsen and Mulliken 1992). Wildlife exports commonly shift to states having the least restrictive legislation or effective enforcement (Figure 12.1, Box 12.1).

Trade in fruit bats, *Pteropus*, between Pacific islands is another example of problems in regulating wildlife trade (Box 12.1). As populations were heavily harvested their size dropped, restrictions were imposed, and trade shifted to another island (Figure 12.1). By the 1990s, the only legal source of bats was Palau, but in 1994 it became independent of the United States and thus subject to CITES. The trade's persistence reflects the cultural value of fruit bats to the Chamorus who have long traditionally eaten fruit bats and with CITES prohibitions of trade between islands, illegal hunting and trade may increase (Wiles 1994).

Since its inception in 1973, CITES has gone through many changes reflected in the 7,000 pages of proceedings and amendments, which Wijnstekers (1990) has ably boiled down into one small reference book. Many of these changes, and the current tension within CITES, arise from a fundamental disagreement between interests that see CITES as being, in its essence, a trade convention and those parties that see CITES as primarily a mechanism to preserve endangered species.

Endangered species in situ are subject to national control; no nation or international organization can compel a sovereign country to protect its biological resources. However, when commodities move from one country to another through trade, exporting countries relinquish the

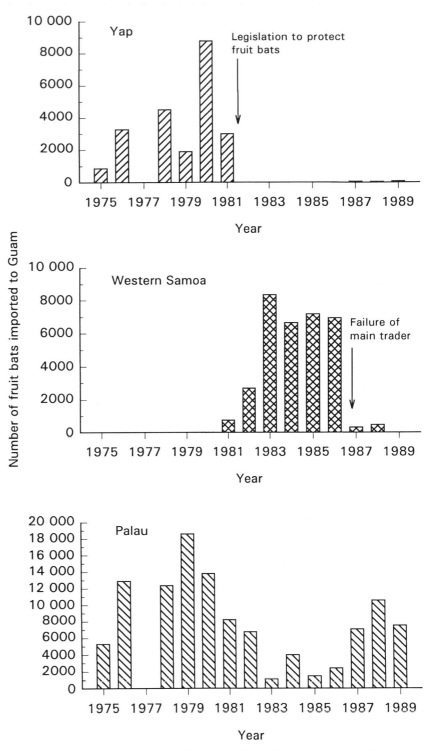

Figure 12.1. The export of fruit bats from Yap, Palau, and western Samoa to Guam. (Modified from Wiles GJ. Recent trends in the fruit bat trade on Guam. In: Wilson DE, Graham GL, eds. Pacific island flying foxes. Washington, DC: U.S. Department of the Interior, 1992:53–67.)

umbrella of sovereign control at the moment of export and are therefore subject to controls levied by importing countries or countries cooperating through the international convention. Trade becomes the lever whereby the international community can influence endangered species conservation within the borders of sovereign nations.

The Berne Criteria state that listed species must not only be at risk but that some of this risk must be attributable to trade. Nations choosing to emphasize the conservation role of CITES argue that there must be no trade in endangered species; whether trade is harmful, beneficial, or neutral is immaterial. Nations choosing to emphasize the trade aspects of the convention argue that trade in well-managed populations of endangered species can be beneficial to species conservation and, in some cases, may represent the only hope for a species long-term survival.

Whether trade is beneficial to conservation of wildlife depends upon whether the market is structured to prevent locking into a positive feedback loop (Sections 11.2, 11.5) and the capabilities and willingness of member nations to implement effective controls. The international track record for agreeing to and conforming to scientifically based management schemes is not encouraging if, for example, the history of the International Whaling Commission's Revised Management Scheme and its precursor is anything to go by. Despite years of wrangling the scheme still was not accepted until 1994 (Cooke and Thomsen 1993,

---

Box 12.1.   CITES and the international trade in fruit bats on the Mariana Islands

The trade in fruit bats
Fruit bat, *Pteropus*, harvests increased in the 1950s after a shift to cash economy and loss of traditional strictures on harvesting combined with availability of firearms and air transport. Guam imported almost 221,000 fruit bats from neighboring islands between 1975 and 1989 after populations collapsed on Guam in the late 1960s. Protection of remaining bats on Guam through local legislation (1981) and the U.S. Endangered Species Act (1984) encouraged further imports from elsewhere. Sources shifted from one Pacific island to another, depending on protective regulations and market preferences. In the 1970s and 1980s most bats came from the Caroline Islands (especially Palau) and other Mariana Islands. In the mid-1980s, bats came from outside the Micronesian islands and countries, especially western Samoa. The main source of bats in the late 1980s reverted to the Caroline Islands.

(Continued)

> **Box 12.1.   (Continued)**
> _____
>
> **Trade restrictions**
> CITES Appendix II listing in 1987 had little impact on the import of bats to Guam. Between 1987 and 1989, 30,000 fruit bats were imported from islands which the United States administered as part of the Trust Territory of the Pacific islands, and as such the trade is considered national and outside CITES. The political administration then changed, and the Caroline Islands (but not Palau) and Commonwealth of the northern Marianas became independent and therefore subject to CITES regulations. Moving seven species of fruit bats from Appendix II to Appendix I in 1989 did not reduce the import of bats to Guam. Between 1990 and 1992, 82% of the 20,000 bats imported to Guam came from Palau. CITES controls did not apply to the import of bats to Guam from Palau as it is considered national trade (and an effort to apply U.S. legislation failed for technical reasons). The level of the trade from Palau exceeds the local availability, and apparently the bats are illegally coming from other islands via Palau. In that case CITES regulations do apply and their apparent failure was lack of effective enforcement.
>
> (Modified from Wiles GJ. Recent trends in the fruit bat trade on Guam. In: Wilson DE, Geraham GL, eds. Pacific island flying foxes: proceedings of an international conservation conference. Washington DC: U.S. Department of the Interior, 1992:53–67; Falanruw MC, Manmaw CJ. Protection of flying foxes on Yap Island. In: Wilson DE, Geraham GL, eds. Pacific island flying foxes: proceedings of an international conservation conference. Washington DC: U.S. Department of the Interior, 1992:150–154.)

Baker and Palumbi 1994). This may be an apt comment on the difficulties that CITES will face in designing management criteria for species whose trade is lucrative and where vested national interests are uppermost. CITES will be essential to support beneficial trade. Coordination in gathering trade statistics and assisting with the transfer of expertise and resources to countries so that they can help with national legislation and train customs inspectors will be among the requisites.

12.1.2   International Union for Conservation of Nature—The World Conservation Union

IUCN does not have the mandate to legislate, but we included it because of its pivotal role in the conservation of animals and plants. On

October 5, 1948, the International Union for Conservation of Nature and Natural Resources (IUCN) was founded and is unquestionably the most weighty international nongovernmental conservation force operating. Eighteen governments and 7 international and 107 national nature protection organizations signed the Constitutive Act of the Union under the auspices of UNESCO. By 1994 those numbers had increased to 100 government agencies and 550 nongovernment organizations distributed through 126 countries (S. Stuart, personal communication, 1994). We quote from the preamble to the 1948 Act:

Whereas the time has come when human standards of living are being depressed because natural resources are becoming inadequate for their maintenance; Whereas this trend may be reversed if people are awakened in time to a full realization of their dependence upon exhaustible natural resources and recognize the need for their protection and restoration as well as for their wise and informed administration in order that the future peace, progress and prosperity of mankind may be assured. . . .

The objectives of IUCN are to encourage and help cooperation between governments and national and international organizations concerned with the protection of nature and especially the preservation of species threatened with extinction. It became quickly apparent that international fundraising would be needed to place conservation on any sort of footing. So in 1961 the World Wildlife Fund was born and has subsequently stayed closely linked with the IUCN, although now retitled World Wide Fund for Nature. In some countries the translation of "wildlife" carried the connotation of meat, which goes to illustrate, in a small way, the difficulties of operating between languages and cultures.

IUCN, operating as it does with a small budget, is influential. Its combination of government and nongovernment members give it the flexibility of the private sector and recognition in the government sector. The second reason for IUCN's influence is that through its commissions (described below) it has access, often free, to a wide range of expert opinions.

Among IUCN's most conspicuous achievements are the Red Data Books started in 1966. Their history relates struggles with the definitions for different categories of endangerment (Fitter and Fitter 1987), and that is not over as IUCN is working to redefine the categories. The lists are updated as more information is obtained and the status of species changes. We give in Table 12.1 the most recent listing of the major taxa by the IUCN category. A problem with formal listings of species is that the basis for the listing—application of the criteria—has rarely if ever been visible, although future listings are meant to be explicit about the criteria. Apart from taxonomic problems and the actual known status of a species, personal interests and biases crop up. The lists are not exhaustive and lean toward the larger, more attractive species from countries where there have been more studies, often by

Table 12.1.   Numbers of species of mammals, birds, reptiles, and amphibians in IUCN designated categories of threat

Vertebrate group	IUCN designated categories of threat[a]						
	Ex	E	V	R	I	K	Total
Mammals	83	177	199	89	68	208	824
Birds	114	188	241	257	176	108	1,084
Reptiles	20	47	88	79	43	59	336
Amphibians	4	32	32	55	14	36	173
Fishes	34	158	226	246	304	45	1,013
Invertebrates	368	582	702	422	941	107	3,122
Total	623	1,184	1,488	1,148	1,546	563	6,552

[a] Ex = extinct, E = endangered, V = vulnerable, R = rare, I = indeterminate, K = insufficiently known. Taxa currently under review and to be categorized.
Modified from Groombridge B. 1994 IUCN red list of threatened animals. Gland, Switzerland: IUCN, 1993.

expatriates but certainly in the English language. The coverage for countries whose faunae were neglected in early lists is, however, being increased. The number of Mexican mammals in the Red Data Book went from 15 species in 1986 to 34 in 1990 (Cuarón 1993).

IUCN cooperates with the International Council for Bird Preservation (ICBP), now called Birdlife International, to publish Red Books on threatened birds of Africa (Collar and Stuart 1985) and the Americas (Collar et al 1992): further volumes are planned. The published volumes are a remarkable compendium of ecological information on species listed by IUCN and reveal the strength of IUCN and ICBP in disseminating information on endangered species. We quote from the introduction to the volume on the Americas:

Books like this get abandoned, not completed. The expectation that any assessment of threatened species can be definitive, or that compilation of data in it can be exhaustive, is illusory. The best that can happen is to aim for closest approximation to these things, for certainly if species are to be saved it is in the *first place* through information.

The compilations impart a sense of urgency from the sheer numbers of species about which we know virtually nothing compared to the handfuls of species that attract extraordinary efforts. They illustrate the magnitude of the task to conserve an array of species and ecological relationships.

The IUCN's Species Survival Commission now encompasses about 102 specialist groups which are mostly responsible for species groupings such as deer, crocodiles, cycads, and ants. The groups are an international network of technical expertise with 5,000 members from

167 countries, and their information is usually available as an action plan or workshop reports.

IUCN and WWF oversee the compilation of statistics on the trade in wildlife and wildlife parts through Trade Record Analysis of Flora and Fauna in Commerce (TRAFFIC). TRAFFIC produces regular bulletins and undertakes the production of specialist reports on particular aspects such as the trade in rhino horns or live wild birds (Thomsen et al 1992). Its trade figures are compiled from a variety of sources and do not just depend on the data supplied by governments which are parties to CITES.

12.2 NATIONAL LEGISLATION

12.2.1 Legislation for take and possession

Most countries have had legislation in one form or another to regulate wildlife hunting, and frequently those regulations included varying degrees of protection for endangered species. In the last few decades many countries are updating their legislation specifically to increase protection for endangered species, and this is a requirement for countries that sign the 1993 Biodiversity Convention. Prohibition of hunting and trading may fail to protect an endangered species unless possession is also covered. Possession of wild-caught parrots in Venezuela is legal, although their hunting, trading, and transportation is illegal (Desenne and Strahl 1991).

Endangered species legislation requires that wildlife be classified in categories of endangerment, but criteria for listing for protection vary between jurisdictions and are not always open and public. New (1991) offers examples for butterflies where listing has been misinformed, overzealous, or uninformed. Categorizing a species as endangered has the merit that it draws the attention of administrators and funding bodies. Public awareness is attracted by designating an animal as endangered. Less wanted consequences of listing are greater desirability to collectors and vulnerability to private land owners wishing to rid themselves of a potential problem from having an endangered species on their property. Removal of species from a list is usually through acts, which tends to impart immutability to the lists. A member of the public may propose a species to be added, but not to be deleted, from the Australian Endangered Species Protection Act 1992. Regardless of the process, the criteria for delisting a species should not be the same as for listing because species' decline and recovery are not mirror images.

Lists of endangered species are influenced by the ebb and flow of taxonomy, which casts up new species and washes others away. The

Hawaiian endemic genus of tree snails, *Achatinella*, had 227 described species in 1888, and after revision in 1914 it had shrunk to 41 species grouped in 11 series. As 16 species are extinct and 23 others may be extinct or nearly so, the genus is on the U.S. Endangered Species List. The listing of 41 species stands, although further taxonomic work dropped 8 species by contracting two series to a single species each. The same study suggested that the other nine series probably each represented a single species, which would reduce the 41 species to 12–16 (U.S. Fish and Wildlife Service 1993).

A more serious problem for endangered species listings arises when the morphologic or geographic criteria for a species or subspecies are contradicted or modified by molecular genetics. It is easy to lose sight of the fact that measures of genetic variation subsample the genes (Section 6.3.2), as do morphometric, biochemical, or any other species attributes used as taxonomic criteria. The controversies that erupted over the red wolf, *Canis rufus*, and the Florida panther after studies of their mtDNA were used to query their taxonomic status signifies the confusion when sight is lost of the biological concepts for species and subspecies. Those concepts are relatively clear (O'Brian and Mayr 1991) and so is their practice with common sense and pragmatism. As Cronin (1993) points out, we would hardly lump polar and brown bears together, but one mitochondrial DNA genotype of brown bears is more similar to polar bears than to four other brown bear genotypes.

Another problem with lists of extinct species is the difficulty of determining when a species is extinct. Assigning a species to the category of extinction depends on the thoroughness of the effort to have found it. Species are being rediscovered after decades of being assumed to be extinct (the cahow is an example; Section 4.3).

A prominent example of national legislation for endangered species is the U.S. Endangered Species Act (for descriptions see Shull 1991, Lowe et al 1990). The 1973 Act applies to species considered to be "endangered" and "threatened." "Endangered" is defined as "in danger of extinction throughout all or a significant portion of its range," and "threatened" is "likely to become an endangered species throughout all or a significant portion of its range."

The U.S. Endangered Species Act has a well-defined procedure for listing, delisting, and modifying the status. A government agency or a private individual or organization can initiate a candidate. The listing procedure is time and dollar consuming and leaves a long list of candidates (3,650 in 1990) compared to the 554 species listed in 1990. Interim protection for candidate species is encouraged. Species traded from other countries, if the United States considers them endangered, can be included. The U.S. Endangered Species Act mandates a recovery plan for each listed national species, and federal government agencies have to ensure that any of their activities or ones that they authorize or fund do not threaten any endangered or threatened species. This is not complete protection, as, for example, the red-cockaded woodpecker lost colonies to military use of land in the "national interest" (Jackson 1986).

The funds spent through the Endangered Species Act that are "reasonably identifiable" are annually reported. In 1991 federal and state agencies spent $176.8 (U.S.) million on 570 of 639 listed species. Seven species (1%) received half the annual funding (bald eagle, *Haliaeetus leucocephalus*; Florida scrub-jay, *Aphelocoma coerulescens*; Florida manatee, *Trichechus manatus*; red-cockaded woodpecker; northern spotted owl; peregrine falcon; and chinook salmon, *Onchorynchus* spp.).

The justification for listing requires a certain amount of information, and for many species, especially invertebrates, this information simply does not exist. Even for islands known for the high levels of endemic species, invertebrates are little featured for protection or even study. Only 4 of the 39 species of the Hawaiian yellow-faced bees, *Nesoprosopis*, are candidates for listing, although 35 of them have not been recorded for 25 years (Hafernik 1992). Lack of information is not the only impediment to listing. Private interests may vigorously resist a listing because once a species is listed its protection can curtail or prevent landowners from any activities considered detrimental to it.

Murphy and Weiss's (1988) account of the 5 years it took to list the bay checkerspot butterfly conveys the controversy that can follow petitions for listing species on the U.S. Endangered Species Act. Just as vested interests can promote a species being listed, lobbying can also halt it. The staggering decline of prairie dogs (Section 4.6) led to the black-footed ferret's extinction in the wild, and two other species (mountain plovers, *Charadrius montanus*, and feruginous hawks, *Buteo regalis*) also dependent on prairie dogs declined to the extent that they are candidates for listing. Yet despite losing 90% of their colonies to a cause that is still extant (government-sanctioned poisoning), black-tailed prairie dogs were rejected as a candidate species in 1994 (Miller et al 1994b).

In contrast to the listing of species in the United States and Australia, Canada relies on a legally nonbinding cooperation between federal, provincial, and territorial governments to list endangered wildlife. The Committee on the Status of Endangered Wildlife in Canada works with government and nongovernment agencies to list vertebrates, invertebrates (at least, Lepidoptera and Mollusca), and plants as extinct, extirpated, endangered, and threatened. However, in 1995, Canada is considering following other countries and establishing endangered species legislation.

The United States has other national legislation that protects species, including the U.S. Marine Mammals Protection Act and the Wild Bird Conservation Act 1992. The latter supports CITES by further restricting the import of wild-caught birds listed by CITES on Appendices I and II. The legislation uses reverse listing. Only birds on an approved list can be imported, and the listing depends on the implementation of CITES in the country of origin. Funds may be available to the countries of origin of the birds through the Act from fines and donations.

Most countries have enacted particular acts to implement their obligations under CITES. An example is Hong Kong's regulation of the import and possession of rhino parts (Milliken 1991). Hong Kong's first steps in 1978 dealt with registration of rhino horn and hide, and a ban on commercial importation followed in 1979. By 1985 all legal international trade of rhino parts from Hong Kong had been stopped. In 1987 after CITES had urged all parties to the convention to ban all trade in rhino parts the potential loophole of noncommercial possession was closed in Hong Kong with the requirement that all recognizable rhino products be covered by possession licenses. The next step, in 1989, was to prohibit the use of rhino products in oriental medications. Subsequently the number of traditional pharmacies selling rhino horn dropped from 73% in 1979 to 5% in 1990 (Leader-Williams 1992). In 1994 Hong Kong passed further legislation which prohibited the sale and possession of medicines containing tiger parts and South Korea's similar steps will have become effective by 1995 (Milliken and Haywood 1994).

12.2.2 *Legislation to protect habitat*

The means to protect habitat is commonly through legislation to set aside areas as reserves (Chapter 10). Other possibilities exist to protect habitats outside government-controlled reserves. Private ownership may protect habitat by attaching conditions to land titles (covenants) or granting the right to use or restrain the use of land owned by someone else (easement). New Zealand applies its Reserves Act 1977 and National Parks Act 1980 to private lands through the use of covenants, conservation leases, conservation cross-leases (trade of land for a specific activity between private and public land), and development embargoes. The legislation for saving habitat through "offset" agreements (trading the protection of one piece of public land for the use of another piece of public land) is covered by endangered species legislation. Reserves can benefit from the different concepts of land proprietorship and the particular relationship with their environment held by many indigenous people. Examples of this in practice are the American Samoa National Park, the Falealupo Rainforest reserve (on Savaii in Western Samoa), and the Tafua Biological reserve (Cox and Elmqvist 1991).

Many countries regulate land use with a permit system to control activities that may degrade the environment. In Britain for example, the Nature Conservancy Council can negotiate for the modification or withdrawal of activities that would damage a Site of Special Scientific Interest (Brotherton 1990).

The species-specific protection of habitat is another approach to establishing reserves of protected habitat. One of the best examples of this approach is in the United States, but, as Groombridge (1992) describes, there are others such as the Flora and Fauna Guarantee Act of the State of Victoria legislation (Australia) and the French Nature Conservation Act 1976. The U.S. Endangered Species Act has provisions for the protection of habitat deemed critical for an endangered species, and the procedure for listing critical habitat is similar to that for listing a species as endangered or threatened. In 1982 the Act was modified to impart some flexibility. The modification allowed a certain level of "incidental taking" of a species if concessions in return benefit it. In southern California, the Coachella Valley fringe-toed lizard, *Uma inornata*, is an endemic in the loose-shifting dunes of a single valley about to be built over. A compromise plan allowed the developers to leave about 10% of dunes for the lizards and to pay an area surcharge to support the reserve (Lowe et al 1990).

The bay checkerspot butterfly was a more contentious case under that provision of the U.S. Endangered Species Act. The butterfly's habitat had declined to a few fragments: a defense contractor that owned land next to one of the habitat fragments contested the listing of the butterfly as endangered. While this was being considered and eventually rejected (Murphy and Weiss 1988), the proposed developer of another fragment entered a management agreement and set up a trust fund to purchase habitat and carry out further ecological studies and monitoring (Murphy 1988). This agreement was drawn up to mitigate and compensate for the "take" through habitat loss from landfill development on the most extensive habitat fragment left.

12.3 RECOVERY PLANS

In the United States and Australia the Endangered Species Acts specifically mandate recovery plans, whereas in Canada it is policy that any terrestrial species listed as endangered, threatened, or extirpated has to have a recovery plan. The first plan under the U.S. Endangered Species Act was for Kirtland's warbler in 1976. By February 1994, 377 recovery plans had been approved in the United States, which is slightly over half of the 632 U.S. species listed as endangered.

Stripped of its jargon, the basic format for the plans is similar. The introduction includes the species ecology and reasons for its decline. The plan's goals usually include downlisting from the category of endangerment. The second part is usually the proposed means for the recovery. The third part is an implantation schedule and estimates of costs.

Operating at a different scale from government species-specific recovery plans are the Action Plans from IUCN's Species Specialist Groups. Since 1986 each specialist group of volunteer experts has started to evaluate status and needs for conservation of species in its group—for example, Chapman and Flux (1990) for pikas, rabbits, and hares and Cumming et al (1990) for elephants and rhinos. By 1993 the groups had published 23 plans on mammals, 2 on reptiles, and 1 on insects. Those action plans, while admirable in their compilation of recent information on status, have some of the shortcomings of species-specific recovery plans. They are longer on treatment and shorter on the science of diagnosis. They are, however, often a convenient summary of information that otherwise would be languishing in files. IUCN's Captive Breeding Specialist Group emphasizes managing captive populations within the context of the wild populations and has spawned an array of plans and processes to prompt regional and international coordination (Seal et al 1994).

An idea of the extent of the implementation of the action plans is the number of projects funded by various agencies, including World Wide Fund for Nature (Gimenez-Dixon and Stuart 1993). African elephants, antelopes, primates, and rhinos account for 89% of the U.S. $57 million raised since the plans were published.

Recovery plans, like many management plans, rarely explicitly distinguish between technical and value judgments (Caughley and Sinclair 1994). It is not the prerogative of conservation biologists to make value judgments to determine conservation goals, although Bell (1983) argues that a biologist should declare them rather than offer them as technical judgments. Bell (1983), while exploring the rationale for culling in national parks, lucidly explains the role of technical and value judgments and the logic traced by his flowcharts applies equally well to other conservation issues. Value judgments, by definition being neither right nor wrong, are subject to a wide range of social, economic, political, and ecological influences. Decision analysis is a technique to reveal the role of those influences (Section 12.4). The degrees of uncertainty (probabilities) associated with the technical judgments should be reported as well as the judgements themselves.

Recovery plans and the IUCN action plans have the merit of bringing together many of those who have or should have a hand in the reversal of a species decline and thus at least prompting exchange of information. At the first level of exchange are the technical people, and they are likely a mixed bag. We quote Wemmer et al (1987), writing in the context of a fictional plan to save tigers in a reserve network: "Ingenious plans are discussed but the population biologists are vexed by the field worker's disregard for theory, and the field workers are appalled by the dogmatic inability of the former to recognize practical problems."

Other levels of interchange are between technical and administrative levels and then within the administrative levels if more than one government is involved. The involvement of nongovernment organizations has to be factored into those consultations. A survey of contacts for distributing IUCN Action Plans (Gimenez-Dixon and Stuart 1993) revealed that the lowest proportion of contacts was with governments (29%) and national and local nongovernment organizations (6%), yet they are essential for implementation of plans.

Even more absent from developing the plans are the local people, who often are impinged by measures to protect species. Frequently they will be on the receiving end of a component of the plan such as an educational program. That is not the same type of consultation as having a say in the plan itself. Without the involvement of local people, conservation projects often falter because they are imposed on, rather than integrated into, the local perceptions and cultural values.

Consultation should act as a safety net to ensure that actions to recover one species do not have a deleterious effect on another. The unintended effects of a recovery action for species on another endangered species (Section 1.4) is not so much a criticism of the species as the appropriate conservation unit (Polhemus 1993) but more the consequences of acting with inadequate ecological information. Recently arguments have been advanced for using ecosystems as the unit of conservation: the reasoning is to ensure compatibility of and coordination between plans when there are several endangered species in one ecosystem. Informally grouping species recovery plans by habitat may be a common sense compromise.

Recovery plans as currently devised do not pivot on diagnosing the causes of the decline. Yet it is this that has to drive effective recovery actions. Recovery actions can and often could be the experimental means to test diagnoses (Box 8.11 gives an example of hypotheses to be tested to identify the factors implemented in the decline of Peary caribou). The ability to determine the success and failure of the recovery treatments is often glossed over. An absence of measurable criteria within a designated time frame can easily leave a recovery program locked into such treatments as control of introduced predators or artificial feeding stations (Box 12.2).

Recovery goals are sometimes given as a single number—a desired size for a population. This can create an impression of certainty which, at the very least, sidesteps the precision of techniques used to estimate population size. Stating the desired population size or number of populations within a time period or stating rates of decrease or increase imparts a sense of dynamics. Static goals can induce complacency once they are reached which may be inappropriate depending on how much guesswork went into establishing them. Just because a population has recovered to a certain size, it is by no means safe.

A sense of urgency and the inpetus to be current in the field often leads to adopting insufficiently tested concepts. Walters (1991) made this point while observing that acceptance of an effective population size for the red-cockaded woodpecker's recovery plan was premature.

Box 12.2.   Identification of feasible options in recovery plans

---

These questions prompt the identification of a feasible option.

1.   Where do we want to go?
2.   Can we get there?
3.   Will we know when we have arrived?
4.   How do we get there?
5.   What disadvantages or penalties accrue?
6.   What benefits are gained?
7.   Will the benefits exceed the penalties?

Question 3 is important: It requires formulating stopping rules which either halt or change the management action. If we cannot determine when the objective has been attained, either for reasons of logic (ambiguous or abstract statement of the objective) or for technical reasons (inability to meansure the state of the system), the option is not feasible.

It follows as a corollary that there must be an easy way of recognizing the failure to attain an objective. The most common is to measure the outcome against that specified by the technical objective. Another is to compare the outcome with a set of *criteria of failure* set before the management action is begun. Those two are not the same. Comparison of outcome with objective can produce assessments like "not quite" or "not yet." Not so with criteria of failure: they take the form "the operation will be judged unsuccessful, and will therefore be terminated, if outcome x has not been attained by time t."

(Modified from Caughley G, Sinclair ARE. Wildlife management and ecology. Cambridge, MA: Blackwell Science, 1994.)

As he aptly observed, it is biologists who bear the blame for uncritically accepting concepts and who fail to clarify limitations attached to newly developed concepts and techniques.

12.4   RISK AND DECISION ANALYSES

Management plans recommend possible options that are uncertain. The risks of implementing those recommendations and then finding that the outcome is an adverse event, ranging from inappropriate expenditure, public dissatisfaction, all the way to a species extinction. A manager will then want to have, as far as possible, quantitative evaluation of those risks. The first step is a risk analysis that entails listing all the events likely to impinge a decision and their consequences, and then assigning probabilities for their occurrence in a specified time interval. The probabilities can be estimated from observed occurrences or can be subjective—and that is where a note of caution should be sounded. The reliability of a risk assessment depends critically upon the accuracy of the estimated probabilities (Speed 1985, Caughley 1994). Once they are estimated, it is easy to lose sight of their origin. Assigning probabilities can be biased and used to favor one option over another (unless they are based on observed occurrences). The second step is a decision analysis. The probabilities are summed and multiplied according to the rules of probability theory. Setting out the probabilities and options is best done as a decision tree which identifies the preferred option (Maguire 1986). The sensitivity of the ranked options can be analyzed by altering the probabilities and determining the outcome.

The key step is the risk assessment: ensuring that all possible events and consequences are considered. It has the undeniable advantage of putting all factors used to make a decision on the table where they are visible. Recommendations for conservation are often controversial, and each special interest group has its own biases. Risk assessment and decision analysis allow all interested parties to see which factors were used and how they were weighted in a decision.

Formal procedures such as decision analysis have been used relatively infrequently despite their advantages (Maguire and Servheen 1992) and then mostly in the context of deciding what to do with small or captive populations. A cautionary tale from the decision analysis for the Sumatran rhino illustrates what happens when not all events and their consequences are included in the risk analysis (Caughley 1994). The decision analysis led to a recommendation for captive breeding (Section 9.5.1). Other data showed it was less expensive to avoid poaching even without considering the poor breeding record of Sumatran rhinos in zoos (Leader-Williams 1993).

12.5 SUMMARY

An array of international conventions impinge on conservation. The Convention on International Trade in Endangered Species of Wild Fauna and Flora is the best known. CITES regulates the international trade of species nominated by the countries that have ratified the convention. The successes of CITES owe much to its support by national legislation that regulates the import, possession, and export of species and their derivative parts. IUCN–The World Conservation Union is not a convention but nonetheless is pivotal in conservation. Straddling as it does government and nongovernment agencies, IUCN plays an unique role in collecting and disseminating information and facilitating research on endangered species.

National legislation to regulate possession and take is moving more toward the protection of endangered species. The first step is usually a listing procedure to designate the status of a species. Endangered species legislation usually requires the production of recovery plans, which are special cases of management plans and have the same requirement for attainable and accountable objectives. The emphasis of the goals should not be tied only to levels of population size but also rates of change. The decisions for the management of an endangered species often span the interests of different audiences. A technique that allows all the criteria for a decision to be disclosed and objectively evaluated is risk and decision analysis. However, like most techniques decision analysis is only as good as the information fed into it.

CHAPTER 13

Summary and Conclusions

This chapter summarizes the more important conclusions of this book. The reasoning and information upon which they are based is given in the appropriate chapters. Throughout the book we have sought to balance conservation biology's two prevailing but contrasted paradigms: the small-population one, which is largely theoretical, deals with the risks facing small populations. The declining-population paradigm is the practicalities of detecting declines and reversing them.

Our emphasis is on how we weigh evidence for why a species is in trouble. The two essential tools are, first, to know the species ecology well enough to propose hypotheses on whether it is declining and why. The second tool is the scientific method: using experimental design to test the hypotheses. Usually we want to avoid accepting a null hypothesis when it is false (Type 2 error). In other words, missing a decline or assuming a treatment has not worked could further a decline and eventually even cause extinction. That sample sizes are small is not a deterrent to experiments, as it can be partially compensated for by the design and computer-intensive sampling techniques.

We rate the most useful unit of conservation status as the species and the population as the unit of management. Although a case can be made for using the ecosystem (or community) as a unit of conservation, it runs into problems of definition. It is particularly ambiguous when written into law.

Extinction is a natural process in geological time, taking the form of a modest background rate which occasionally accelerates to provide a mass extinction. There have been several of these over the last 50,000 years. The most recent, in New Zealand, Madagascar, and the islands of the Pacific, are clearly associated with the initial arrival of people. The causes of the mass extinctions during the Pleistocene in Australia and at the end of the Pleistocene in the Americas are still controversial. We have restricted ourselves to outlining the evidence for and against their climatic change interpretation and their human impact interpretation.

We chose sets of mammals and birds for surveying historic causes of extinction, but we were left to conclude how little is known about most past extinctions. In only a few cases can the cause of extinction be deduced with confidence. Perhaps it is those apparently secure instances that have contributed more than is warranted to the certainty that man was directly or indirectly responsible for all the others.

With an eye to what can be learned from a detailed look at which factors drive declines, we selected case histories for 17 species in trouble. The only secure generalization is that factors must be detected by the disciplined application of scientific method and ecological acumen rather than by assumptions, modeling, or arguing from "general principles." A recurring theme in conservation biology is that, since time and resources are often short and data are scarce, simple models or assumptions rather than ecological research can generate management recommendations. Our case studies suggest that this is not so, that often the wrong hypotheses will be selected. Abandoning research methodology is likely to provide the wrong answer as a consequence of asking the wrong question. In addition to being sloppy science, it can lose us yet another species.

The case histories and other examples indicate that rarely do single factors drive a decline. Problems tend to be nested insofar that treating one reveals another which sometimes reflects flaws in the initial diagnosis. Seven species from the case histories ended up in refuges where they persisted because the original cause of their decline was excluded. The refuge may be on the periphery of the species range, and it may be suboptimal habitat lacking specific resources (food, shelter, nest sites), and these, too, have to be part of the diagnosis.

Our analysis suggests that evidence for a species in trouble is a decline in numbers or a contraction in range. Even just the knowledge of habitat modification on a large scale can be sufficient to tip us off to investigate further. Introduced predators or overharvesting are relatively easy to diagnose. But when it comes to habitat changes it is insufficient to simply rest the case as that. Instead diagnosis depends on knowing which specific resources have changed and that may require greater ecological skills. Factors are not always the conspicuous ones such as loss of nest holes in trees. Dealing at the level of specific resources holds whether the habitat modification is uniform or if land use changes are piecemeal and fragment the habitats. Changes in habitat dispersion have unequal effects on species, depending on their ecology and life histories.

To extend our understanding of declines we need more than case-by-case listings of causes. The trajectory of a declining population and its demographic components are rarely described, and, as well as data, we lack theories with which to predict and test ideas about how species decline. Much attention has focused on theoretical risks at the final stage of low numbers, which is the essence of the small population paradigm. The population may die out according to the various demographic or environmental problems imposed by the sampling probabilities—the role of bad luck. This is a tenuous conclusion: extinctions usually happen out of sight and we rarely know the fate of the last individuals of a species.

Despite oft-cited concerns, there is no convincing evidence that genetic malfunction has been responsible for the final stages of the extinction. Nonetheless, that could simply reflect a lack of information. The major questions that need to be addressed by conservation geneticists are the following:

- How is fitness influenced by its equilibrium level of heterozygosity and over what time scale or environmental changes?
- Why does inbreeding differ so widely among species, and what are its effects?

Concern over loss of genetic variation is one reason why meta-population dynamics is at the forefront of conservation biology. The reasoning is that a metapopulation encompasses more genetic variation than any one of its individual populations. The other predicted advantage of a metapopulation structure is safety in numbers: having several populations raises the chance that at least one will be left unscathed by an environmental catastrophe. That is why meta-population dynamics feature in designing reserve networks and managing captive populations. However, whether populations should be kept isolated or linked, such as by corridors or translocations between reserves, are as yet unanswered questions. If a population is already small, dividing it further to create other populations may simply increase the risk of extinction. Our caveat for application of meta-population dynamics is that there is a continuum of population structures, habitat dispersion, and dispersive behavior, which limit generalizations or using a single model.

The answer to the question of how long a population is likely to persist if present conditions obtain (population viability analysis) is seldom useful in diagnosing why the population declined to its present level or how it might be rescued from it. Population viability analyses as presently constituted call for detailed information on the population's lower-order demographic parameters (survivorship and fecundity by age, or those summarized by birth rate and death rate). These are fiddly to estimate and almost impossible to do so to the level of accuracy required to drive a viability analysis. The parameter that can be estimated in the field is r, the realized rate of increase, which for a species in trouble is usually negative. It is estimated from the trend of population size or density or from an index of these. Expected time to extinction can then be estimated directly from the trend.

Treatment of a decline can test a hypothesis whether the diagnosis was correct. Resources supplementation (food, nest sites) or reducing mortality (overharvesting or predators) are treatments that lend themselves to experiments. When it comes to treating a small population, we would like to find out why the population declined to low numbers, to identify and eliminate the agent of decline, and then to reverse the process. The treatment of low numbers as such, often leading to a recommendation that a significant proportion of the population be put in a zoo, is to treat a symptom as if it were a cause. Supplementation of low numbers through captive breeding or captive raising is a useful strategy to complement the treatment of a species declining in the wild. Translocation to supplement population size or establish separate populations is popular, but popularity and success do not, however, go hand in hand. Care in preparing the translocated animals for release helps, but the release will often be of individuals changed by their sojourn in captivity, and they are released into an ecological community subtly changed from the original one.

The design of reserves is well served by theory, but, in practice, reserves as a treatment for species in trouble often falter, which may start with failure to clearly state the reserve objectives, be they to conserve states or processes. Theory has dwelt largely on area and its related themes of shape, size, and number, but we are starting to see a return to determining why some species persist and others do not. Merely treating habitat loss or modification by setting aside a reserve for that habitat is not always going to be enough. Conspicuous failures in reserves have come when the reserve did not meet the ecological requirements of the species, often through a lack of appreciation for the dynamics of communities. And discounting the needs of people displaced from or living near the reserve has hampered the effectiveness of reserves.

International treaties and national legislation play their part in helping species in trouble, but their success is mixed when trying to reduce overharvesting, whether or not it includes international trade. The problem is that legislation to reduce harvesting does not get to the root of the matter, and to do that we turn to three aspects of economics. First is the positive feedback loop between value and rarity; then there is the fact that sustained-yield harvesting is economically inefficient: it pays to rapidly deplete natural capital (animals or plants) and reinvest the money profitably elsewhere. Finally, some market structures leave harvesters no incentive to conserve because the return on their effort is too low. These problems are not insurmountable, but just the same as ecological problems, treatment without diagnosis is difficult at best. Treatment has to take into account the harvesters' social institutions and customs as well as individual aspirations, and this applies to almost any conservation issue or problem. If people through their shared need for resources are part of the problem for a species in trouble, then they are also part of the solution.

More countries are developing national legislation for endangered species, and, as often as not, the legislation requires species recovery plans. The plans are an opportunity for all interested parties to deal openly with issues underlying species recovery, and here risk analysis can help. On the other hand, recovery plans sometimes are short on diagnosis of factors determining the decline. We also note the temptation in the rush to save a species to accept a theory (corridors, minimum viable population sizes, population viability analyses) whose testing is incomplete, especially considering the diverse life-styles and ecology.

Throughout the text we have described using common species as surrogates rather than risk the last individuals of a species. Most examples are the obvious ones to test techniques to capture, handle, breed, and release, but surrogates could contribute much more to saving species. We have commented on the lack of theory in the declining-population paradigm and the small-population paradigm's scarcely tested theory. We suggest that the way to cope with this is to experimentally test the paradigm's predictions. A paradigm which is not questioned is an ideology. And that is where experiments with common species could be used to test existing or new theories on how populations decline—what combinations of events start a decline, and so forth. We already have numerous examples of experimental extinctions under the guise of pest control.

The time to tackle a conservation problem is when a decline is detected. The major aim of conservation biology should be to detect the problem well before the final stage of low numbers. Thus, it is trends in numbers and not absolute population sizes that cause worry. Waiting until a population is small runs the risk of changing conservation management into crisis management, with all the panic and shortcuts that are thereby generated. As for the species now in trouble, we judge that help for them will largely come from the well-tested toolboxes of experimental design and ecological research.

References

Aguilera-Klink F. Some notes on the misuse of classic writings in economics on the subject of common property. Ecol Econ 1994;9:221–228.

Akçakaya HR, Ginzberg LR. Ecological risk analysis for single and multiple populations. In: Seitz A, Loeschcke V, eds. Species conservation: a population-biological approach. Basel: Birkhauser Verlag, 1991:78–87.

Ali S. Asiatic wild ass. In: Encyclopedia of Indian natural history: centenary publication of the Bombay Natural History Society. Delhi: Oxford University Press, 1986:25–26.

Ali S, Santapau H, Abdulali H (eds). The hispid hare, *Caprolagus hispidus* (Pearson). J Bombay Nat Hist Soc 1960;57:400–402.

Allen GM. Extinct and vanishing mammals of the western hemisphere with the marine species of all the oceans. New York: ACIWP, 1942.

Allen GM. Extinct and vanishing mammals of the western hemisphere with the marine species of all the oceans. New York: Cooper Square, 1972.

Allen H. Nineteenth century faunal change in western New South Wales and north-western Victoria. Working Papers in Anthropology, Archaeology, Linguistics and Maori Studies, No. 64. Department of Anthropology, University of Auckland, Auckland, NZ, 1983.

Allendorf FW. Isolation, gene flow, and genetic differentiation among populations. In: Schonewald-Cox CM, Chambers SM, MacBryde B, Thomas L, eds. Genetics and conservation: a reference for managing wild animal and plant populations. Menlo Park, CA: Benjamin/Cummings, 1983.

Alvarez VB, González AC. The critical condition of hutias in Cuba. Oryx 1991;25:206–208.

Amarasekare P. Potential impact of mammalian nest predators on endemic forest birds of western Mauna Kea, Hawaii. Conserv Biol 1993;7:316–324.

Anderson A. Prodigious birds. Cambridge: Cambridge University Press, 1989

Anderson A, McGovern-Wilson R. The pattern of prehistoric Polynesian colonisation in New Zealand. J R Soc NZ 1990;20:41–63.

Anderson JL. Management of translocated white rhino in South Africa. In: Ryder OA, ed. Rhinoceros biology and conservation. Proceedings of an international conference, May 9–11, 1991, San Diego, CA. San Diego: Zoological Society of San Diego, 1993:287–293.

Anderson RM, May RM. Vaccination and herd immunity to infectious diseases. Nature 1985;318:323–329.

Andresen S. The effectiveness of the International Whaling Commission. Arctic 1993;46:108–115.

Andrewartha HG, Birch LC. The distribution and abundance of animals. Chicago: Chicago University Press, 1954.

Anon. Marine Mammal Commission. Annual report to U.S. Congress 1992.

Apps P. Cats on Dassen Island. Acta Zool Fenn 1984;172:115–116.

Ardern SL, Mclean IG, Anderson S, Maloney R, Lambert DM. The effects of blood sampling on the behavior and survival of the endangered Chatham Island black robin (*Petroica traversi*). Conserv Biol 1994;8:857–862.

Armbruster P, Lande R. A population viability analysis for African elephant (*Loxodonta africana*): how big should reserves be? Conserv Biol 1993;7:602–610.

Armstrong S. Dining with vultures. New Sci 1993;140(1899):41–43.

Arnold GW, Weeldenburg JR. The distributions and characteristics of remnant vegetation in parts of the Kellerberrin, Tammin, Trayning and Wyalkatchem Shires of Western Australia. CSIRO Division of Wildlife and Ecology Technical Memorandum No. 33, 1991.

Askins RA. Population trends in grassland, shrubland, and forest birds in eastern North America. Curr Ornithol 1993;11:1–34.

Atkinson IAE. Spread of the ship rat (*Rattus r. rattus* L.) in New Zealand. J R Soc NZ 1973;3:457–472.

Atkinson IAE. A reassessment of factors, particularly *Rattus rattus* L., that influenced the decline of endemic forest birds in the Hawaiian Islands. Pacific Sci 1977;31:109–133.

Atkinson IAE. The spread of commensal species of *Rattus* to oceanic islands and their effect on island avifaunas. ICBP Tech Publ 1985;3:35–81.

Atkinson IAE, Greenwood RM. Relationships between moas and plants. NZ J Ecol 1989;12(Suppl):67–96.

Atkinson IAE, Moller H. Kiore, Polynesian rat. In: King CM, ed. The handbook of New Zealand mammals. Oxford: Oxford University Press, 1990:174–192.

Baker AJ. A review of New Zealand ornithology. Curr Ornithol 1991;8:1–67.

Baker CS, Palumbi SR. Which whales are hunted? A molecular genetic approach to monitoring whaling. Science 1994;265:1538–1539.

Baker RC, Wilke F, Baltzo CH. The northern fur seal. United States Department of the Interior, Circular 336, Washington, DC, 1970.

Balakrishnan M, Ndhlovu DE. Wildlife utilization and local people: a case-study in Upper Lupande Game Management Area, Zimbabwe. Environ Conserv 1992;19:135–145.

Ballance AP. *Paryphanta* at Pawakatutu. J Auckland Univ Field Club 1986;31:13–18.

Ballou J. Calculating inbreeding coefficients from pedigrees. In: Schonewald-Cox CM, Chambers SM, MacBryde B, Thomas L, eds. Genetics and conservation: a reference for managing wild animal and plant populations. Menlo Park, CA: Benjamin/Cummings, 1983:509–520.

Balouet JC, Olson SL. Fossil birds from the late Quaternary deposits in New Caledonia. Smithsonian Contrib Zool 1989;469:1–38.

Banko PC. Constraints on productivity of wild nene or Hawaiian geese, *Branta sandwicensis*. Wildfowl 1992;43:99–106.

Banks B, Beebee TJC, Cooke AS. Conservation of the natterjack toad *Bufo calamita* in Britain over the period 1970–1990 in relation to site protection and other factors. Biol Conserv 1994;67:111–118.

Bannerman DA. Birds of the Atlantic islands. Edinburgh: Oliver and Boyd, 1963.

Bannerman DA. A probable sight record of a Canarian black oystercatcher. Ibis 1969;111:257.

Barbour MS, Litvaitis JA. Niche dimensions of New England cottontails in relation to habitat patch size. Oecologia 1993;95:321–327.

Barinaga M. Where have all the froggies gone? Science 1990;247:1033–1034.

Bart J. Impact of human visitations on avian nesting success. Living Bird 1978;16:187–192.

Bart J, Robson DS. Estimating survivorship when the subjects are visited periodically. Ecology 1982;63:1078–1090.

Barton M. Black-footed ferret rides again. BBC Wildlife 1992a;10(9):63.

Barton M. They're going wild again in Wyoming. BBC Wildlife 1992b;10(1):56.

Barzdo J. Elephant crisis. Oryx 1989;23:181–183.

Batisse M. Biosphere reserves: an overview. Nature Resour 1993;29:3–5.

Bayliss P. Kangaroo dynamics. In: Caughley G, Shepherd N, Short J, eds. Kangaroos: their ecology and management in the sheep rangelands of Australia. Cambridge: Cambridge University Press, 1987:119–134.

Beauchamp AJ, Worthy TH. Decline in distribution of the takahe *Porphyrio* (= *Notornis*) *mantelli*: a re-examination. J R Soc NZ 1988;18:103–112.

Beck BB, Kleiman DG, Castro I, Rettberg-Beck B, Carvalho C. Preparation of captive-born golden lion tamarins for release into the wild. In: Kleiman DG, Rosenberger AL, eds. A case study in conservation: the golden lion tamarin. Washington, DC: Smithsonian Institution Press, in press.

Beck BB, Rapaport LG, Stanley Price MR, Wilson AC., Reintroduction of captive-born animals. In: Olney PJS, Mace GM, Feistner ATC, eds. Creative conservation: interactive management of wild and captive animals. London: Chapman and Hall, 1994:265–286.

Beebe W. Rediscovery of the Bermuda cahow. Bull NY Zool Soc 1935;38:189–196.

Beissinger SR, Bucher EH. Sustainable harvesting of parrots for conservation. In: Beissinger ER, Snyder NFR, eds. New world parrots in crisis. Solutions from conservation biology. Washington, DC: Smithsonian Institution, 1992:73–115.

Belbin L. Environmental representativeness: regional partitioning and reserve selection. Biol Conserv 1993;66:223–230.

Bell DJ. A study of the hispid hare (*Caprolagus hispidis*) in Royal Suklaphanta Wildlife Reserve, western Nepal: a summary report. Dodo, J Jersey Wildlife Preserv Trust 1986;23:24–31.

Bell DJ, Hartley JRM. An investigation into current captive breeding problems with the pink pigeon *Nesoenas mayeri:* a report on the first year's results. Dodo, J Jersey Wildlife Preserv Trust 1987;24:86–102.

Bell DJ, Oliver WLR. Northern Indian tall grasslands: management and species conservation with special reference to fire. In: Singh KP, Singh JS, eds. Tropical ecosystems: ecology and management. New Delhi: Wiley, 1992:109–123.

Bell DJ, Oliver WLR, Ghose RK. The hispid hare, *Caprolagus hispidis*. In: Chapman JA, Flux JEC, eds. Rabbits, hares and pikas; status survey and conservation action plan. Gland, Switzerland: IUCN/SSC Lagomorph Specialist Group, International Union Conservation of Nature, 1990:128–136.

Bell RHV. Decision-making in wildlife management with reference to problems of overpopulation. In: Owen-Smith RN, ed. Management of large mammals in African conservation areas. Pretoria: HAUM, 1983:145–172.

Bellwood P. Man's conquest of the Pacific. The prehistory of southeast Asia and Oceania. Auckland: Collins, 1978.

Belovsky GA. Extinction models and mammalian persistence. In: Soulé ME, ed. Viable populations for conservation. Cambridge: Cambridge University Press, 1987:35–57.

Berger AJ. Hawaiian wildlife. Honolulu: University Press of Hawaii, 1981.

Berger J. Rhino conservation tactics. Nature 1993;361(6408):121.

Berger J. Science, conservation and black rhinos. J Mammal 1994;75:298–308.

Berry RJ. Conserving genetical variety. In: Warren A, Goldsmith FB, eds. Conservation in practice. London: Wiley, 1974:99–115.

Beverton RHJ. Small marine pelagic fish and the threat of fishing; are they endangered? J Fish Biol 1990;37(Suppl A):5–16.

Bhattacharya A. The status of the Kaziranga rhino population. Tiger Paper 1993;20:1–6.

Bibikov DI. The steppe marmot—its past and future. Oryx 1991;25:45–49.

Biggins DE, Crete RA. Black-footed ferret recovery. In: Bjugstad AJ, Uresk DW, Hamre RH, eds. Proceedings of the Ninth Great Plains Wildlife Damage Control Workshop. USDA Forest Service General Technical Report RM-171, Fort Collins, CO, 1989:59–63.

Black FL. Why did they die? Science 1992;258:1739–1740.

Black JM. Reintroduction and restocking: guidelines for bird recovery programmes. Bird Conserv Int 1991;1:329–334.

Black JM, Banko PC. Is the Hawaiian goose (*Branta sandvicensis*) saved from extinction? In: Olney PJS, Mace GM, Feistner ATC, eds. Creative conservation: interactive management of wild and captive animals. London: Chapman and Hall, 1994:394–410.

Black JM, Duvall F, Hoshide H, et al. The current status of the Hawaiian goose *Branta sandvicensis* and its recovery program. Wildfowl 1991;42:149–154.

Bloomer JP, Bester MN. Control of feral cats on sub-Antarctic Marion Island, Indian Ocean. Biol Conserv 1992;60:211–219.

Bodmer RE, Fang TG, Moya L, Gill R. Managing wildlife to conserve Amazonian forests: population biology and economic considerations of game hunting. Biol Conserv 1994;67:29–35.

Bogan AE. Freshwater bivalve extinctions (Mollusca: Unionoida): a search for causes. Am Zool 1993;33:599–609.

Bonner WN. Seals and man. A study of interactions. Seattle: University of Washington Press, 1982.

Bourn NAD, Thomas JA. The ecology and conservation of the brown argus butterfly, *Arica agrestis*, in Britain. Biol Conserv 1993;63:67–74.

Bourne WRP. FitzRoy's foxes and Darwin's finches. Arch Nat Hist 1992;19:29–37.

Bourne WRP. The story of the great auk, *Pinguinis impennis*. Arch Nat Hist 1993;20:257–278.

Bowers J. Economics of the environment: the conservationists' response to the Pearce Report. Telford, Shropshire: British Association of Nature Conservationists, 1990.

Boyce MS. Population viability analysis. Ann Rev Ecol System 1992;23:481–506.

Brash AR. The history of avian extinction and forest conversion on Puerto Rico. Biol Conserv 1987;42:97–111.

Brechtel SH, Carbyn LN, Hjertaas D, Mamo C. Canadian swift foxes reintroduction feasibility study: 1989 to 1992. Report and recommendations of the National Recovery team. Edmonton, Alberta: Alberta Fish and Wildlife Department, 1993.

Brett D. Sea birds in the trees. Ecos 1989;61:4–8.

Briggs DEG, Crowther PR (eds). Palaeobiology: a synthesis. Oxford: Blackwell, 1990.

Brock MK, White BN. Application of DNA fingerprinting to the recovery program of the endangered Puerto Rican parrot. Proc Nat Acad Sci 1992; 89:11121–11125.

Brotherton I. On loopholes, plugs and inevitable leaks: a theory of SSSI protection in Great Britain. Conserv Biol 1990;52:187–203.

Brown JH. Mammals on mountaintops: nonequilibrium insular biogeography. Am Natural 1971;105:467–478.

Bryden HA. The decline and fall of the South African elephant. Fortnight Rev 1903;79:100–108. [Not seen: quoted in Douglas-Hamilton, 1987.]

Buechner HK, Dawkins HC. Vegetation change induced by elephants and fire in Murchison Falls National Park, Uganda. Ecology 1961;42:752–766.

Burgman M, Cantoni D, Vogel P. Shrews in suburbia: an application of Goodman's extinction model. Biol Conserv 1992;61:117–123.

Burgman MA, Ferson S, Akçakaya HR. Risk assessment in conservation biology. London: Chapman and Hall, 1993.

Busch BC. The war against the seals. A history of the North American seal fishery. Montreal: McGill and Queen's University Press, 1985.

Butler D, Merton D. The black robin: saving the world's most endangered bird. Auckland: Oxford University Press, 1992.

Butler PJ. Parrots, pressures, people, and pride. In: Beissinger ER, Snyder NFR, eds. New world parrots in crisis. Solutions from conservation biology. Washington DC: Smithsonian Institution, 1992:25–46.

Cade TJ, Henderson JH, Thelander CG, White CM (eds). Peregrine falcon populations: their management and recovery. Boise, ID: Peregrine Fund, 1988.

Cade TJ, Jones CJ. Progress in restoration of the Mauritius kestrel. Conserv Biol 1993;7:169–175.

Calaby JH. Observations on the banded ant-eater *Myrmecobius fasciatus* Waterhouse (Marsupialia), with particular reference to its food habits. Proc Zool Soc London 1960;135:183–207.

Calaby JH, Flannery TF. Marsupials of Australia, Vol III. Sydney: Geo, in press.

Campbell DG, Lowell KS, Lightbourn ME. The effect of introduced hutias (*Geocapromys ingrahami*) on the woody vegetation of Little Wax Cay, Bahamas. Conserv Biol 1991;5:536–541.

Carey C. Hypothesis concerning the causes of the disappearance of the boreal toads from the mountains of Colorado. Conserv Biol 1993;7:355–362.

Carlton JT. Neoextinctions of marine invertebrates. Am Zool 1993;33:499–509.

Caro TM, Laurenson MK. Ecological and genetic factors in conservation: a cautionary tale. Science 1994;263:485–486.

Carrington R. Elephants: a short account of their natural history, evolution and influence on mankind. London: Chatto and Windus, 1958.

Case TJ, Bolger DT, Richman AD. Reptilian extinctions: the last ten thousand years. In: Fiedler PL, Jain SK, eds. Conservation biology: the theory and practice of nature conservation, preservation and management. New York: Chapman and Hall, 1992:91–125.

Caton A, Williams MJ, Ramirez C. The Australian 1989–90 to 1992–93 southern bluefin tuna seasons. Twelfth Australian, Japanese and New Zealand Trilateral Scientific Meeting on Southern Bluefin Tuna, CSIRO Marine Laboratories, Hobart. Working Paper No. SBFWS/93/7, 1993.

Caughley G. Mortality patterns in mammals. Ecology 1966;47:906–918.

Caughley G. Growth, stabilisation and decline of New Zealand populations of the Himalayan thar (*Hemitragus jemlahicus*). Ph.D. (Zoology) thesis, University of Canterbury, Christchurch, New Zealand, 1967a.

Caughley G. Parameters for seasonally breeding populations. Ecology 1967b;48:834–839.

Caughley G. Wildlife and recreation on the Trisuli watershed and other areas in Nepal. HMG/FAO/UNDP Trisuli Watershed Development Project. Project Report No. 6, 1969.

Caughley G. Eruption of ungulate populations, with special emphasis on Himalayan thar in New Zealand. Ecology 1970;51:53–72.

Caughley G. Interpretation of age ratios. J Wildlife Man 1974;38:557–562.

Caughley G. Plant herbivore systems. In: May RM, ed. Theoretical ecology: principles and applications. London: Blackwell, 1976a:94–113.

Caughley G. The elephant problem: an alternative hypothesis. E Afr Wildlife J 1976b;14:265–283.

Caughley G. Analysis of vertebrate populations. Chichester: Wiley, 1977.

Caughley G. Dynamics of large animals and their relevance to culling. In: Owen-Smith RN, ed. Management of large mammals in African conservation areas. Pretoria: HAUM, 1983:115–126.

Caughley G. Rangelands, livestock and wildlife, the ecological equivalent of sulphur, saltpetre and charcoal. In: Joss PJ, Lynch PW, Williams OB, eds. Rangelands: a resource under seige. Canberra: Australian Academy of Science, 1986:545.

Caughley G. Ecological relationships. In: Caughley G, Shepherd N, Short J, eds. Kangaroos: their ecology and management in the sheep rangelands of Australia. Cambridge: Cambridge University Press, 1987:159–187.

Caughley G. The colonisation of New Zealand by the Polynesians. J R Soc NZ 1988;18:245–270.

Caughley G. New Zealand plant-herbivore systems: past and present. NZ J Ecol 1989;12(Suppl):3–10.

Caughley G. Directions in conservation biology. J Animal Ecol 1994;63:215–244.

Caughley G, Dublin H, Parker I. Projected decline of the African elephant. Biol Conserv 1990;54:157–164.

Caughley G, Goddard J. Abundance and distribution of elephants in the Luangwa Valley, Zambia. E Afr Wildlife J 1975;13:39–48.

Caughley G, Grice D, Barker R, Brown B. The edge of the range. J Animal Ecol 1988;57:771–785.

Caughley G, Grigg GC, Short J. How many kangaroos? Search 1983:14:151–152.

Caughley G, Gunn A. Dynamics of large herbivores in deserts: kangaroos and caribou. Oikos 1993;67:47–55.

Caughley G, Krebs CJ. Are big mammals simply little mammals writ large? Oecologia 1983;59:7–17.

Caughley G, Sinclair ARE. Wildlife management and ecology. Cambridge, MA: Blackwell Science, 1994.

Caughley G, Walker BH. Working with ecological ideas: ecological concepts. In: Ferrar AA, ed. Guidelines for the management of large mammals in African conservation areas. South African National Scientific Programmes Report No. 69, 1983:13–34.

Ceballos G, Navarro D. Diversity and conservation of Mexican mammals. In: Mares MA, Schmidly DJ, eds. Latin American mammalogy. History, biodiversity and conservation. Norman: University of Oklahoma Press, 1991:167–198.

Chapin JP. The birds of the Belgian Congo. Part 1, Bull Am Museum Nat Hist 1938;68:1–766.

Chapman JA, Flux JEC (eds). Rabbits, hares and pikas: status survey and conservation action plan. IUCN/SSC Lagomorph Specialist Group, Gland, Switzerland: IUCN, 1990.

Chapuis JL, Boussès P, Barnaud G. Alien mammals, impact and management in the French subantarctic islands. Biol Conserv 1994;67:97–104.

Chase A. Playing god in Yellowstone: the destruction of America's first national park. San Diego: Harcourt Brace Jovanovich, 1987.

Cheke AS. An ecological history of the Mascarene Islands, with particular reference to extinctions and introductions of land vertebrates. In: Diamond AW, ed. Studies of Mascarene Island birds. Cambridge: Cambridge University Press, 1987a:5–89.

Cheke AS. The legacy of the dodo—conservation in Mauritius. Oryx 1987b; 21:29–36.

Cheke AS, Dahl JF. The status of bats on the western Indian Ocean islands, with special reference to *Pteropus*. Mammalia 1981;45:205–238.

Chepko-Sade BD, Shields WM, Berger J, et al. The effects of dispersal and social structure on effective population size. In: Chepko-Sade BD, Halpin ZT, eds. Mammalian dispersal patterns: the effects of social structure on population genetics. Chicago: The University of Chicago Press, 1987:287–321.

Cherfas J. The hunting of the whale: a tragedy that must end. London: The Bodley Head, 1988.

Chisholm AG, Moran AJ (eds). The price of preservation. Melbourne: Tasman Institute, 1993.

Chittleborough RG. Dynamics of two populations of the humpback whale, *Megaptera novaeangliae* (Borowski). Austral J Marine Fresh Res 1965;16:33–128.

Clarke B, Murray J, Johnson MS. The extinction of endemic species by a program of biological control. Pacific Sci 1984;38:97–104.

Clark CW. Mathematical bioeconomics: the optimum management of renewable resources. New York: Wiley Interscience, 1976.

Clough GC. Current status of two endangered Caribbean rodents. Biol Conserv 1976;10:43–47.

Clutton-Brock J. Man-made dogs. Science 1977;197:1340–1342.

Cochran WG. Sampling techniques, 2nd ed. New York: Wiley, 1963.

Cocks KD. Use with care: managing Australia's natural resources in the 21st century. Sydney: University of New South Wales Press, 1992.

Cohen B. Multiple use and nature conservation in South Australia's arid zone. Rangelands J 1992;14:205–214.

Cole BP. Recovery planning for endangered and threatened species. In: Seal US, Thorne ET, Bogan MA, Anderson SH, eds. Conservation biology and the black-footed ferret. New Haven: Yale University Press, 1989:201–209.

Collar NJ, Gonzaga LP, Krabbe N, et al. Threatened birds of the Americas. Washington: Smithsonian Institution Press, London: ICBP, 1992.

Collar NJ, Stuart SN. Threatened birds of Africa and related islands. The ICBP/IUCN red data book, Part 1. Cambridge: International Council for Bird Preservation and International Union for the Conservation of Nature, 1985.

Conner RN, Rudolph DG. Forest habitat loss, fragmentation, and red-cockaded woodpecker populations. Wilson Bull 1991;103:446–457.

Conrad J, Bjørndal T. On the resumption of commercial whaling: the case of the minke whale in the northeast Atlantic. Arctic 1993;46:164–171.

Conway AJ, Goodman PS. Population characteristics and management of black rhinoceros *Diceros bicornis minor* and white rhinoceros *Ceratotherium simum simum* in Ndumu Game Reserve, South Africa. Biol Conserv 1989;47:109–122.

Cooke J, Thomsen J. International Whaling Commission: a report of the 45th annual meeting. TRAFFIC 1993;14:21–24.

Cooper A, Atkinson IAE, Lee WG, Worthy TH. Evolution of the moa and their effect on the New Zealand flora. Trends Ecol Evol 1993;8:433–437.

Cooper J, Fourie A. Improved breeding success of great-winged petrels *Pterodroma macroptera* following control of feral cats *Felis catus* at subantarctic Marion Island. Bird Conserv Int 1991;1:171–175.

Costanza R, Kemp WM, Boynton WR. Predictability, scale, and biodiversity in coastal and estuarine ecosystems: implications for management. Ambio 1993;22:88–96.

Coulson JC, Potts GR, Deans IR, Fraser SM. Exceptional mortality of shags and other seabirds caused by paralytic shellfish poisoning. Br Birds 1968;61:381–404.

Coulter MC. Seabird conservation in the Galápagos Islands, Ecuador. Int Council Bird Preserv Tech Publ 1984;2:237–244.

Cowles GS. The fossil record. In: Diamond AW, ed. Studies of Mascarene Island birds. Cambridge: Cambridge University Press, 1987:90–100.

Cox PA, Elmqvist T. Indigenous control of tropical rainforest reserves: an alternative strategy for conservation. Ambio 1991;20:317–321.

Craig JL. Meta-populations: is management as flexible as nature? In: Olney PJS, Mace GM, Feistner ATC, eds. Creative conservation: interactive management of wild and captive animals. London: Chapman and Hall, 1994:50–66.

Cronin MA. Mitochondrial DNA in wildlife taxonomy and conservation biology: cautionary notes. Wildlife Soc Bull 1993;21:339–348.

Cruz JB, Cruz F. Conservation of the dark-rumped petrel *Pterodroma phaeopygia* in the Galápagos Islands, Ecuador. Biol Conserv 1987;42:303–311.

Cuarón AD. Extinction rate estimates. Nature 1993;366:118.

Cumming DHM, Du Toit RF, Stuart SN. African elephants and rhinos. Status survey and conservation action plan. IUCN/SSC African elephant and rhino specialist group. Gland, Switzerland: IUCN, 1990.

Curry PJ, Hacker RB. Can pastoral grazing management satisfy endorsed conservation objectives in arid Western Australia? J Environ Man 1990:30:295–320.

Daniel MJ. Greater short-tailed bat. In: King CM, ed. The handbook of New Zealand mammals. Auckland: Oxford University Press, 1990;131–136.

Davis JR, Guynn DC Jr, Hyder BD. Feasibility of using tribromoethanol to recapture wild turkeys. Wildlife Soc Bull 1994;22:496–500.

Davis PE, Newton I. Population and breeding of red kites in Wales over a 30-year period. J Animal Ecol 1981;50:759–772.

Day D. The doomsday book of animals. New York: Viking Press, 1981.

Derrickson SR, Snyder NFR. Potentials and limits of captive breeding in parrot conservation. In: Beissinger ER, Snyder NFR, eds. New world parrots in crisis. Solutions from conservation biology. Washington: Smithsonian Institution, 1992:133–163.

Desenne P, Strahl SD. Trade and the conservation status of the family Psittacidae in Venezuela. Bird Conserv Int 1991;1:153–169.

Dewar RE. Extinctions in Madagascar: the loss of the subfossil fauna. In: Martin PS, Klein RG, eds. Quaternary extinctions. Tucson: University of Arizona Press, 1984:574–593.

Diamond AW (ed). Studies of Mascarene Island birds. Cambridge: Cambridge University Press, 1987.

Diamond JM. The island dilemma: lessons of modern biogeographic studies for the design of nature reserves. Biol Conserv 1975;7:129–146.

Diamond JM. Island biogeography and conservation: strategy and limitations. Science 1976;193:1027–1029.

Diamond JM. "Normal" extinction of isolated populations. In: Nitecki MH, ed. Extinctions. Chicago: Chicago University Press, 1984a:191–246.

Diamond JM. Historic extinctions: a Rosetta Stone for understanding prehistoric extinctions. In: Martin PS, Klein RG, eds. Quaternary extinctions. Tucson: University of Arizona Press, 1984b:824–862.

Diamond JM. Twilight of the Hawaiian birds. Nature 1991;353:505–506.

Diamond JM, Bishop KD, van Balen S. Bird survival in an isolated Javan woodland: island or mirror. Conserv Biol 1987;1:132–142.

Dinerstein E, Jnawali SR. Greater one-horned rhinoceros populations in Nepal. In: Ryder OA, ed. Rhinoceros biology and conservation. Proceedings of an international conference, May 9–11, 1991, San Diego, CA. San Diego: Zoological Society of San Diego, 1993:196–207.

Dodd CK, Seigel RA. Relocation, repatriation, and translocation of amphibians and reptiles: are they conservation strategies that work? Herpetologica 1991;47:336–350.

Donovan SK. Mass extinctions. Processes and evidence. New York: Columbia University Press, 1989.

Douglas-Hamilton I. African elephant population study. Pachyderm 1987a; 8:1–10.

Douglas-Hamilton I. African elephants: population trends and their causes. Oryx 1987b;21:11–24.

Dublin H. In the eye of the beholder: our image of the African elephant. Endanger Species Tech Bull 1994;19(1):6–7.

Dukto R. Protected at last: the Hibernia mine. Bats 1994;12(3):3–5.

Durán JC, Cattan PE, Yañez JL. Food habits of foxes (*Canis* sp.) in the Chilean National Chinchilla Reserve. J Mammal 1987;68:179–181.

Durant SM, Harwood J. Assessment of monitoring and management strategies for local populations of the Mediterranean monk seal *Monachus monachus*. Biol Conserv 1992;61:81–92.

Eberhardt LL, Thomas JM. Designing environmental field studies. Ecol Mono 1991;61:53–73.

Edgington ES. Randomization tests. New York: Marcel Dekker, 1987.

Edwards EP. A coded workbook of the birds of the world. Vol. 1: Non-passerines, 2nd ed. Sweet Briar, Vancouver: E.P. Edwards, 1982.

Edwards EP. A coded workbook of the birds of the world. Vol 2: Passerines, 2nd ed. Sweet Briar, Vancouver: E.P. Edwards, 1986.

Eldredge N. Life pulse. Episodes from the story of the fossil record. New York: Facts on File, 1987.

Eldredge N. The miner's canary: unravelling the mysteries of extinction. London: Virgin Books, 1991:214.

Ellsworth DL, Honeycutt RL, Silvy NS, Smith MH, Bickham JW, Klimstra WD. White-tailed deer restoration to the southeastern United States: evaluating genetic variation. J Wildlife Man 1994;58:686–697.

Elmes GW, Thomas JA. Complexity of species conservation in managed habitats: interaction between *Maculinea* butterflies and their ant hosts. Biodiv Conserv 1992;1:155–169.

Erwin DH. The great Paleozoic crisis: life and death in the Permian. New York: Columbia University Press, 1993.

Etheridge R. General zoology. In: Ramsay EP (curator), Lord Howe Island. Its zoology, geology and physical characteristics. Mem Austral Museum No. 2, 1889:1–42.

Evans IM, Love JA, Galbraith CA, Pienkowski MW. Population and range restoration of threatened raptors in the United Kingdom. In: Chancellor B-U, Chancellor RD, eds. Raptor conservation today. World Working Group on Birds of Prey and Owls. Berlin: Pica Press, 1994:447–457.

Evans PGH. Electrophoretic variability of gene products. In: Cooke F, Buckley PA, eds. Avian genetics: a population and ecological approach. London: Academic Press, 1987:105–162.

Ewens WJ, Brockwell PJ, Gani JM, Resnick SI. Minimum viable population size in the presence of catastrophes. In: Soulé ME, ed. Viable populations for conservation. Cambridge: Cambridge University Press, 1987:59–68.

Falanruw MC, Manmaw CJ. Protection of flying foxes on Yap Islands. In: Wilson DE, Graham GL, eds. Pacific Island flying foxes: proceedings of an international conservation conference. Washington: U.S. Department of the Interior, Biological Report 90, 1992:150–154.

Fay FH. History and present status of the Pacific walrus population. Trans N Am Wildlife Conf 1957;22:431–445.

Fay FH, Kelly BP, Sease JL. Managing the exploitation of Pacific walruses: a tragedy of delayed response and poor communication. Marine Mammal Sci 1989;5:1–16.

Field CR. Elephant ecology in the Queen Elizabeth National Park, Uganda. E Afr Wildlife J 1971;9:99–123.

Fitter R, Fitter M (eds). The road to extinction. A symposium held by the Species Survival Commission, Madrid, November 7–9, 1984. Gland, Switzerland: IUCN, 1987.

Fitzgerald BM. House cat. In: King CM, ed. The handbook of New Zealand mammals. Oxford: Oxford University Press, 1990:330–348.

Fitzgerald LA. *Tupinambis* lizards and people: a sustainable use approach to conservation and development. Conserv Biol 1994;8:12–16.

Fitzgerald LA, Chani JM, Donadio OE. *Tupinambis* lizards in Argentina: implementing management of a traditionally exploited resource. In: Robinson JG, Redford KH, eds. Neotropical wildlife use and conservation. Chicago: Chicago University Press, 1991:303–316.

Fitzgerald S. International wildlife trade: whose business is it? Baltimore: World Wildlife Fund, 1989.

Fjeldså J. The decline and probable extinction of the Colombian grebe *Podiceps andinus*. Bird Conserv Int 1993;3:221–234.

Fleischer LA. Guadalupe fur seal, *Arctocephalus townsendi*. In: Croxall JP, Gentry RL, eds. Status, biology, and ecology of fur seals. Proceedings of an International Symposium and Workshop, Cambridge, England. NOAA Technical Report NMFS 51, 1987:43–48.

Fleischner TL. Ecological costs of livestock grazing in western North America. Conserv Biol 1994;8:629–644.

Flux JEC. Relative effect of cats, myxomatosis, traditional control, or competitors in removing rabbits from islands. NZ J Zool 1993;20:13–18.

Forbes LS. A note on statistical power. Auk 1990;107:438–439.

Fowler CW. Northern fur seals of Pribilof Islands. In: The northern fur seal. Species of the USSR and contiguous countries. USSR Academy of Sciences, Moscow: Nauka, 1993.

Frankel OH, Soulé ME. Conservation and evolution. Cambridge: Cambridge University Press, 1981.

Franklin IR. Evolutionary change in small populations. In: Soulé ME, Wilcox BA, eds. Conservation biology: an evolutionary-ecological perspective. Sunderland, MA: Sinauer Associates, 1980:135–149.

Frazer NB. Sea turtle conservation and halfway technology. Conserv Biol 1992;6:179–184.

Friend JA. The numbat *Myrmecobius fasciatus* (Myrmecobiidae): history of decline and potential for recovery. Proc Ecol Soc Austral 1990;16:369–377.

Friend JA, Whitford D. Maintenance and breeding of the numbat (*Myrmecobius fasciatus*) in captivity. In: Roberts M, Carnio J, Crawshaw G, Hutchins M, eds. The biology and management of Australasian carnivorous marsupials. Bulletin No. 3 of the Monotreme and Marsupial Advisory Group of the American Association of Zoological Parks and Aquariums. Toronto: Metropolitan Toronto Zoo and AAZPA, 1993:103–124.

Friend M, Thomas NJ. Disease prevention and control in endangered avian species: special considerations and needs. Acta XX Congr Int Ornithol 1991;4:2331–2337.

Fritts TH, Scott NJ Jr, Smith BE. Trapping *Boiga irregularis* on Guam using bird odors. J Herpetol 1989;23:189–192.

Fullagar PJ. The woodhens of Lord Howe Island. Avicult Mag 1985;91:15–30.

Gadgil M, Berkes F, Folke C. Indigenous knowledge for biodiversity conservation. Ambio 1993;22:151–156.

Gakahu CK. African rhinos: current numbers and distribution. In: Ryder OA, ed. Rhinoceros biology and conservation. Proceedings of an international conference, May 9–11, 1991, San Diego, CA. San Diego: Zoological Society of San Diego, 1993:160–165.

Galbreath RA. Walter Buller: the reluctant conservationist. Wellington: GP Books, 1989.

Gardner AS. The biogeography of the lizards of the Seychelles Islands. J Biogeog 1986;13:237–253.

Gaski AL. Species in danger. Bluefin tuna: an examination of the international trade with emphasis on the Japanese market. Cambridge: TRAFFIC International, 1993.

Gee EP. The Indian wild ass. Oryx 1963;7:9–21.

Gee EP. The wildlife of India. London: Collins, 1964.

Gibbons JRH, Watkins IF. Behavior, ecology, and conservation of south Pacific banded iguanas, *Brachylophus*, including a newly discovered species. In: Burghardt GM, Rand AS, eds. Iguanas of the world. Park Ridge, NJ: Noyes, 1982:418–441.

Gilpin ME, Soulé ME. Minimum viable populations: the processes of population extinction. In: Soulé ME, ed. Conservation biology: the science of scarcity and diversity. Sunderland, MA: Sinauer Associates, 1986:13–34.

Gimenez-Dixon M, Stuart S. Action plans for species conservation, an evaluation of their effectiveness. Species (Newsletter IUCN/SSC) 1993; 20:6–10.

Gipps JHW (ed). Beyond captive breeding: re-introducing endangered mammals to the wild: the proceedings of a symposium held at the Zoological Society of London on 24th and 25th November 1989. Symp Zool Soc London No. 62, 1991.

Glover J. The elephant problem at Tsavo. E Afr Wildlife J 1963;1:30–39.

Goldschmidt T, Witte F, Wanink J. Cascading effects of the introduced Nile perch on the detritivorous/phyloplanktivorous species in the sublittoral areas of Lake Victoria. Conserv Biol 1993;7:686–700.

Goodman D. The demography of chance extinction. In: Soulé ME, ed. Viable population for conservation. Cambridge: Cambridge University Press, 1987:11–34.

Graetz D, Fisher R, Wilson M. Looking back: the changing face of the Australian continent, 1972–1992. Canberra: CSIRO Office of Space Science and Applications, 1992.

Graham RW, Lundelius EL Jr. Coevolutionary disequilibrium and Pleistocene extinctions. In: Martin PS, Klein RG, eds. Quaternary extinctions. Tucson: University of Arizona Press, 1984:223–249.

Grau J. Chinchilla. In: Mason IL, ed. Evolution of domesticated animals. London: Longman Group, 1984:260–262.

Graveland J, van der Wal R, van Balen JH, van Noordwijk AJ. Poor reproduction in forest passerines from decline in snail abundance on acidified soils. Nature 1994;368:446–448.

Grayson DK, Livingston SD. Missing mammals on Great Basin Mountains: Holocene extinctions and inadequate knowledge. Conserv Biol 1993;7:527–532.

Green MJB. Exploiting the musk deer for its musk. Traffic Bull 1987;8:59–61.

Green RH. Sampling design and statistical methods for environmental biologists. New York: Wiley, 1979.

Green RH, Young RC. Sampling to detect rare species. Ecol Appl 1993;3:351–356.

Greenway JC Jr. Extinct and vanishing birds. Special Publication No. 13. New York: American Committee for International Wildlife Protection, 1958.

Gretton A, Komdeur J, Komdeur M. Saving the magpie robin. World Birdwatch 1991;13:10–11.

Groombridge B (ed). Global biodiversity: status of the Earth's living resources. A report compiled by the World Conservation Monitoring Centre. London: Chapman and Hall, 1992.

Groombridge B (ed). 1994 IUCN red list of threatened animals. Gland, Switzerland: IUCN, 1993.

Grossblatt N (ed). Decline of the sea turtles: causes and prevention. Washington DC: National Academy Press, 1990.

Grossman D, Ferrar TA, du Plessis PC. Socio-economic factors influencing conservation in South Africa. Traffic Bull 1992;13:29–31.

Groves CP. Horses, asses and zebras in the wild. London: David and Charles, 1974.

Grumbine RE. Viable populations, reserve size, and federal lands management: a critique. Conserv Biol 1990;4:127–134.

Grynberg R. The tuna dilemma. Pacific Islands Monthly May 1993:9–11.

Gudynas E. The conservation status of South American rodents: many questions but few answers. In: Lidicker WZ Jr, ed. Rodents—a world survey of species of conservation concern. International Union of the Conservation of Nature/Species Survival Commission Occasional Paper No. 4, 1989:20–25.

Guthrie RD. Mosaics, allelochemics, and nutrients: an ecological theory of late Pleistocene megafaunal extinctions. In: Martin PS, Klein RG, eds. Quaternary extinctions. Tucson: University of Arizona Press, 1984:259–298.

Hachisuka M. The dodo and kindred birds or the extinct birds of the Mascarene Islands. London: H.F. & G. Witherby, 1953.

Hadfield MG, Miller SE, Carwile AH. The decimation of endemic Hawai'ian tree snails by alien predators. Am Zool 1993;33:610–622.

Hadfield MG, Mountain BS. A field study of a vanishing species, *Achatinella mustelina* (Gastropoda, Pulmonata), in the Waianae Mountains of Oahu. Pacific Sci 1980;34:345–358.

Hafernik JH Jr. Threats to invertebrate biodiversity: implications for conservation strategies. In: Fiedler PL, Jain SK, eds. Conservation biology: the theory and practice of nature conservation, preservation and management. London: Chapman and Hall, 1992:171–195.

Hairston NG Sr. Ecological experiments purpose, design and execution. Cambridge: Cambridge University Press, 1989.

Hall ER. The mammals of North America. New York: Wiley, 1981.

Hall-Martin A. Elephant survivors. Oryx 1980;15:355–362.

Hamilton WD. Inbreeding in Egypt and in this book: a childish perspective. In: Thornhill WM, ed. The natural history of inbreeding and outbreeding: theoretical and empirical perspectives. Chicago: University of Chicago Press, 1993:429–452.

Hammer CU, Clausen HB, Friedrich WL, Tauber H. The Minoan eruption of Santorini in Greece dated to 1645 BC? Nature 1987;328:517–519.

Hansen AJ, Garman SL, Marks B, Urban DL. An approach for managing vertebrate diversity across multiple-use landscapes. Ecol Appl 1993;3:481–496.

Hanski I. Single-species metapopulation dynamics: concepts, models and observations. Biol J Linnean Soc 1991;42:17–38.

Hanski I, Gilpin M. Metapopulation dynamics: brief history and conceptual domain. Biol J Linnean Soc 1991;42:3–16.

Hanski I, Thomas CD. Metapopulation dynamics and conservation: a spatially explicit model applied to butterflies. Biol Conserv 1994;68:167–180.

Hardin G. The tragedy of the commons. Science 1968;162:1243–1248.

Harper F. Extinct and vanishing mammals of the Old World. New York: ACIWP, 1945.

Harris MP. The biology of an endangered species, the dark-rumped petrel (*Pterodroma phaeopygia*) in the Galápagos Islands. Condor 1970;72:76–84.

Harris RB, Allendorf FW. Genetically effective population size of large mammals: an assessment of estimators. Conserv Biol 1989;3:181–191.

Harrison S. Local extinction in a metapopulation context: an empirical evaluation. Biol J Linnean Soc 1991;42:73–88.

Hartman G. Long-term population development of a reintroduced beaver (*Castor fiber*) population in Sweden. Conserv Biol 1994;8:713-717.

Hassell MP, Lawton JH, May RM. Patterns of dynamical behaviour in single-species population. J Animal Ecol 1976;45:471–486.

Haynes AM. Human exploitation of seabirds in Jamaica. Biol Conserv 1987; 41:99–124.

Hedrick PW, Miller PS. Conservation genetics: techniques and fundamentals. Ecol Appl 1992;2:30–46.

Heinen JT. Park–people relations in Kosi Tappu Wildlife Reserve, Nepal: a socio-economic analysis. Environ Conserv 1993;20:25–34.

Heisterberg JF, Nunez-Garcia F. Controlling shiny cowbirds in Puerto Rico. In: Crabb AC, Marsh RE, eds. Proceedings of the Thirteenth Vertebrate Pest Conference. University of California, Davis, 1988:295–300.

Henry JD. Home again on the range. Equinox July/August 1994:47–53.

Herrero S, Mamo C, Carbyn L, Scott-Brown M. Swift fox reintroduction into Canada. Occasional Paper. Provincial Museum of Alberta, Nat Hist 1991; 15:246–252.

Hickey JJ (ed). Peregrine falcon populations: their biology and decline. Madison: University of Wisconsin Press, 1969.

Hickson RE, Moller H, Garrick AS. Poisoning rats on Stewart Island. NZ J Ecol 1986;9:111–121.

Hill ES. Lord Howe Island. Official visit of the Water Police Magistrate, and the Director of the Botanic Gardens, Sydney; together with a description of the Island. Sydney: Government Printer, 1870.

Hoagland DB, Horst GR, Kilpatrick CW. Biogeography and population biology of the mongoose in the West Indies. Biogeog West Indies 1989:611–634.

Hobbs RJ, Saunders DA, Lobry de Bruyn LA, Main AR. Changes in biota. In: Hobbs RJ, Saunders DJ, eds. Reintegrating fragmented landscapes towards sustainable production and nature conservation. New York: Springer-Verlag, 1992:65–106.

Hockey PAR. The influence of coastal utilisation by man on the presumed extinction of the Canarian black oystercatcher, *Haematopus meadewaldoi* Bannerman. Biol Conserv 1987;39:49–62.

Holdaway RN. New Zealand's pre-human avifauna and its vulnerability. NZ J Ecol 1989;12(Suppl):11–25.

Holden C. Tuna stocks: east meets west. Science 1994;265:1525–1526.

Holden P. Along the dingo fence. Sydney: Hodder and Stoughton, 1993.

Hone J. Analysis of vertebrate pest control. New York: Cambridge University Press 1994:270.

Hornaday WT. Our vanishing wildlife. New York: New York Zoological Society, 1913.

Horton DR. Red kangaroos: the last of the Australian megafauna. In: Martin PS, Klein RG, eds. Quaternary extinctions. Tucson: University of Arizona Press, 1984:639–680.

Howard R, Moore A. A complete checklist of the birds of the world. Oxford: Oxford University Press, 1980.

Howell PG. An evaluation of the biological control of the feral cat *Felis catus* (Linnaeus, 1758). Acta Zool Fenn 1984;172:111–113.

Hubbard AL, McOrist S, Jones TW, Boid R, Scott R, Esterby N. Is survival of European wildcats *Felis silvestris* in Britain threatened by interbreeding with domestic cats? Biol Conserv 1992;61:203–208.

Hubbs CL. Back from oblivion. Guadalupe fur seal: still a living species. Pacific Disc 1956;9(6):14–21.

Hubbs CL, Norris KS. Original teeming abundance, supposed extinction, and survival of the Juan Fernandez fur seal. In: Burt WH, ed. Antarctic Pinnipedia. American Geophysical Union, Antarctic Research Series 18, 1971.

Hurlbert SH. Pseudoreplication and the design of ecological field experiments. Ecol Mono 1984;54:187–211.

ICBP. Report of the 1986 University of East Anglia Martinique Oriole expedition. Study Report 23. Cambridge: Birdlife International, 1987.

ICCAT. Report by the International Commission for the Conservation of Atlantic Tunas (ICCAT) on the status of the bluefin tuna populations and on the related conservation initiatives in the Atlantic. Madrid: ICCAT, 1994.

Inglis G, Underwood AJ. Comments on some designs proposed for experiments on the biological importance of corridors. Conserv Biol 1992;6:581–586.

Iñigo-Elias EI, Ramos MA. The Psittacine trade in Mexico. In: Robinson JG, Redford KH, eds. Neotropical wildlife use and conservation. Chicago: Chicago University Press, 1991:383–392.

Irwin LL, Wigley TB. Toward an experimental basis for protecting forest wildlife. Ecol Appl 1993;3:213–217.

IUCN. The IUCN red list of threatened animals. Gland, Switzerland: IUCN, 1990.

Ivlev VS. Experimental ecology of the feeding of fishes. New Haven: Yale University Press, 1961.

Jackson JA. Biopolitics, management of federal lands, and the conservation of the red-cockaded woodpecker. Am Birds 1986;40:1162–1168.

Jackson P. Cat Specialist Group report. Species 1994;20:59–60.

Jacobson SK, Lopez AF. Biological impacts of ecotourism: tourists and nesting turtles in Tortuguero National Park, Costa Rica. Wildlife Soc Bull 1994;22:414–419.

James HF, Olson SL. Description of thirty-two new species of birds from the Hawaiian Islands: Part II. Passeriformes. Ornithol Mono 1991;46:1–88.

Jamieson GS. Marine invertebrate conservation: evaluation of fisheries over-exploitation concerns. Am Zool 1993;33:551–567.

Jarman P. The red fox—an exotic, large predator. In: Kitching RL, ed. The ecology of exotic animals and plants; some case histories. Brisbane: Wiley, 1986:44–61.

Järvinen O. Conservation of endangered plant populations: single large or several small reserves. Oikos 1982;38:301–307.

Jiménez JA, Hughes KA, Alaks G, Graham L, Lacy RC. An experimental study of inbreeding depression in a natural habitat. Science 1994;5183:271–273.

Jiménez JE, Feinsinger P, Jaksic FM. Spatiotemporal patterns of an irruption and decline of small mammals in northcentral Chile. J Mammal 1992;73:356–364.

Johnson AM. The status of northern fur seal populations. In: Ronald K, Mansfield AW, eds. Biology of the seal. Proceedings of a symposium. Rapports et Procès-verbaux des Réunions 169, 1975:263–266.

Johnson RR. Release and translocation strategies for the Puerto Rican crested toad, *Peltophryne lemur*. Endanger Species Update 1990;8:54–57.

Johnson RR. Model programs for reproduction and management: ex situ and in situ conservation of toads and the family *Bufonidae*. In: Murphy JB, Adler K, Collins JT, eds. Captive management and conservation of amphibians and reptiles. Contributions to Herpetology, 11. Ithaca, NY: Society for the Study of Amphibians and Reptiles, 1994:243–254.

Jones CG. The larger land-birds of Mauritius. In: Diamond AW, ed. Studies of Mascarene Island birds. Cambridge: Cambridge University Press, 1987:208–300.

Jones CG, Jeggo DF, Hartley J. The maintenance and captive-breeding of the pink pigeon *Nesoenas mayeri*. Dodo, J Jersey Wildlife Preserv Trust 1983;20:16–26.

Jones CG, Swinnerton KJ, Taylor CJ, Mungroo Y. The release of captive-bred pink pigeons *Columba mayeri* in native forest on Mauritius. A progress report July 1987–June 1992. Dodo, J Jersey Wildlife Preserv Trust 1992;28:92–125.

Jones CG, Todd DM, Mungroo Y. Mortality, morbidity and breeding success of the pink pigeon (*Columba (Nesoenas) mayeri*). Int Council Bird Preserv Tech Publ 1989;10:89–113.

Jorgenson JP, Jorgenson AB. Imports of CITES-regulated mammals into the United States from Latin America: 1982–84. In: Mares MA, Schmidly DJ, eds. Latin American mammalogy: history, biodiversity, and conservation. Norman: University of Oklahoma Press, 1991:322–335.

Kear J, Berger AJ. The Hawaiian goose. An experiment in conservation. Calton, Staffordshire, England: T. & A.D. Poyser, 1980.

Keegan DW, Coblentz BE, Winchell CS. Feral goat eradication on San Clemente Island, California. Wildlife Soc Bull 1994;22:56–61.

Kennedy M. Australasian marsupials and monotremes. An action plan for their conservation. IUCN/SSC Australasian Marsupial and Monotreme Specialist Group. Gland, Switzerland: IUCN, 1992.

Key G, Heredia EM. Distribution and current status of rodents in the Galápagos. Not Galápagos 1994;53:21–24.

Khan MK, Nor BHM, Yusof E, Rahman MA. In-situ conservation of the Sumatran rhinoceros (*Dicerhinus sumatrensis*) a Malaysian experience. In: Ryder OA, ed. Rhinoceros biology and conservation. Proceedings of an international conference, May 9–11, 1991, San Diego, California. San Diego: Zoological Society of San Diego, 1993:238–247.

King CM. Stoat. In: King CM, ed. The handbook of New Zealand mammals. Oxford. Oxford University Press, 1990:288–319.

King WB, Lincer JE. DDE residues in the endangered Hawaiian dark-rumped petrel (*Pterodroma phaeopygia sandwichensis*). Condor 1973;75:460–461.

Kinnear JE, Onus ML, Bromilow RN. Fox control and rock-wallaby population dynamics. Austral Wildlife Res 1988;15:435–450.

Kirk DA, Racey PA. Effect of the introduced black-naped hare *Lepus nigricollis nigricollis* on the vegetation of Cousin Island, Seychelles and possible implications for avifauna. Biol Conserv 1992;61:171–179.

Kirkpatrick JB. An iterative method for establishing priorities for the selection of nature reserves: an example from Tasmania. Biol Conserv 1983;25:127–134.

Kirkpatrick JB, Harwood CE. Conservation of Tasmanian macrophytic wetland vegetation. Proc R Soc Tasmania 1983;117:5–20.

Kirmse W. Raptors' plasticity of nest site selection. In: Chancellor B-U, Chancellor RD, eds. Raptor conservation today. World Working Group on Birds of Prey and Owls. Berlin: Pica Press, 1994:143–145.

Kitchener A. On the external appearance of the dodo, *Raphus cucullatus*. Arch Nat Hist 1993;20:279–301.

Kleiman DG, Beck BB, Dietz JM, Dietz LA. Costs of re-introduction and criteria for success: accounting and accountability in the golden lion tamarin conservation program. Symp Zool Soc London 1991;62:125–144.

Kleiman DG, Stanley Price MR, Beck BB. Criteria for reintroductions. In: Olney PJS, Mace GM, Feistner ATC, eds. Creative conservation: interactive management of wild and captive animals. London: Chapman and Hall, 1994:287–303.

Klein DR. The introduction, increase and crash of reindeer on St. Matthew Island. J Wildlife Man 1968;32:350–357.

Komdeur J. Influence of territory quality and habitat saturation on dispersal options in the Seychelles warbler: an experimental test of the habitat saturation hypothesis for cooperative breeding. Acta XX Congr Int Ornithol 1990;3:1325–1333.

Komdeur J. Importance of habitat saturation and territory quality for evolution of cooperative breeding in the Seychelles warbler. Nature 1992;358:493–495.

Komdeur J. Conserving the Seychelles warbler, *Acrocephalus sechellensis*, by translocation from Cousin Island to the Islands of Aride and Cousine. Biol Conserv 1994;67:143–152.

Komdeur J, Bullock ID, Rands MRW. Conserving the Seychelles warbler *Acrocephalus sechellensis* by translocation: a transfer from Cousin Island to Aride Island. Bird Conserv Int 1991;1:177–185.

Krebs CJ. Ecological methodology. New York: Harper and Row, 1989.

Kremsater LL, Bunnell FL. Testing responses to forest edges: the example of black-tailed deer. Can J Zool 1992;70:2426–2435.

Kuchling G, Dejose JP, Burbidge AA, Bradshaw SD. Beyond captive breeding: the western swamp tortoise *Pseudemydura umbrina* recovery programme. Int Zoo Yearb 1992;31:37–41.

Kurtén B. Pleistocene mammals of Europe. Chicago: Aldine, 1968.

Lacy RC. VORTEX: a computer simulation model for population viability analysis. Wildlife Res 1993;20:45–65.

Lacy RC, Kreeger T. VORTEX users manual. A stochastic simulation of the extinction process. Chicago: Chicago Zoological Society, 1992.

Lahti T, Ranta E. The SLOSS principle and conservation practice: an example. Oikos 1985;44:369–370.

Lambert FR. Trade, status and management of three parrots in the North Molluccas, Indonesia: white cockatoo *Cacatua alba*, chattering lory *Lorius garrulus* and violet-eared lory *Eos squamata*. Bird Conserv Int 1993;3:145–168.

Lande R. Extinction thresholds in demographic models of territorial populations. Am Natural 1987;130:624–635.

Lande R. Demographic models of the northern spotted owl (*Strix occidentalis caurina*). Oecologia 1988a;75:601–607.

Lande R. Genetics and demography in biological conservation. Science 1988b;241:1455–1460.

Lande R. Risks of population extinction from demographic and environmental stochasticity, and random catastrophes. Am Natural 1993;142:911–927.

Lande R, Barrowclough GF. Effective population size, genetic variation, and their use in population management. In: Soulé ME, ed. Viable populations for conservation. Cambridge: Cambridge University Press, 1987:87–123.

Lander RH, Kajimura H. Status of northern fur seals. Mammals Seas 1982;4:319–345. FAO Fisheries Series No. 5.

Laurenson MK, Caro TM. Monitoring the effects of non-trivial handling in free-living cheetahs. Animal Behav 1994;47:547–557.

Laws RM. Elephants as agents of habitat and landscape change in East Africa. Oikos 1970;21:1–15.

Laws RM, Parker ISC, Johnstone RCB. Elephants and their habitats. The ecology of elephants in North Bunyoro, Uganda. Oxford: Clarendon Press, 1975.

Leader-Williams N. Reindeer on South Georgia: the ecology of an introduced species. Cambridge: Cambridge University Press, 1988.

Leader-Williams N. Black rhinos and African elephants: lessons for conservation funding. Oryx 1990;24:23–29.

Leader-Williams N. The world trade in rhino horn: a review. Cambridge: TRAFFIC International, 1992.

Leader-Williams N. Theory and pragmatism in the conservation of rhinos. In: Ryder OA, ed. Rhinoceros biology and conservation. Proceedings of an international conference, May 9–11, 1991, San Diego, California. San Diego: Zoological Society, 1993:69–81.

Leader-Williams N, Albon SD. Allocation of resources for conservation. Nature 1988;336:533–535.

Leader-Williams N, Milner-Gulland EJ. Policies for the enforcement of wildlife laws: the balance between detection and penalties in Luanga Valley, Zambia. Conserv Biol 1993;7:611–617.

Leader-Williams N, Walton DHW, Prince PA. Introduced reindeer on South Georgia—a management dilemma. Biol Conserv 1989;47:1–11.

Leberg PL, Stangel PW, Hillestad HO, Marchinton RL, Smith MW. Genetic structure of reintroduced wild turkey and white-tailed deer populations. J Wildlife Man 1994;58:698–711.

Lehman N, Wayne RK, Stewart BS. Comparative levels of genetic variability in harbour seals and northern elephant seals as determined by genetic fingerprinting. Symp Zool Soc London 1993;66:49–60.

Leopold A. Game management. New York: Charles Scribners, 1933.

Levins R. Evolution in changing environments: some theoretical explorations. Monograph in Population Biology, Princeton, NJ: Princeton University Press, 1968.

Levins R. Some demographic and genetic consequences of environmental heterogeneity for biological control. Bull Entomol Soc Am 1969;15:237–240.

Levins R. Extinction. In: Gesternhaber M, ed. Some mathematical problems in biology. Providence, RI: American Mathematical Society, 1970:77–107.

Lewis D, Kaweche GB, Mwcnya A. Wildlife conservation outside protected areas—lessons from an experiment in Zambia. Conserv Biol 1990;4:171–180.

Lind CR. The effects of multiple clutching on the size, fertility and hatchability of the eggs of the pink pigeon *Nesoenas mayeri* at Jersey Wildlife Preservation Trust. Dodo, J Jersey Wildlife Preserv Trust 1989;26:93–98.

Lindenmayer DB, Clark TW, Lacy RC, Thomas VC. Population viability analysis as a tool in wildlife conservation policy: with reference to Australia. Environ Man 1993;17:745–758.

Lindenmayer DB, Cunningham RB, Tanton MT, Smith AP. The conservation of arboreal marsupials in the montane ash forests of the Central Highlands of Victoria, south-east Australia. II: The loss of trees with hollows and its implications for the conservation of Leadbeater's possum *Gymnobelideus leadbeateri* McCoy (Marsupialia: Petauridae). Biol Conserv 1990a;54:133–145.

Lindenmayer DB, Cunningham RB, Tanton MT, Smith AP, Nix HA. The conservation of arboreal marsupials in the montane ash forests of the Central Highlands of Victoria, south-east Australia. I: Factors influencing the occupancy of trees with hollows. Biol Conserv 1990b;54:111–131.

Ling F. Nene alert. Art Center Gaz Aug–Sept 1992:11.

Litvaitis JA. Response of early successional vertebrates to historic changes in land use. Conserv Biol 1993;7:866–873.

Loh J, Loh K. Rhino horn in Taipei, Taiwan. Traffic 1994;14:55–58.

Lokemoen JT, Woodward RO. An assessment of predator barriers and predator control to enhance duck nest success on peninsulas. Wildlife Soc Bull 1993;21:275–282.

Lotka AJ. Relation between birth rates and death rates. Science 1907a;26:21–22.

Lotka AJ. Studies on the mode of growth of material aggregates. Am J Sci (4) 1907b;24:199–216.

Lovegrove R. The kite's tail: the story of the red kite in Wales. Sandy, Bedfordshire, UK: Royal Society for the Protection of Birds, 1990.

Lovejoy TE. Accelerating rates of extinction. Dodo, J Jersey Wildlife Preserv Trust 1974;11:11–12.

Lovejoy TE, Bierregaard RO Jr, Rylands AB Jr, et al. Edge and other effects of isolation on Amazon forest fragments. In: Soulé ME, ed. Conservation biology: the science of scarcity and diversity. Sunderland, MA: Sinauer Associates, 1986:257–285.

Lovejoy TE, Rankin JM, Bierregaard RO Jr, Brown KS Jr, Emmons LH, van der Voort ME. Ecosystem decay of Amazon forest remnants. In: Nitecki MH, ed. Extinctions. Chicago: Chicago University Press, 1984:295–326.

Lowe DW, Matthews JR, Moseley CJ (eds). The official World Wildlife Fund guide to endangered species of North America. Washington: Beacham, 1990.

Lowe PR. A naturalist on desert islands. New York: Charles Scribners, 1911.

Luxmoore R. An examination of the ranching criteria. Part 10 (6 pages) in Applying new criteria for listing species on the CITES appendices. London: IUCN/SSC Workshop, 1992.

MacArthur RH, Wilson EO. An equilibrium theory of insular zoogeography. Evolution 1963;17:373–387.

MacArthur RH, Wilson EO. The theory of island biogeography. Princeton, NJ: Princeton University Press, 1967.

Mackey B, Nix HA, Stein JA, Cork SE, Bullen FT. Assessing the representativeness of the wet tropics of Queensland World Heritage property. Biol Conserv 1989;50:279–303.

MacKinnon K, MacKinnon J. Habitat protection and re-introduction programs. Symp Zool Soc London 1991;62:173–198.

Maguire LA. Using decision analysis to manage endangered species populations. J Environ Man 1986;22:345–360.

Maguire LA, Servheen C. Integrating biological and sociological concerns in endangered species management: augmentation of grizzly bear populations. Conserv Biol 1992;6:426–434.

Manly BFJ. The design and analysis of research studies. Cambridge: Cambridge University Press, 1992.

Manly BFJ, McDonald L, Thomas D. Resource selection by animals: statistical design and analysis of field studies. London: Chapman and Hall, 1993.

Margules C, Higgs AJ, Rafe RW. Modern biogeographic theory: are there any lessons for nature reserve design? Biol Conserv 1982;24:115–128.

Margules CM. Wog Wog habitat fragmentation experiment. Environ Conserv 1992;19:316–325.

Margules CR, Cresswell ID, Nicholls AO. A scientific basis for establishing networks of protected areas. In: Florey PL, Humphries CJ, Vane-Wright RI, eds. Systematics and conservation evaluation. Systematics Association Special Volume No. 50. Oxford: Clarendon Press, 1994b:327–350.

Margules CR, Milkovits GA. Smith GT. Contrasting effects of habitat fragmentation on the scorpion *Cercophonius squama* and an amphipod. Ecology 1994a;75:2033–2042.

Margules CR, Nicholls AO. Assessing conservation values of remnant "islands": mallee patches on the western Eyre Peninsula, South Australia. In: Saunders DA, Arnold GW, Burbidge AA, Hopkins AJM, eds. Nature conservation: the role of remnant vegetation. Sydney: Surrey Beatty, 1987:89–102.

Margules CR, Nicholls AO, Pressey RL. Selecting networks of reserves to maximize biological diversity. Biol Conserv 1988;43:63–76.

Margules CR, Nicholls AO, Usher MB. Apparent species turnover, probability of extinction and the selection of nature reserves: a case study on the Ingleborough limestone pavements. Conserv Biol 1994c;8:398–409.

Marshall AP. Censusing Laysan ducks *Anas laysanensis*: a lesson in the pitfalls of estimating threatened species populations. Bird Conserv Int 1992;2:239–251.

Marshall AP, Black JM. The effect of rearing experience on subsequent behavioural traits in Hawaiian geese *Branta sandvicensis*: implications for the recovery programme. Bird Conserv Int 1992;2:131–147.

Marshall FHA. On the change over in the oestrous cycle in animals after transfer across the equator, with further observations on the incidence of the breeding seasons and the factors controlling sexual periodicity. Proc R Soc London Ser B 1937;122:413–428.

Marshall LG. The great American interchange: an invasion induced crisis for South American mammals. In: Nitecki MH, ed. Biotic crisis in ecological and evolutionary time. London: Academic Press, 1981:133–229.

Martin C. Interspecific relationships between Barasingha (*Cervus duvauceli*) and Axis deer (*Axis axis*) in Kanha National Park, India, and relevance to management. In: Wemmer CM, ed. Biology and management of the Cervidae. Washington DC: Smithsonian Institution Press, 1987:299–306.

Martin EB. The poisoning of rhinos and tigers in Nepal. Oryx 1992;26:82–86.

Martin EB. The present-day trade routes and markets for rhinoceros products. In: Ryder OA, ed. Rhinoceros biology and conservation. Proceedings of an international conference, May 9–11, 1991, San Diego, California. San Diego: Zoological Society, 1993:1–9.

Martin PS. Prehistoric overkill: the global model. In: Martin PS, Klein RG, eds. Quaternary extinctions. Tucson: University of Arizona Press, 1984:354–403.

Martin PS, Klein RG (eds). Quarternary extinctions. Tucson: University of Arizona Press, 1984.

May CA, Wetton JH, Davis PE, Brookfield JFY, Parkin DT. Single-locus profiling reveals loss of variation in inbred populations of the red kite (*Milvus milvus*). Proc R Soc London B, 1993;251:165–170.

May RM. Stability and complexity in model ecosystems. Princeton, NJ: Princeton University Press, 1973.

May RM. Patterns of species abundance and diversity. In: Cody ML, Diamond JM, eds. Ecology and evolution of communities. Cambridge, MA: Belknap Press, 1975:81–120.

May RM. Harvesting whale and fish populations. Nature 1976;263:91–92.

May RM. Taxonomy as destiny. Nature 1990;347:129–130.

May RM. The role of ecological theory in planning re-introduction of endangered species. Symp Zool Soc London 1991;62:145–163.

May RM, Oster GF. Bifurcations and dynamic complexity in simple ecological models. Am Natural 1976;110:573–599.

Mayfield HF. Brood parasitism: reducing interactions between Kirtland's warblers and brown-headed cowbirds. In: Temple SA, ed. Endangered birds: management techniques for preserving threatened species. Madison: University of Wisconsin Press, 1978:85–92.

Mayr E. Avifauna: turnover on islands. Science 1965;150:1587–1588.

McCallum H. Quantifying the impact of disease on threatened species. Pacific Conserv Biol 1994;1:107–117.

McCarthy MA, Franklin DC, Burgman MA. The importance of demographic uncertainty: an example from the helmeted honeyeater *Lichenostomus melanops cassidix*. Biol Conserv 1994;67:135–142.

McIntyre S, Barrett GW, Kitching RL, Recher HF. Species triage—seeing beyond wounded rhinos. Conserv Biol 1992;6:604–606.

McKone MJ, Lively CM. Statistical analysis of experiments conducted at multiple sites. Oikos 1993;67:184–186.

McNay RS, Morgan JA, Bunnell FL. Characterising independence of observations in movements of black-tailed deer. J Wildlife Man 1994;58:422-429.

McNeely JA. Economic incentives for conserving biodiversity: lessons for Africa. Ambio 1993;22:144–150.

Mead JI, Meltzer DJ. North American late Quaternary extinctions and the radiocarbon record. In: Martin PS, Klein RG, eds. Quaternary extinctions. Tucson: University of Arizona Press, 1984:440–450.

Meadows R. Conservation at a snail's pace. Zoogoer March 1989:30–33.

Merton DB. Success in re-establishing a threatened species: the saddleback—its status and conservation. Bull Int Council Bird Preserv 1975;12:150–157.

Merton D. Eradication of rabbits from Round Island, Mauritius: a conservation success story. Dodo, J Jersey Wildlife Preserv Trust 1987;24:19–43.

Merton D. Current status of kakapo populations: Fiordland. In: Rasch G, ed. Proceedings of a workshop on the conservation of kakapo with recommendations for the 1989–1994 kakapo recovery plan. Wellington: Department of Conservation, 1989:5–6.

Mestel R. Reckless young condors moved away from temptation. New Sci 1993;140(1901):4.

Meyers JM. Improved capture techniques for psittacines. Wildlife Soc Bull 1994;22:511–516.

Milberg P, Tyrberg T. Naïve birds and noble savages—a review of man-caused prehistoric extinctions of island birds. Ecography 1993;16:229–250.

Millener PR. Evolution, extinction and the subfossil record of New Zealand's avifauna. In: Gill BJ, Heather BD, eds. A flying start. Commemorating fifty years of the Ornithological Society of New Zealand 1940–1990. Auckland: Random Century, 1990:93–100.

Miller B, Biggins D, Hanebury L, Vargas A. Reintroduction of the black-footed ferret (*Mustela nigripes*). In: Olney PJS, Mace GM, Feistner ATC, eds. Creative conservation: interactive management of wild and captive animals. London: Chapman and Hall, 1994a:455–464.

Miller B, Ceballos G, Reading R. The prarie dog and biotic diversity. Conserv Biol 1994b;8:677–681.

Miller B, Mullette KJ. Rehabilitation of an endangered Australian bird: the Lord Howe Island woodhen *Tricholimnas sylvestris* (Sclater). Biol Conserv 1985; 34:55–95.

Miller BJ, Anderson SH. Ethology of the endangered black-footed ferret *Mustela nigripes*. Adv Ethol 1993;No.31.

Miller FL. Interactions between men, dogs and wolves on western Queen Elizabeth Islands, Northwest Territories, Canada. Musk-ox 1978;22: 70–72.

Miller FL. Peary caribou status report. Report prepared for The Committee on the Status of Endangered Wildlife in Canada (COSEWIC). Ottawa: Canadian Wildlife Service, 1990.

Miller SD, Rottmann J, Raedeke KJ, Taber RD. Endangered mammals of Chile: status and conservation. Biol Conserv 1983;25:335–352.

Miller TJ. Husbandry and breeding of the Puerto Rican toad (*Peltophryne lemur*) with comments on its natural history. Zoo Biol 1985;4:281–286.

Milliken T. The evolution of legal controls on rhinoceros products in Hong Kong—an Asian model worth considering. Oryx 1991;25:209–214.

Milliken T, Du Toit R. Rhino dehorning in Zimbabwe: an update. TRAFFIC 1994;14:45–46.

Milliken T, Haywood M. Recent data on trade in rhino and tiger products. TRAFFIC 1994;14:21–24.

Milliken T, Nowell K, Thomsen JB. The decline of the black rhino in Zimbabwe: implications for future rhino conservation. Cambridge: TRAFFIC International, 1993.

Mills JA, Lavers RB, Lee WG. The takahe: a relict of the Pleistocene grassland avifauna of New Zealand. NZ J Ecol 1984;7:57–70.

Mills JA, Lavers RB, Lee WG. The post-Pleistocene decline of takahe (*Notornis mantelli*): a reply. J Roy Soc NZ 1988;18:112–118.

Mills LS, Soulé ME, Doak DF. The keystone-species concept in ecology and conservation. Bioscience 1993;43:219–224.

Mishra HR, Wemmer C, Smith JLD. Tigers in Nepal: management conflicts with human interests. In: Tilson RL, Seal US, eds. Tigers of the world. The biology, biopolitics, management, and conservation of an endangered species. Park Ridge, NJ: Noyes, 1987:449–463.

Misra A. Leatherback turtle—breeding and behaviour. Tiger Paper 1993;20:15–21.

Mitton JB. Theory and data pertinent to the relationship between heterozygosity and fitness. In: Thornhill NW, ed. The natural history of inbreeding and outbreeding; theoretical and empirical perspectives. Chicago: University of Chicago Press, 1993:17–41.

Moore DE III, Smith R. The red wolf as a model for carnivore re-introductions. Symp Zool Soc London 1991;62:263–278.

Morgan S. Land settlement in early Tasmania. Cambridge: Cambridge University Press, 1992.

Morrell B. A narrative of four voyages to the South Sea, North and South Pacific Ocean, Chinese Sea, Ethiopic and Southern Atlantic Ocean, Indian and Antarctic Ocean from the year 1822 to 1831. Comprising critical surveys of coasts and islands, with sailing directions, and an account of some new and valuable discoveries, including the Mascarene Islands, where thirteen of the author's crew were massacred and eaten by cannibals, to which is prefixed a brief sketch of the author's early life. New York: J & J Harper, 1832.

Morton AC. The effects of marking and handling on recapture frequencies of butterflies. In: Vane-Wright RI, Ackery PR, eds. The biology of butterflies. London: Academic Press, 1984:55–58.

Morton SR. The impact of European settlement on the vertebrate animals of arid Australia: a conceptual model. Proc Ecol Soc Austral 1990;16:201–213.

Muggleton J, Benham BR. Isolation and the decline of the Large Blue butterfly (*Maculinea arion*) in Great Britain. Biol Conserv 1975;7:119–128.

Mulliken TA, Broad SR, Thomsen JB. The wild bird trade—an overview. In: Thomsen JB, Edwards SR, Mulliken TA, eds. Perceptions, conservation and management of wild birds in trade. Cambridge: TRAFFIC International, 1992:1–42.

Murphy DD. The Kirby Canyon Conservation Agreement: a model for the resolution of land-use conflicts involving threatened vertebrates. Environ Conserv 1988;15:45–48, 57.

Murphy DD, Noon BR. Integrating scientific methods with habitat conservation planning: reserve design for northern spotted owls. Ecol Appl 1992; 2:3–17.

Murphy DD, Weiss SB. A long-term monitoring plan for a threatened butterfly. Conserv Biol 1988;2:367–374.

Murphy RC, Mowbray LS. New light on the cahow, *Pterodroma cahow*. Auk 1951;68:266–280.

Murray P. Extinctions downunder: a bestiary of extinct Australian late Pleistocene monotremes and marsupials. In: Martin PS, Klein RG, eds. Quaternary extinctions. Tucson: University of Arizona Press, 1984:600–628.

Murray TE, Bartle JA, Kalish SR, Taylor PR. Incidental capture of seabirds by Japanese southern bluefin tuna longline vessels in New Zealand waters, 1988–1992. Bird Conserv Int 1993;3:181–210.

Nakamura K. An essay on the Japanese sea lion, *Zalophus californianus japonicus*, living on the Seven Islands of Izu. Bull Kanagawa Prefect Mus (Nat Sci) 1991;20:59–66.

Nelson EW. Smaller mammals of North America. Nat Geogr Mag 1918;33:371–493.

Nelson EW, Goldman EA. Six new racoons of the Procyon lotor group. J Mammal 1930;11:453–459.

Nelson JB. The biology of Abbott's booby, *Sula abbotti*. Ibis 1971;113:429–467.

Nelson JB, Powell D. The breeding ecology of Abbott's booby *Sula abbotti*. Emu 1986;86:33–46.

Nelson K, Sullivan TA (eds). New criteria for listing species in the CITES appendices. Species 1993;20:15–17.

New TR. Butterfly conservation. Oxford: Oxford University Press, 1991.

Newmark WD. Legal and biotic boundaries of western North American national parks: a problem of congruence. Biol Conserv 1985;33:197–208.

Newton E. On the land birds of the Seychelles Archipelago. Ibis 1867;3:335–361.

Nicholls AO, Margules CR. The design of studies to demonstrate the biological importance of corridors. In: Saunders DA, Hobbs RJ, eds. Nature conservation 2: the role of corridors. Sydney: Surrey Beatty, 1991:49–61.

Nicholls AO, Margules CR. An upgraded reserve selection algorithm. Biol Conserv 1993;64:165–169.

Nicolaus LK, Herrera J, Nicolaus JC, Dimmick CR. Carbachol as a conditioned taste aversion agent to control avian predation. Agric Ecosys Environ 1989;26:13–21.

Nicolaus LK, Nellis DW. The first evaluation of the use of conditioned taste aversion to control predation by mongooses upon eggs. Appl Animal Behav Sci 1987;17:329–346.

Nishida T. Status report on Japanese southern bluefin tuna (*Thunnus maccoyii*) fisheries. Twelfth Australian, Japanese and New Zealand Trilateral Scientific Meeting on Southern Bluefin Tuna, CSIRO Marine Laboratories, Hobart. Working Paper No. SBFWS/93/5, 1993.

Norman DR. Man and tegu lizards in eastern Paraguay. Biol Conserv 1987;41:39–56.

North SG, Bullock DJ, Dulloo ME. Changes in the vegetation and reptile populations on Round Island, Mauritius, following eradication of rabbits. Biol Conserv 1994;67:21–28.

Noss RF. Corridors in real landscapes: a reply to Simberloff and Cox. Conserv Biol 1987;1:159–164.

Nour N, Matthysen E, Dhondt AA. Artificial nest predation and habitat fragmentation: different trends in bird and mammal predators. Ecography 1993;16:111–116.

Nowak RM. Walker's Mammals of the World, 5th ed. Baltimore: Johns Hopkins University Press, 1991.

Nowell K, Wei-Lien C, Chia-Jai P. The horns of a dilemma: the market for rhino horn in Taiwan. Cambridge: TRAFFIC International, 1992.

Oakleaf B, Luce B. First spring: ferrets return to the wild in Wyoming's Shirley Basin. Wyoming Wildlife 1992;56:21–29.

Oakleaf B, Luce B, Thorne ET, Torbit S. Black-footed ferret: reintroduction in Shirley Basin, 1991 Completion Report. Cheyenne, WY: Wyoming Game and Fish Department, 1992.

Oakleaf B, Luce B, Thorne ET, Williams B. Black-footed ferret: reintroduction in Shirley Basin, 1992 Completion Report. Cheyenne, WY: Wyoming Game and Fish Department, 1993.

O'Brien SJ, Martenson JS, Eichelberger MA, Thorne ET, Wright F. Genetic variation and molecular systematics of the black-footed ferret. In: Seal US, Thorne ET, Bogan MA, Anderson SH, eds. Conservation biology and the black-footed ferret. New Haven: Yale University Press, 1989:21–36.

O'Brien SJ, Mayr E. Bureaucratic mischief: recognizing endangered species and subspecies. Science 1991;251:1187–1188.

Ogle CC. The incidence and conservation of animal and plant species in remnants of native vegetation in New Zealand. In: Saunders DA, Arnold GW, Burbidge AA, Hopkins AIM, eds. Nature conservation: the role of remnants of native vegetation. Sydney: Surrey Beatty, 1987:79–87.

Oli MK, Taylor IR, Rogers ME. Snow leopard *Panthera uncia* predation of livestock: an assessment of local perceptions in the Annapurna Conservation Area, Nepal. Biol Conserv 1994;68:63–68.

Oliver WLR. The continuing decline of the pigmy hog and hispid hare. Dodo, J Jersey Wildlife Preserv Trust 1977;14:80–85.

Oliver WLR. The doubtful future of the pigmy hog and the hispid hare. J Bombay Nat Hist Soc 1979;75:341–372.

Oliver WLR. Pigmy hog and hispid hare: further observations of the continuing decline (or, a lament for Barnadi and a good cause for scepticism). Dodo, J Jersey Wildlife Preserv Trust 1981;18:10–20.

Oliver WLR. The coney and the yellow snake: the distribution and status of the Jamaican hutia, *Geocapromys brownii*, and the Jamaican boa, *Epicrates subflavus*. Dodo, J Jersey Wildlife Preserv Trust 1982;19:6–33.

Oliver WLR. The distribution and status of the hispid hare *Caprolagus hispidis*: the summarized findings of the 1984 pigmy hog/hispid hare field survey in northern Bangladesh, southern Nepal and northern India. Dodo, J Jersey Wildlife Preserv Trust 1984;21:6–32.

Oliver WLR. The pigmy hog and the hispid hare: case histories of conservation problems and related considerations in north-eastern India. In: Allchin B, Allchin FR, Thapar BK, eds. The conservation of the Indian heritage. New Delhi: Cosmo, 1989:67–82.

Oliver WRB. New Zealand birds. Wellington: A.H. & A.W. Reed, 1955.

Olney PJS, Ellis P. 1990 International zoo yearbook, Vol 30. London: Zoological Society, 1991.

Olney PJS, Mace GM, Feistner ATC (eds). Creative conservation: interactive management of wild and captive animals. London: Chapman and Hall, 1994.

Olson SL. Extinction on islands: man as a catastrophe. In: Western D, Pearl MC, eds. Conservation for the twenty-first century. New York: Oxford University Press, 1989:50–53.

Olson SL, James HF. Description of thirty-two new species of birds from the Hawaiian Islands. Part I: Non-passeriformes. Ornithol Mono 1991;45:1–88.

Opler PA. The parade of passing species: a survey of extinctions in the U.S. Sci Teach 1976;43:30–34.

Osemeobo GJ. Effects of land-use and collection on the decline of African giant snails in Nigeria. Environ Conserv 1992;19:153–159.

Owen-Smith RN (ed). Management of large mammals in African conservation areas. Pretoria: HAUM, 1983.

Packer C, Pusey AE. Dispersal, kinship, and inbreeding in African lions. In: Thornhill NW, ed. The natural history of inbreeding and outbreeding; theoretical and empirical perspectives. Chicago: University of Chicago Press, 1993:375–391.

Paddle RN. Thylacines associated with the Royal Zoological Society of New South Wales. Austral Zool 1993;29:97–101.

Paine FL, Duval JJ. The resurrected toad. Animal King 1985;88(5):33–38.

Paine FL, Miller JD, Crawshaw G, et al. Status of the Puerto Rican crested toad, *Peltophryne lemur*. Int Zoo Yearb 1989;28:53–58.

Panwar HS. Decline and restoration success of the central Indian barasinga *Cervus duvauceli branderi*. In: Threatened deer. Proceedings of a working meeting of the Deer Specialist Group of the Survival Service Commission on the IUCN Threatened Deer Programme and a dossier on the planning of restoration programmes for threatened mammals with special reference to deer held at Longview, Washington State, USA, 26 September–10 October 1977. Morges, Switzerland: IUCN, 1978:143–158.

Parker ISC. The ivory trade. Washington: U.S. Department of the Interior, Fish and Wildlife Service, 1979.

Parker ISC. The Tsavo story: an ecological case history. In: Owen-Smith RN, ed. Management of large mammals in African conservation areas. Pretoria: Haum, 1983:37–49.

Parker ISC, Amin. Ivory crisis. London: Chatto and Windus, 1983.

Parker ISC, Graham AD. Elephant decline (Part I) downward trends in African elephant distribution and numbers (Part 1). Int J Environ Stud 1989a;34:287–305.

Parker ISC, Graham AD. Elephant decline: downward trends in African elephant distribution and numbers (Part II). Int J Environ Stud 1989b;35:13–26.

Pascual MA, Adkison MD. The decline of the Steller sea lion in the northeast Pacific: demography, harvest or environment? Ecol Appl 1994;4:393–403.

Paton PWC. The effect of edge on avian nest success: how strong is the evidence? Conserv Biol 1994;8:17–26.

Patel T. Halfway home for rare horses. New Sci 1993;140:8.

Patterson BD. Mammalian extinction and biogeography in the southern Rocky Mountains. In: Nitecki MH, ed. Extinctions. Chicago: Chicago University Press, 1984:247–293.

Payne MR. Growth in the Antarctic fur seal, *Arctocephalus gazella.* J Zool Soc London 1979;187:1–20.

Pearce D, Markandya A, Barbier EB. Blueprint for a green economy. London: Earthscan, 1989.

Pemberton JM, Albon SD, Guinness FE, Clutton-Brock TH, Berry RJ. Genetic variation and juvenile survival. Evolution 1988;42:921–934.

Perkins RCL. Vertebrata, Aves. In: Fauna Hawaiiensis, Vol 1, Part IV. Cambridge: Cambridge University Press for the Bernice Pauahi Bishop Museum, 1903:386–466.

Peterson R. Isle Royale: eight pups bring wolf numbers to 15. Wolf 1994;12(2):38.

Petranka JW, Eldridge ME, Haley KE. Effects of timber harvesting on southern Appalachian salamanders. Conserv Biol 1993;7:363–370.

Pettigrew JD. A burst of feral cats in the Diamantina—a lesson for the management of pest species? In: Siepen G, Owens C, eds. Proceedings of a cat management workshop. Brisbane and Canberra: Queensland Department of Environment and Heritage and Australian Nature Conservation Agency, 1993:25–32.

Philips JC. A natural history of the ducks. New York: Dover, 1986.

Pielou EC. Mathematical ecology. New York: Wiley, 1977.

Pimm SL. The balance of nature. Chicago: Chicago University Press, 1991.

Pimm SL, Gilpin ME. Theoretical issues in conservation biology. In: Roughgarden J, May RM, Levin SA, eds. Perspectives in ecological theory. Princeton, NJ: Princeton University Press, 1989:287–305.

Pletschet S, Kelly JF. Breeding biology and nesting success of palila. Condor 1990;92:1012–1021.

Polhemus DA. Conservation of aquatic insects: worldwide crisis or localized threats? Am Zool 1993;33:588–598.

Possingham HP, Lindenmayer DB, Norton TW. A framework for the improved management of threatened species based on population viability analysis. Pacific Conserv Biol 1993;1:39–45.

Powell AN, Cuthbert FJ. Augmenting small populations of plovers: an assessment of cross-fostering and captive-rearing. Conserv Biol 1993;7:160–168.

Pregill G. Cranial morphology and the evolution of the West Indian toads (Salientia: Bufonidae): resurrection of the genus *Peltophryne* Fitzinger. Copeia 1981;2:273–285.

Pregill GK. Prehistoric extinction of giant iguanas in Tonga. Copeia 1989;2:505–508.

Pressey RL, Bedward M, Nicholls AO. Reserve selection in mallee lands. In: Noble JC, Joss PJ, Jones GK, eds. The mallee lands: a conservation perspective. Melbourne: CSIRO, 1990:167–178.

Pressey RL, Humphries CJ, Margules CR, Vane-Wright RI, Williams PH. Beyond opportunism: key principles for systematic reserve selection. Trends Ecol Evol 1993;8:124–128.

Pressey RL, Nicholls AO. Efficiency in conservation evaluation: scoring versus iterative approaches. Biol Conserv 1989a;50:199–218.

Pressey RL, Nicholls AO. Application of a numerical algorithm to the selection of reserves in semi-arid New South Wales. Biol Conserv 1989b;50:263–278.

Pressey RL, Nicholls AO. Reserve selection in the Western Division of New South Wales: development of a new procedure based on land system mapping. In: Margules CR, Austin MP, eds. Nature conservation: cost effective biological surveys and data analysis. Melbourne: CSIRO, 1991:98–105.

Prins HHT. The pastoral road to extinction: competition between wildlife and traditional pastoralism in East Africa. Environ Conserv 1992;19:117–123.

Quattro JM, Vrijenhoek RC. Fitness differences among remnant populations of the endangered Sonoran topminnow. Science 1989;245:976–978.

Quinn H, Quinn H. Estimated number of snake species that can be managed by species survival plans in North America. Zoo Biol 1993;12:243–255.

Rahbek C. Captive breeding—a useful tool in the preservation of biodiversity? Biodiv Conserv 1993;2:426–437.

Ralls K, Ballou JD, Templeton A. Estimates of lethal equivalents and the cost of inbreeding in mammals. Conserv Biol 1988;2:185–193.

Ralls K, Brugger K, Ballou J. Inbreeding and juvenile mortality in small population of ungulates. Science 1979;206:1101–1103.

Ralls K, Meadows R. Breeding like flies. Nature 1993;361:689–690.

Ramono WS, Santiapillai C, Mackinnon K. Conservation and management of Javan rhino (*Rhinoceros sondaicus*) in Indonesia. In: Ryder OA, ed. Rhinoceros biology and conservation. Proceedings of an international conference, May 9–11, 1991, San Diego, California. San Diego: Zoological Society, 1993:265–273.

Rasker R, Martin MV, Johnson RL. Economics: theory versus practice in wildlife management. Conserv Biol 1992;6:338–349.

Raupach MR, Bradley EF, Ghadiri H. Wind tunnel investigation into the aerodynamic effect of forest clearing on the nesting of Abbott's booby on Christmas Island. Technical Report No T12. Canberra: CSIRO Centre for Environmental Mechanics, 1987.

Reading RP, Kellert SR. Attitudes toward a proposed reintroduction of black-footed ferrets (*Mustela nigripes*). Conserv Biol 1993;7:569–580.

Reading RP, Beissinger SR, Grensten JJ, Clark TW. Attributes of black-tailed prairie dog colonies in northcentral Montana, with management recommendations for the conservation of biodiversity. In: Clark TW, Hinckley D, Rich T, eds. Montana BLM Wildlife Tech Bull No 2;1989:13–23.

Rebelo AG, Siegfried WR. Where should nature reserves be located in the Cape Floristic Region, South Africa? Models for the spatial configuration of a reserve network aimed at maximizing the protection of floral diversity. Conserv Biol 1992;6:243–252.

Recher HF, Clark SS. A biological survey of Lord Howe Island with recommendations for the conservation of the island's wildlife. Biol Conserv 1974;6:263–273.

Reed CEM, Nilsson RJ, Murray DP. Cross-fostering New Zealand's black stilt. J Wildlife Man 1993;57:608–611.

Reed JM, Doerr PD, Walters JR. Determining minimum population size for birds and mammals. Wildlife Soc Bull 1986;14:255–261.

Reijnders P, Brasseur S, van der Toorn J, et al. Seals, fur seals, sea lions, and walrus. Status Survey and Conservation Plan. Gland, Switzerland: IUCN, 1993.

Reville BJ, Tranter JD, Yorkston HD. Impact of forest clearing on the endangered seabird *Sula abbotti*. Biol Conserv 1990;51:23–38.

Rice EK. The chinchilla: endangered whistler of the Andes. Animal King 1988;91(1):6–7.

Richter AR, Humphrey SR, Cope JB, Brack V Jr. Modified cave entrances: thermal effect on body mass and resulting decline of endangered Indiana bats (*Myotis sodalis*). Conserv Biol 1993;7:407–415.

Richter WE. Handbook of computations for biological statistics of fish population. Fish Res Board Can Bull 1958;119:1–300.

Richter-Dyn N, Goel NS. On the extinction of a colonizing species. Theoret Popul Biol 1972;3:406–433.

Ripley SD. Rails of the world. Boston: David R. Godine, 1977.

Robertson G. Plant dynamics. In: Caughley G, Shepherd N, Short J, eds. Kangaroos: their ecology and management in the sheep rangelands of Australia. Cambridge: Cambridge University Press, 1987:50–68.

Robertson G, Short J, Wellard G. The environment of the Australian sheep rangelands. In: Caughley G, Shepherd N, Short J, eds. Kangaroos: their ecology and management in the sheep rangelands of Australia. Cambridge: Cambridge University Press, 1987:14–34.

Robinson GR, Quinn JF. Habitat fragmentation, species diversity, extinction, and the design of nature reserves. In: Jain SK, Botsford LW, eds. Applied population biology. Dordrecht: Kluwer, 1992:223–248.

Rodda GH, Fritts TH. The impact of the introduction of the colubrid snake *Boiga irregularis* on Guam's lizards. J Herpetol 1992;26:166–174.

Rowley I, Russell E, Brooker M. Inbreeding: benefits may outweigh costs. Animal Behav 1986;34:939–941.

Rowley I, Russell E, Brooker M. Inbreeding in birds. In: Thornhill WM, ed. The natural history of inbreeding and outbreeding: theoretical and empirical perspectives. Chicago: University of Chicago Press, 1993:429–452.

Sadler PM. Sediment accumulation rates and the completeness of stratigraphic sections. J Geol 1981;98:569–584.

Safina C. Bluefin tuna in the west Atlantic: negligent management and the making of an endangered species. Conserv Biol 1993;7:229–234.

Salvador DJI. Socio-economic incentives for the conservation of the rainforest habitat of the Philippine eagle *Pithecophaga jefferi*. In: Chancellor B-U, Chancellor RD, eds. Raptor conservation today. World Working Group on Birds of Prey and Owls. Berlin: Pica Press, 1994:277–282.

Salwasser H, Schonewald-Cox C, Baker R. The role of interagency cooperation in managing viable populations. In: Soulé ME, ed. Viable population for conservation. Cambridge: Cambridge University Press, 1987:159–173.

Sankhala K. Tiger! The story of the Indian tiger. New York: Simon and Schuster, 1977.

Santiapillai C, Giao PM, Dung VV. Conservation and management of Javan rhino (*Rhinoceros sondaicus annamiticus*) in Vietnam. In: Ryder OA, ed. Rhinoceros biology and conservation. Proceedings of an international conference, May 9–11, 1991, San Diego, California. San Diego: Zoological Society, 1993:248–256.

Santiapillai C, Mackinnon K. Conservation and management of Sumatran rhino (*Dicerorhinus sumatrensis*) in Indonesia. In: Ryder OA, ed. Rhinoceros biology and conservation. Proceedings of an international conference, May 9–11, 1991, San Diego, California. San Diego: Zoological Society, 1993:257–264.

Santiso C. The Maya Biosphere reserves: an alternative for the sustainable use of resources. Nature Resour 1993;29:6–11.

Sanyal P. Managing the man-eaters in the Sundarbans Tiger Reserve of India—a case study. In: Tilson RL, Seal US, eds. Tigers of the world. The biology, biopolitics, management, and conservation of an endangered species. Park Ridge, NJ: Noyes, 1987:427–433.

Sarre S, Dearn JM, Georges A. The application of fluctuating asymmetry in the monitoring of animal populations. Pacific Conserv Biol 1993;1:118–122.

Saunders DA, Hobbs RJ, Margules CR. Biological consequences of ecosystem fragmentation: a review. Conserv Biol 1991;5:18–32.

Savidge JA. Extinction of an island forest avifauna by an introduced snake. Ecology 1987;68:660–668.

Scheffer VB. Seals, sea lions and walruses—a review of the Pinnipedia. Stanford: Stanford University Press, 1958.

Schonewald-Cox CM. Conclusions: guidelines to management: a beginning attempt. In: Schonewald-Cox CM, Chambers SM, MacBryde B, Thomas WL, eds. Genetics and conservation: a reference for managing wild animal and plant populations. Menlo Park, CA: Benjamin/Cummings, 1983:414–445.

Schorger AW. The passenger pigeon: its natural history and extinction. Madison: University of Wisconsin Press, 1955.

Scott JM, Csuti B, Estes JE, Caicco S. Beyond endangered species: an integrated conservation strategy for the preservation of biological diversity. Endanger Species Update 1988;5:43–48.

Scott JM, Mountainspring S, van Riper C III, et al. Annual variation in the distribution, abundance, and habitat response of the palila (*Loxioides bailleui*). Auk 1984;101:647–664.

Seal US. Life after extinction. Symp Zool Soc London 1991;62:39–55.

Seal US, Foose TJ, Ellis S. Conservation assessment and management plans (CAMPs) and global captive action plans (GCAPs). In: Olney PJS, Mace GM, Feistner ATC, eds. Creative conservation: interactive management of wild and captive animals. London: Chapman and Hall, 1994:312–327.

Seal US, Thorne ET, Bogan MA, Anderson SH (eds). Conservation biology and the black-footed ferret. New Haven: Yale University Press, 1989.

Setchell KDR, Gosselin SJ, Welsh MB, et al. Dietary estrogens—a probable cause of infertility and liver disease in captive cheetahs. Gastroenterology 1987;93:225–233.

Shaffer ML. Determining minimum viable population sizes for the grizzly bear. Int Conf Bear Res Man 1983;5:133–139.

Shaw JH. The outlook for sustainable harvests of wildlife in Latin America. In: Robinson JG, Redford KH, eds. Neotropical wildlife use and conservation. Chicago: University of Chicago Press, 1991:24–34.

Sheail J. The extermination of the muskrat (*Ondatra zibethicus*) in interwar Britain. Arch Nat Hist 1988;15:155–170.

Sheets-Pyenson S. How to "grow" a natural history museum: the building of colonial collections, 1850–1900. Arch Nat Hist 1988;15:121–147.

Shepherd N. Condition and recruitment of kangaroos. In: Caughley G, Shepherd N, Short J, eds. Kangaroos: their ecology and management in the sheep rangelands of Australia. Cambridge: Cambridge University Press, 1987:135–158.

Sherwin WB, Murray ND, Marshall Graves JA, Brown PR. Measurement of genetic variation in endangered populations: bandicoots (Marsupialia: Peramelidae) as an example. Conserv Biol 1991;5:103–108.

Shields WM. The natural and unnatural history of inbreeding and outbreeding. In: Thornhill NW, ed. The natural history of inbreeding and outbreeding; theoretical and empirical perspectives. Chicago: University of Chicago Press, 1993:143–172.

Short J. Factors affecting food intake of rangeland herbivores. In: Caughley G, Shepherd N, Short J, eds. Kangaroos: their ecology and management in the sheep rangelands of Australia. Cambridge: Cambridge University Press, 1987:84–99.

Short J, Bradshaw SD, Giles J, Prince RIT, Wilson GR. Reintroduction of macropods (Marsupialia: Macropodoidea) in Australia—a review. Biol Conserv 1992;62:189–204.

Short J, Turner B. A test of the vegetation mosaic hypothesis: a hypothesis to explain the decline and extinction of Australian mammals. Conserv Biol 1994;8:439–449.

Shrader-Frechette, McCoy ED. Method in ecology: strategies for conservation. Cambridge: Cambridge University Press, 1993.

Shull AM. The role of the United States Fish and Wildlife Service in preserving endangered species. In: Mares MA, Schmidly DJ, eds. Latin American mammalogy. Norman: University of Oklahoma Press, 1991:336–351.

Silva JL, Strahl SD. Human impact on populations of chachalacas, guans, and curassows (Galliformes: Cracidae) in Venezuela. In: Robinson JG, Redford KH, eds. Neotropical wildlife use and conservation. Chicago: University of Chicago Press, 1991:37–52.

Simberloff D. The proximate causes of extinction. In: Raup DM, Jablonski D, eds. Patterns and processes in the history of life. Berlin: Springer-Verlag, 1986:259–276.

Simberloff D. The spotted owl fracas: mixing academic, applied, and political ecology. Ecology 1987;68:766–772.

Simberloff D. The contribution of population and community biology to conservation science. Ann Rev Ecol System 1988;19:473–511.

Simberloff DS, Abele LG. Island biogeography theory and conservation practice. Science 1976;191:285–286.

Simberloff D, Cox J. Consequences and costs of conservation corridors. Conserv Biol 1987;1:63–71.

Simberloff D, Farr JA, Cox J, Mehlman DW. Movement corridors: conservation bargains or poor investments? Conserv Biol 1992;6:493–504.

Simons TR. Biology and behavior of the endangered Hawaiian dark-rumped petrel. Condor 1985;87:229–245.

Sinha SP, Sawarkar VB. Management of the reintroduced great one horned rhinoceros (*Rhinoceros unicornis*) in Dudhwa National Park, Uttar Pradesh, India. In: Ryder OA, ed. Rhinoceros biology and conservation. Proceedings of an international conference, May 9–11, 1991, San Diego, California. San Diego: Zoological Society, 1993:218–227.

Smith FDM, May RM, Pellew R, Johnson TH, Walter KR. How much do we know about the current extinction rate? Trends Ecol Evol 1993;8:375–378.

Smith K, Smith F. Conserving rhinos in Garamba National Park. In: Ryder OA, ed. Rhinoceros biology and conservation. Proceedings of an international conference, May 9–11, 1991, San Diego, California. San Diego: Zoological Society, 1993:166–177.

Snyder NFR. Puerto Rican parrots and nest-site scarcity. In: Temple SA, ed. Endangered birds management techniques for preserving threatened species. Madison: University of Wisconsin Press, 1978:47–53.

Snyder NFR, Wiley JW, Kepler CB. The parrots of Luquillo: natural history and conservation of the Puerto Rican parrot. Los Angeles: Western Foundation of Vertebrate Zoology, 1987.

Soulé ME. What do we really know about extinctions? In: Schonewald-Cox CM, Chambers SM, MacBryde B, Thomas WL, eds. Genetics and conservation: a reference for managing wild animal and plant populations. Menlo Park, CA: Benjamin Cummings, 1983:111–125.

Soulé ME, Kohm KM (eds). Research priorities for conservation biology. Washington DC: Island Press, 1989.

Sowerby A de C. Notes upon the sika-deer of North China. Ann Mag Nat Hist 1918;9:119–122.

Sparrow HR, Sisk TD, Ehrlich PR, Murphy DD. Techniques and guidelines for monitoring neotropical butterflies. Conserv Biol 1994;8:800–809.

Speed TP. Probabilistic risk assessment in the nuclear industry: WASH-1400 and beyond. In: Le Cam LM, Olshen RA, eds. Proceedings of the Berkeley Conference in Honor of Jerzy Neyman and Jack Kiefer, Vol I. Monterey, CA: Wadsworth, 1985:173–200.

Springer PF, Byrd GV, Woolington DW. Reestablishing Aleutian Canada geese. In: Temple SA, ed. Endangered birds management techniques for preserving threatened species. Madison: University of Wisconsin Press, 1978:331–338.

Stacey PB, Taper M. Environmental variation and the persistence of small populations. Ecol Appl 1992;2:18–29.

Stanley SM. Macroevolution: pattern and process. San Francisco: W.H. Freeman, 1979.

Stanley SM. Extinction. New York: Scientific American Library, 1987.

Stanley Price MR. Animal re-introductions; the Arabian oryx in Oman. Cambridge: Cambridge University Press, 1989.

Stanley Price MR. A review of mammal reintroductions, and the role of the re-introduction specialist group of IUCN/SSC. Symp R Soc London 1991; 62:9–26.

Steadman DW. Two new species of rails (Aves: Rallidae) from Mangaia, southern Cook Islands. Pacific Sci 1987;40:27–43.

Steadman DW. Extinction of birds in eastern Polynesia: a review of the record, and comparisons with other Pacific Island groups. J Archaeol Sci 1989; 16:177–205.

Steadman DW. Extinction of species: past, present and future. In: Wyman RL, ed. Global climate change and life on earth. New York: Routledge, Chapman and Hall, 1991:156–169.

Steadman DW. Extinct and extirpated birds from Rota, Mariana Islands. Micronesia 1992;25:71–84.

Steadman DW, Olson SL. Bird remains from an archaeological site on Henderson Island, South Pacific: man-caused extinctions on an "uninhabited" island. Proc Nat Acad Sci 1985;82:6191–6195.

Steel WS. The technology of wildlife management cropping in the Luangwa Valley, Zambia. E Afr Agric For J 1968;32:266–270.

Stern C. The Hardy-Weinberg law. Science 1943;97:137–138.

Stirling IG. Research and management of polar bears *Ursus maritimus*. Polar Rec 1986;23:167–176.

Stoll JR, Johnson LA. Concepts of value, nonmarket valuation, and the case of the whooping crane. Trans N Am Wildlife Nat Resour Conf 1984;49:382–393.

Stork NE, Lyal CHC. Extinction or "co-extinction" rates? Nature 1993;366:307.

Stuart SN. Re-introductions: to what extent are they needed? Symp Zool Soc London 1991;62:27–37.

Swanson TM. Economics and animal welfare: the case of the live bird trade. In: Thomsen JB, Edwards SR, Mulliken TA, eds. Perceptions, conservation and management of wild birds in trade. Cambridge: TRAFFIC International, 1992:43–57.

Swanson TM, Barbier EB (eds). Economics for the wilds. Wildlife, wildlands, diversity and development. London: Earthscan, 1992.

Swart R de L, Ross PS, Vedder LJ, et al. Impairment of immune function in harbor seals (*Phoca vitulina*) feeding on fish from polluted waters. Ambio 1994;23:155–159.

Talbot LM. A look at threatened species. A report on some animals of the Middle East and southern Asia which are threatened with extermination. London: Fauna Preservation Society for IUCN, 1960.

Taylor BL, Gerrodette T. The uses of statistical power in conservation biology: the vaquita and northern spotted owl. Biol Conserv 1993;7:489–500.

Taylor R. Radiocarbon dating of Pleistocene bone: towards criteria for selection of samples. Radiocarbon 1980;22:969–979.

Taylor RH. How the Macquarie Island parakeet became extinct. NZ J Ecol 1979;2:42–45.

Taylor RH, Thomas BW. Rats eradicated from rugged Breaksea Island (170 ha), Fiordland, New Zealand. Biol Conserv 1993;65:191–198.

Terborgh J, Winter B. Some causes of extinction. In: Soulé ME, Wilcox BA, eds. Conservation biology. An evolutionary-ecological perspective. Sunderland, MA: Sinauer Associates, 1980:119–134.

Thomas CD. Extinction, colonization, and metapopulations: environmental tracking by rare species. Conserv Biol 1994;8:373–378.

Thomas JA. Ecology and conservation of the Large Blue butterfly—second report. Monks Wood, Abbots Ripton, UK: Institute of Terrestrial Ecology, 1977.

Thomas JA. Why did the Large Blue become extinct in Britain? Oryx 1980; 15:243–247.

Thomas JA. The conservation of butterflies in temperate countries: past efforts and lessons for the future. In: Vane-Wright RI, Ackery PR, eds. The biology of butterflies. Symposium of the Royal Entomological Society No. 11. London: Academic Press, 1984a:333–353.

Thomas JA. The behaviour and habitat requirements of *Maculinea nausithous* (the dusky Large Blue butterfly) and *M. teleius* (the scarce Large Blue) in France. Biol Conserv 1984b;28:325–347.

Thomas JA. Ecological lessons from the re-introduction of Lepidoptera. The Entomol 1989;108:56–68.

Thomas JA. Rare species conservation: case studies of European butterflies. Br Ecol Soc Symp 1990;31:149–197.

Thomas JA, Wardlaw JC. The effect of queen ants on the survival of *Maculinea arion* larvae in *Myrmica* ant nests. Oecologia 1990;85:87–91.

Thompson HV, King CM (eds). The European rabbit. The history and biology of a successful colonizer. Oxford: Oxford University Press, 1994.

Thompson PM, Hall AJ. Seals and epizootics—what factors might affect the severity of mass mortalities? Mammal Rev 1993;23:147–152.

Thomsen JB, Brautigam A. Sustainable use of neotropical parrots. In: Robinson JG, Redford KH, eds. Neotropical wildlife use and conservation. Chicago: University of Chicago Press, 1991:359–379.

Thomsen JB, Edwards SR, Mulliken TA. Perceptions, conservation and management of wild birds in trade. Cambridge: TRAFFIC International, 1992.

Thomsen JB, Mulliken TA. Trade in neotropical Psittacines and its conservation implications. In: Beissinger SR, Snyder NFR, eds. New World parrots in crisis. Solutions from conservation biology. Washington: Smithsonian Institution Press, 1992:221–240.

Thornback J, Jenkins MD. The IUCN mammal red data book. Gland, Switzerland: IUCN, 1982.

Thorne ET, Oakleaf B. Species rescue for captive breeding: black-footed ferret as an example. Symp Zool Soc London 1991;62:241–261.

Thorne ET, Williams ES. Disease and endangered species: the black-footed ferret as a recent example. Conserv Biol 1988;2:66–74.

Thornhill WM (ed). The natural history of inbreeding and outbreeding: theoretical and empirical perspectives. Chicago: University of Chicago Press, 1993.

Thouless CR, Chongqui L, Loudon ASI. The milu or Père David's deer *Elaphurus davidianus* reintroduction project at Da Feng. Int Zoo Yearb 1988;27:223–230.

Todd DM. The release of pink pigeons (*Columba* (*Nesoenas*) *mayeri*) at Pamplemousses, Mauritius—a progress report. Dodo, J Jersey Wildlife Preserv Trust 1984;21:43–57.

Tomkins RJ. Breeding success and mortality of dark-rumped petrels in the Galápagos, and control of their predators. Int Council Bird Preserv Tech Publ 1985;3:159–175.

Tonge S, Bloxam Q. A review of the captive-breeding programme for Polynesian tree snails *Partula* spp. Int Zoo Yearb 1991;30:51–59.

TRAFFIC. Southern bluefin tuna TAC unchanged. TRAFFIC 1994;14(2):51.

Trauger DL, Bocetti CI. Kirtland's warbler recovery team effectively coordinates interagency research and management. Endanger Species Tech Bull 1993;18:4–17.

Trotter MM. Tai Rua: a moa-hunter site in North Otago. In: Anderson A, ed. Birds of a feather. British Archaeological Reports International Series 62, and New Zealand Archaeological Association Monograph II, 1979:203–230.

Turbott EG (Convener). Birds of New Zealand. Auckland: Whitcombe & Tombs, 1967.

Turbott EG (Convener). Checklist of the birds of New Zealand and the Ross Dependency, Antarctica. Auckland: Ornithological Society of New Zealand and Random Century, 1990.

Turcek FJ. Effect of introductions on two game populations in Czechoslovakia. J Wildlife Man 1951;15:113–114.

Tustin K. History of introduction and exploitation. In: The future of New Zealand's wild animals? Seminar 2000: Wild Animal Management, 21–22 November 1988. A seminar organized by the New Zealand Deerstalkers Association, Christchurch, New Zealand, 1988:40–42.

Underwood AJ. Physical factors and biological interactions. In: Moore PG, ed. The ecology of rocky coasts. New York: Columbia University Press, 1986a: 372–390.

Underwood AJ. What is a community? In: Raup DM, Jablonski D, eds. Patterns and processes in the history of life. Berlin: Springer-Verlag, 1986b:351–367.

Underwood AJ. Experiments in ecology and management: their logics, functions and interpretations. Austral J Ecol 1990;15:365–389.

Underwood AJ. The logic of ecological experiments: a case history from studies of the distribution of macro-algae on rocky intertidal shores. J Marine Biol Assoc UK 1991;71:841–866.

Underwood AJ. The mechanics of spatially replicated sampling programmes to detect environmental impacts in a variable world. Austral J Ecol 1993;18:99–116.

US Fish and Wildlife Service. Black-footed ferret recovery plan. Denver, CO: US Fish and Wildlife Service, 1988.

US Fish and Wildlife Service. Native forest birds of Guam and Rota of the Commonwealth of the Northern Mariana Islands recovery plan. Portland, OR: US Fish and Wildlife Service, 1990.

US Fish and Wildlife Service. Recovery plan for the O'ahu tree snails of the genus *Achatinella*. Portland, OR: US Fish and Wildlife Service, 1993.

van Aarde RJ. The diet and feeding behaviour of feral cats, *Felis catus* at Marion Island. S Afr J Wildlife Res 1980;10:123–128.

van Balen B, Gepak VH. The captive breeding and conservation programme of the Bali starling (*Leucopsar rothschildi*). In: Olney PJS, Mace GM, Feistner ATC, eds. Creative conservation: interactive management of wild and captive animals. London: Chapman and Hall, 1994:420–430.

van Noordwijk AJ, Scharloo W. Inbreeding in an island population of the great tit. Evolution 1981;35:674–688.

van Rensburg PJJ, Skinner JD, van Aarde RJ. Effects of feline panleucopaenia on the population characteristics of feral cats on Marion Island. J App Ecol 1987;24:63–73.

Vane-Wright RI, Humphries CJ, Williams PH. What to protect? Systematics and the agony of choice. Biol Conserv 1991;55:235–254.

Verboom J, Lankester K, Metz JAJ. Linking local and regional dynamics in stochastic metapopulation models. Biol J Linnean Soc 1991;42:39–55.

Verrill AE. The "cahow" of the Bermudas, an extinct bird. Ann Mag Nat Hist 1902;7:26–32.

Vesey-Fitzgerald D. The birds of the Seychelles. I: The endemic birds. Ibis 1940;4:480–489.

Viggers KL, Lindenmayer DB, Spratt DM. The importance of disease in reintroduction programmes. Wildlife Res 1993; 20:687–698.

Vigne J-D. Zooarchaeology and the biogeographical history of the mammals of Corsica and Sardinia since the last ice age. Mammal Rev 1992;22:87–96.

Vigne L, Martin EB. Assam's rhinos face new poaching threats. Oryx 1991; 25:215–221.

Walker AC. Patterns of extinction among the subfossil Madagascar lemuroids. In: Martin PS, Wright HE Jr, eds. Pleistocene extinctions. New Haven: Yale University Press, 1976:425–432.

Walker BH. Biodiversity and ecological redundancy. Conserv Biol 1992;6:18–23.

Walker BH. Rangeland ecology: understanding and managing change. Ambio 1993;22:80–87.

Wallace M, Toone W. Captive management for the long term survival of the California condor. In: McCullough DR, Barrett RH, eds. Wildlife 2001: populations. New York: Elsevier, 1992:766–774.

Wallace MP, Wiley W. Use of Andean condors to enhance success of the California condor reintroduction to the wild. Acta XX Congr Int Ornithol 1991;4:2417–2422.

Walters CJ, Holling CS. Large-scale management experiments and learning by doing. Ecology 1990;71:2060–2068.

Walters JR. Application of ecological principles to the management of endangered species: the case of the red-cockaded woodpecker. Ann Rev Ecol Syst 1991; 22:505–523.

Walters JR, Copeyon CK, Carter JH III. Test of ecological basis of cooperative breeding in red-cockaded woodpeckers. Auk 1992;109:90–97.

Warner RE. Recent history and ecology of the Laysan duck. Condor 1963;65: 3–23.

Warner RE. The role of introduced diseases in the extinction of the endemic Hawaiian avifauna. Condor 1968;70:101–120.

Warren MS. A review of butterfly conservation in central southern Britain: I: Protection, evaluation and extinction on prime sites. Biol Conserv 1993;64:25–35.

Warren MS. The UK status and suspected metapopulation structure of a threatened European butterfly, the marsh fritillary *Eurodryas aurinia*. Biol Conserv 1994;67:239–249.

Watson J, Warman C, Todd D, Laboudallon V. The Seychelles magpie robin, *Copsychus sechellarum*: ecology and conservation of an endangered species. Biol Conserv 1992;61:93–106.

Watson RM, Bell RHV. The distribution, abundance and status of the elephant in the Serengeti region of northern Tanzania. J Appl Ecol 1969;6: 115–132.

Wayne RK, Bruford MW, Girman D, Rebholz WER, Sunnucks P, Taylor AC. Molecular genetics of endangered species. In: Olney PJS, Mace GM, Feistner ATC, eds. Creative conservation: interactive management of wild and captive animals. London: Chapman and Hall, 1994: 92–117.

Wayne RK, Lehman N, Gogan PJP, et al. Conservation genetics of the endangered Isle Royale gray wolf. Conserv Biol 1991;5:41–51.

Webb SD. Ten million years of mammalian extinctions in North America. In: Martin PS, Klein RG, eds. Quaternary extinctions. Tucson: University of Arizona Press, 1984:189–210.

Wells MP, Brandon KE. The principles and practice of buffer zones and local participation in biodiversity conservation. Ambio 1993;22:157–162.

Wemmer C, Smith JLD, Mishra HR. Tigers in the wild: the biopolitical challenges. In: Tilson RL, Seal US, eds. Tigers of the world. The biology, biopolitics, management, and conservation of an endangered species. Park Ridge, NJ: Noyes, 1987:396–405.

Whitaker AH. Lizard populations on islands with and without Polynesian rats, *Rattus exulans* (Peale). Proc NZ Ecol Soc 1973;20:121–130.

Whitcomb RF, Lynch JF, Opler PA, Robbins CS. Island biogeography and conservation: strategy and limitations. Science 1976;193:1030–1032.

Whitehead GK. Deer of the world. London: Constable, 1972.

Whiteley PL, Yuill TM. Interactions of environmental contaminants and infectious disease in avian species. Acta XX Congr Int Ornithol 1991;4:2338–2342.

Whitten AJ, Muslimin M, Henderson GS. The ecology of Sulawesi. Gadjah Mada University Press, 1986.

Wijnstekers W. The evolution of CITES: a reference to the Convention on International Trade in Endangered Species. Lausanne, Switzerland: CITES, 1990.

Wilcox BA. Insular ecology and conservation. In: Soulé ME, Wilcox BA, eds. Conservation biology. An evolutionary-ecological perspective. Sunderland, MA: Sinauer Associates, 1980:95–118.

Wiles GJ. The Pacific island flying fox trade: a new dilemma. Bats 1994; 12(3):15–18.

Wiles GJ. Recent trends in the fruit bat trade on Guam. In: Wilson DE, Graham GL, eds. Pacific island flying foxes: proceedings of an international conservation conference. Biological Report No 90. Washington: US Department of the Interior, 1992:53–67.

Wiley JW, Post W, Cruz A. Conservation of the yellow-shouldered blackbird *Agelaius xanthomus*, an endangered West Indian species. Biol Conserv 1991;55:119–138.

Williams ES, Mills K, Kwiatowski DR, Thorne ET, Boerger-fields A. Plague in a black-footed ferret (*Mustela nigriceps*). J Wildlife Diseases 1994;30.

Williams ES, Thorne ET, Kwiatowski DR, Oakleaf B. Overcoming disease problems in the black-footed ferret recovery program. Trans N Am Wildlife Nat Res Conf 1992;57:474–485.

Williams PH, Gaston KJ. Measuring more of biodiversity: can higher-taxon richness predict wholesale species richness? Biol Conserv 1994;67:211–217.

Wilson AC, Stanley Price MR. Reintroduction as a reason for captive breeding. In: Olney PJS, Mace GM, Feistner ATC, eds. Creative conservation: interactive management of wild and captive animals. London: Chapman and Hall, 1994:243–264.

Wilson EO, Willis EO. Applied biogeography. In: Cody ML, Diamond JM, eds. Ecology and evolution of communities. Cambridge, MA: Belknap Press, 1975:522–534.

Wilson J, Wilson R. Observations on the Seychelles magpie-robin *Copsychus seychellarum*. Br Ornithol Club Bull 1978;98:15–21.

Wilson MH, Kepler CB, Snyder NFR, et al. Puerto Rican parrots and potential limitations of the metapopulation approach to species conservation. Conserv Biol 1994;8:114–123.

Wingate DB. Excluding competitors from Bermuda petrel nesting burrows. In: Temple SA, ed. Endangered birds: management techniques for preserving threatened species. Madison: University of Wisconsin Press, 1978:93–102.

Wingate DB. The restoration of Nonsuch Island as a living museum of Bermuda's pre-colonial terrestrial biome. ICBP Tech Publ 1985;3:225–238.

Woodford MH, Rossiter PB. Disease risks associated with wildlife translocation projects. In: Olney PJS, Mace GM, Feistner ATC, eds. Creative conservation: interactive management of wild and captive animals. London: Chapman and Hall, 1994.

Woods CA. Endemic rodents of the West Indies: the end of a splendid isolation. In: Lidicker WZ Jr, ed. Rodents—a world survey of species of conservation concern. International Union of the Conservation of Nature/Species Survival Commission Occasional Paper No. 4;1989:11–19.

Worthy TH. An illustrated key to the main leg bones of moas (Aves: Dinornithiformes). Nat Museum NZ Misc Ser 17:1988.

Worthy TH, Edwards AR, Millener PR. The fossil record of moas (Aves: Dinornithiformes) older than the Otira (last) Glaciation. J R Soc NZ 1991; 21:101–118.

Wright S. Evolution in Mendelian populations. Genetics 1931;16:97–159.

Wright S. Size of population and breeding structure in relation to evolution. Science 1938;87:430–431.

Wright S. Breeding structure of populations in relation to speciation. Am Natural 1940;74:232–248.

Wurster CF, Wingate DJ. DDT residues and declining reproduction in the Bermuda petrel. Science 1968;159:979–981.

York AE. Northern fur seal, *Callorhinus ursinus*, eastern Pacific population (Pribilof Islands, Alaska, and San Miguel Island, California). In: Croxall JP, Gentry RL, eds. Status, biology, and ecology of fur seals. Proceedings of an international symposium and workshop, Cambridge, England. NOAA Technical Report NMFS 51;1987:9 21.

Zimmerman DR. To save a bird in peril. New York: Coward McCann and Geoghegan, 1975.

Index